HISTOLOGY OF THE VERTEBRATES
a comparative text

HISTOLOGY OF THE VERTEBRATES

a comparative text

WARREN ANDREW

Department of Anatomy
Indiana University Medical Center
Indianapolis, Indiana

CLEVELAND P. HICKMAN

Department of Zoology
DePauw University
Greencastle, Indiana

with 537 illustrations

THE C. V. MOSBY COMPANY

SAINT LOUIS 1974

Library of Congress Cataloging in Publication Data

Andrew, Warren.
    Histology of the vertebrates.

    1. Histology. 2. Anatomy, Comparative.
I. Hickman, Cleveland Pendleton, 1895-   joint
author. II. Title. [DNLM: 1. Histology,
Comparative. 2. Vertebrates—Anatomy and histology.
QL807 A563h 1974]
QM551.A52   596'.08'24   73-17378
ISBN 0-8016-0247-5

CB/CB/B 9 8 7 6 5 4 3 2 1

PREFACE

Microscopy as a science is now somewhat more than three centuries old. Its rate of progress has varied, but the past century has seen tremendous strides. As with other rapidly expanding fields, it has come to have its own special areas, each of which may require a lifetime or more to master. A recent encyclopedic work on the subject lists twenty-six kinds of microscopy subdivisions, with overlapping boundaries fitting together to form the mosaic of the entire field. Among these are phase, fluorescence, x-ray, field emission, and electron microscopy, to name only a few.

From the high point of the many technical advances in this area, it is not always easy to look back and to appreciate the exhilaration of observation and discovery that the early microscopists enjoyed. To Anton van Leeuwenhoek, grinding and mounting his own lenses, there was no lengthy "literature search" to bring him up to date on the subject of his current study, whether it was blood, saliva, or a drop of pond water. Indeed, well into the nineteenth century, books were written on the almost purely exploratory aspect of the use of magnifying instruments. Such pioneer works as *The Microscope and Its Revelations,* by Carpenter, and *Evenings at the Microscope,* by Gosse, can still be read with pleasure.

Although the subject of this book is a scientific one and our approach is calculated to maintain the scientific outlook and method, we hope that we have captured some of the flavor of romance and enjoyment that should accompany the study of the subject. It is difficult, indeed, to think of a field with greater breadth and depth than that of the microscopic structure of the varied forms of animal life.

In contrast to books on human or general histology, those of a comparative nature are not numerous and have been difficult to review, partly because of the breadth of the subject covered and partly because of the long span between publication of comprehensive works in this field. But the student of life sciences and those preparing for a career in medicine or some other health profession deserve a book that will give them the advantages of a broad survey of the histological structure of vertebrate animals. Such a book should fit in well with other subjects in a paramedical curriculum, and its study will not run the risk of being repeated in medical school. Such repetition has become a major problem where human histology is concerned. In addition, the student continuing with advanced work in the biological sciences, the preprofessional student who looks forward to a career in dentistry, and the student who simply is seeking a wider knowledge of animal life as a part of his picture of this varied world will profit from use of this text.

The breadth of view that can be gained from comparative studies is seen in the writings of the great philosophers of biology. Such breadth of view forms an extremely favorable substrate for the action of new facts and new ideas on the mind. In this book, the broad comparative aspects of each system have been reviewed, chiefly from the standpoint of probable evolutionary directions, but attention also has been given to special features of diverse members of the different classes.

We have concentrated here on structure visible with the light microscope. The light microscope is still the "workhorse" of the microscopist and the instrument readily available to the beginning student, which he can learn to use well without too great an expenditure of time and effort. The general knowledge of microscopic structure, at levels from low magnification up to 1000 diameters, gives the student basic knowledge needed in anatomy, physiology, pathology, neurology, endocrinology, and other fields. The number of electron micrographs and the references to recent studies by numerous authors on fine structure will serve to show what important contributions to knowledge of structure are being made with the remarkable powers of resolution of the electron microscope.

We are indebted to scientists in many laboratories in the United States, Europe, and Asia for encouragement and helpful suggestions and for generously furnishing illustrations from their own research and writing. A special debt is acknowledged to Mr. W. C. Gamble, President, Ward's Natural Science Establishment, and to Miss Marita E. Heymann, who have prepared hundreds of tissue and organ specimens. Many of the photomicrographs in this book have been made in our laboratories from these preparations. Other photomicrographs and electron micrographs have been supplied to us by investigators whose kindness is acknowledged in the legends of the figures in this book.

Warren Andrew
Cleveland P. Hickman

CONTENTS

1 THE VERTEBRATES

In the discussion of the evolutionary development of microscopic anatomy throughout the vertebrate world, examples will be drawn from various animals in all the vertebrate classes. However, much of the information and ideas presented will be based on specific animals in each group. These include the sea lamprey (Cyclostomata), shark (Chondrichthyes), brook trout (Osteichthyes), frog (Amphibia), lizard (Reptilia), pigeon (Aves), rabbit (Mammalia), and man (Mammalia).

The evolutionary advances of the different vertebrate classes are discussed briefly in this chapter as background for the phylogenetic surveys of the microanatomy of organ systems. Since the mechanism of evolution has an active and functional basis, the life-styles of these representative vertebrates are also described in an attempt to make the anatomic differences more meaningful.

CYCLOSTOMATA

The lampreys and hagfishes are notable in two respects: (1) they are considered to be the most primitive living vertebrates and (2) they are very destructive to higher fishes. They have caused many millions of dollars of damage to the fisheries of the Great Lakes.

Petromyzon marinus, the sea lamprey (Fig. 1-1), is a swift-swimming fish of primitive nature. Its body is long and eel-like. As do other cyclostomes, the lamprey lacks jaws and has no paired fins. It has three fins, one caudal and two dorsal, all of which are small and inconspicuous. The head bears two well-formed eyes and, midway between them, a **single** nostril. Along the cervical region extends a series of seven openings that give access to the gills. The skeleton is cartilaginous. The mouth is a rounded one, adapted for sucking; the sea lamprey lives by sucking the blood and body fluids from other fishes to which it attaches itself.

There are more than two dozen different species of lampreys. Some are permanent inhabitants of fresh water, living in lakes and ascending streams to breed. The sea lampreys, when adult and 2 to 3 feet in length, enter freshwater streams to build nests and reproduce. The young lampreys are called **ammocoete larvae** and are unlike their parents. They live for several years in the bottom mud, sifting out their food from the ooze. The larval form closely resembles *Amphioxus,* a primitive chordate that lacks a backbone.

A form closely related to the lamprey is the hagfish, *Myxine glutinosa.* It has been described as "the most primitive living vertebrate" (Brodal and Fänge, 1963), and as one might expect, there is a great deal of information available on *Myxine.* However, *Petromyzon* is usually discussed as a

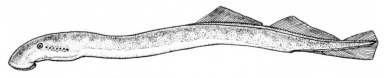

Fig. 1-1. *Petromyzon marinus,* the sea lamprey.

representative of the Cyclostomata because *Myxine* possesses a number of characteristics considered to be **degenerate,** including its atrophic eyes, its habit of burrowing into the body cavity of other fish, and its generally low state of activity. The fact is, however, that many resemblances exist in the fundamental microscopic anatomy of *Petromyzon* and *Myxine*. Therefore some description of microscopic structure will be given for both, always with the specification as to which form is meant.

The lampreys and hagfishes, although commonly thought of as "fishes," are placed in a division known as the Agnatha ("jawless"), separate from all the other Craniata, which are called Gnathostoma (jaw-mouthed). Other primitive features are the lack of paired appendages and the persistence of a notochord. Some species of cyclostomes are hermaphroditic. It is important from the evolutionary standpoint that several ancient groups of fishes, found now only as fossils, are also in the division Agnatha. These include the Cephalaspidae, Anaspidae, and Pteraspidae, all of which lived in the Paleozoic Era and apparently had become extinct before the Mesozoic Era. Superficially, these early fishes looked very unlike the cyclostomes, but in their pouched gills, single nostril, and other features they show an important relationship to the cyclostomes.

CHONDRICHTHYES

The class Chondrichthyes is commonly divided into two subclasses: Elasmobranchii (sharks, skates, and rays) and Holocephali (chimaeras, or ghostfishes).

Sharks belong to the order Selachii. Some of of them are the largest of all fishes, but the ones studied in college laboratories, the dogfish sharks, are of a more modest size, rarely exceeding 3 feet in length. Familiar examples of common dogfish are the Atlantic spiny dogfish *(Squalus acanthias,* Fig. 1-2), the Pacific spiny dogfish *(Squalus suckleyi),* the smooth dogfish *(Mustelus canis),* and the small dogfish *(Scyliorhinus caniculus).* The last two species are mainly bottom-dwellers.

Sharks differ from other fishes in several ways. Their skeleton is composed entirely of cartilage,

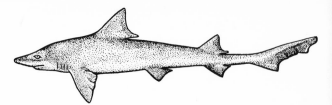

Fig. 1-2. *Squalus acanthias,* the dogfish shark.

which may be stiffened by mineral deposits. They have no swim bladder, and their body is heavier than water. Their gill slits open separately to the outside. Another unique feature is the separation of their suprarenal and interrenal glands. The head terminates anteriorly in a blunt snout, and the crescent-shaped mouth is usually located ventrally, a short distance from the anterior end. The paired pectoral fins have little flexibility and are spread out like airfoils or hydroplanes to provide lift as they swim, in contrast to the bony fishes that hold their pectoral fins close to the body when swimming rapidly through the water. The shark's fins cannot serve as "brakes" because they have only restricted flexure in a vertical plane, and so these fish must turn to avoid collision with objects. Their heterocercal (unequally divided) tail tends to drive the head downward; this tendency is corrected by the flattened shape of the head and by the action of the pectoral fins (Norman, 1963).

Cartilaginous fishes have often been considered primitive, but their lack of bone may be due to specialization or fetalization. Moreover, their fossil record does not extend backward as far as that of bony fishes. The oldest known shark fossil is *Cladoselache* from the Upper Devonian Era (Norman, 1963). Many of their characters were no doubt derived from their extinct placoderm ancestors. The earliest sharks were probably adapted for living on the bottom as are their relatives, the rays and skates, and even some of the present-day sharks.

Although chiefly marine forms, they can, and some do, enter fresh water, but few of them take up a permanent freshwater existence. For example, one of these *(Carcharhinus nicaraguensis)* appears to be permanently established in Lake Nicaragua.

Adaptive nature of cartilaginous tissue. Sharks

have no bony armor, but their purely cartilaginous skeleton has made them lighter for swift predation. Their long snout and ventral mouth position have added to the streamlining of their bodies.

Since they have no hydrostatic organ or air bladder, sharks in midwater keep their position above bottom by staying constantly on the move. In fact, constant movement in many sharks is necessary to keep them alive. Without movement they are unable to keep a sufficient flow of water over the gills. A sluggish circulatory system also requires muscular movement to aid the heart in the propulsion of the blood. Buoyancy is aided by the presence of large livers containing a large amount of oil. Although they have no defensive armor, they compensate for this by swift swimming, fairly efficient sense organs and brain, and powerful jaws for seizing their prey.

Sharks exhibit a generalized anatomic arrangement of their systems—a kind of blueprint of the architectural plan of the vertebrate body. The cartilaginous fishes also show a basic vertebrate plan in their histology. Although the freshwater origin of vertebrates is well established, sharks have been marine forms since the beginning of their long evolution, and their tissues in general indicate a structural and functional adaptation to a saline environment, in many ways different from that of marine bony fishes. Their body fluids, for example, contain about 2.25% urea, which offsets the high osmotic effect of sea water and makes water available by direct absorption. Bony fishes have no such adaptations and are constantly confronted with the necessity of ridding their body fluids of excess salt by active transport systems through the gills, a process requiring the expenditure of much energy.

OSTEICHTHYES

The class Osteichthyes (bony fishes) consists of the Actinopterygii (ray-finned fishes) and the Choanichthyes (lobe-finned, air-breathing forms). The latter include the lungfish, etc., apparent predecessors of air-breathing tetrapods. The most advanced of the ray-finned fishes are the teleosts (superorder Teleostei). They have completely bony skeletons and include most of the more familiar species of fishes.

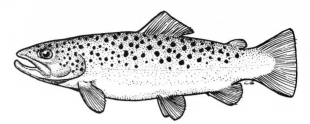

Fig. 1-3. *Salvelinus fontinalis,* the brook trout.

Salvelinus fontinalis, the American brook trout (Fig. 1-3), is common in streams of North America. It bears a generally close resemblance to its close relatives such as the salmon, lake trout, and brown trout. They are all bony fishes, or teleosts. The body is elongated, compressed from side to side, and tapered toward head and tail. The brook trout is a swift and strong swimmer, although it does not face the tremendous demands made on the salmon for such abilities.

The trout of streams and rivers are believed to represent descendants of once migratory fishes. It is thought that as the glaciers retreated northward at the return of milder climatic conditions in the Northern Hemisphere, the migratory fishes in general moved back with them but that some colonies were left behind, not being willing perhaps to leave the cool waters of the streams and lakes. These permanently freshwater forms would have arisen in various lakes of Scandinavia, Switzerland, the British Isles, North America, and other regions. This isolation of forms led to the means for evolution in various directions in the same way that isolation on "land islands" made such evolution possible. Thus the family Salmonidae, typically migratory and anadromous (running up rivers to spawn), has members such as the grayling that are strictly river- and lake-dwellers.

AMPHIBIA

The vertebrate class Amphibia is divided into 3 orders or subclasses: Gymnophiona, or Apoda —limbless, blind, burrowing amphibians (caecelians); Urodela, or Caudata—tailed amphibians (newts and salamanders, Fig. 1-4); and Anura (Salientia)—tailless amphibians (toads and frogs). The amphibians as a group have worldwide distribution, but there are restricted distributions

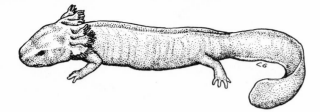

Fig. 1-4. *Necturus,* the salamander.

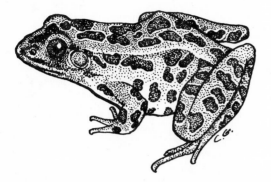

Fig. 1-5. *Rana pipiens,* the leopard frog.

among the different orders or subclasses. The Apoda are found only in certain tropical equatorial regions of Central America, South America, Central Africa, and Southeast Asia. The Urodela are chiefly residents of the Northern Hemisphere and do not cross the equator (Bentley, 1966). The Anura have a much wider distribution, being found from the arctic regions to the deserts of Africa, America, and Australia. Although suited mostly to moist, cool areas, many have become adapted to hot, dry regions.

The amphibians were the first terrestrial vertebrates. With the reptiles, birds, and mammals they constitute the superclass Tetrapoda, or four-legged animals. It is thought that amphibians originated during the latter part of the Devonian geological period from crossopterygian (lobe-finned) fishes whose crawling from pool to pool put a premium on natural selection for terrestrial locomotion.

The most successful of all amphibians are the members of the order or subclass Anura (frogs and toads), and the most successful of all the frogs belong to the genus *Rana* (Fig. 1-5). This genus is distributed in most parts of the world except southern South America, New Zealand, and oceanic islands. These frogs usually show a preference for damp places such as marshes and ditches, where they live chiefly on insects that they catch by their tongue. As tadpoles, they are omnivorous. They avoid salt water, although Ruibal (1959) found a brackish water population of *Rana pipiens* in California.

Except in certain minor particulars, the structure of the *Rana* members is much the same. The most familiar *Rana* species are *Rana temporaria* and *Rana esculenta* in England and the European continent; *Rana pipiens, Rana catesbeiana, Rana clamitans,* and *Rana sylvatica* in North America; and *Rana goliath* in Africa. *R. pipiens* is the common leopard frog widely used in biology laboratories. *R. catesbeiana,* whose body length may reach 9 inches, is the largest American frog, but it is exceeded by *R. goliath,* which is more than 12 inches long.

Moore (1967) considers *R. pipiens* to be the most widespread and common frog in North America. He also stresses the striking evolutionary divergence of local populations, which results in many differentiated allopatric populations. Fowler (1961) indicates that the degree of racial abnormalities among the different hybrids of *R. pipiens* is measured in a North-South direction; East-West directions do not affect hybrids in their interbreeding and production of viable offspring. Thus a Florida frog cannot produce viable offspring when crossed with one from Vermont but can do so when interbred with a Texas frog.

Adaptive nature of amphibian tissue. The larvae or tadpoles of frogs show many fishlike characteristics, but the adults have distinct differences from any of the fishes. The frog skull is simpler and has fewer bones than that of the bony fishes. Frogs also have more cartilage. Their notochord may be entirely lost, and they have well-formed vertebrae. Since they are tetrapods, they have limbs with digits and the skeletal elements that accompany such appendages. The highly specialized frogs meet the requirements of true terrestrial life by having limbs that lift the body off the ground instead of dragging the belly as most of the urodeles do. Alternate and independent movements of the frog's limbs have removed the necessity of the segmental arrangement of muscles characteristic of fishes and primitive amphibians.

Because frog eggs are not amniotic, they have only a jellylike covering for protection instead of the membranes and shell of true terrestrial animals. Normally the frog has to lay its eggs in the water, and since the eggs are small, the tadpoles hatched from them are small and must grow up in water. This "double life" of the frog is reflected in many of its tissues, especially the skin. The skin, devoid of scales or other armor, serves as an accessory breathing organ and must be kept moist by damp surroundings or by glands.

Frogs in water do not drink, but take water through the skin. As a survival adaptation, 25% of their body water is kept in large lymph spaces; this water is lost first when frogs are exposed to arid environments, and thus the tissue water is spared. To conserve water, they also store it in the urinary bladder, which first appears in frogs. In the Australian form, *Chiroleptes,* the bladder may hold water equal to 50% of the body weight (Bentley, 1966). Since the tadpoles require aquatic surroundings, some frogs may postpone reproduction for long periods of time.

Internally, amphibians may have certain resemblances to fishes such as the Dipnoi and Crossopterygii. No adult frogs have gills, but their lung air sacs are similar to those of lungfishes. Many internal tissues and organs in frogs had to be reorganized to meet the requirements of their double life, especially in the vascular and urogenital systems.

REPTILIA

There are at present four existing orders of reptiles. These represent only a fraction of those that existed in the past when more than 14 orders of reptiles flourished in the Mesozoic Era. The reptiles were the first vertebrates to become entirely terrestrial. In that remote geologic period, more than 200 million years ago, their many types moved into several ecologic niches. They underwent an amazing adaptive radiation that was bushlike in its evolution, following the amphibians but preceding the birds and mammals. One of the greatest secrets of their success in colonizing the land was their development of the amniotic egg.

This type of egg freed the reptiles from aquatic dependence in rearing their offspring.

The 4 modern orders of reptiles are the Squamata (lizards and snakes), the Rhynchocephalia (tuatara), the Crocodilia (crocodiles, alligators, and caimans), and the Chelonia (tortoises and turtles). These types are all that remain of the many groups flourishing during the Mesozoic, which culminated in the largest of all land animals, the giant dinosaurs. The causes of their decline are not known in all cases, but climatic changes may have been partially responsible. The colder climate of the Tertiary was far less suitable for reptiles than the warmer Mesozoic.

Although at present reptiles are found in both the temperate and tropical zones, they flourish chiefly in the latter. Reptiles are restricted in their distribution because they are cold-blooded and cannot physiologically maintain a temperature. However, some reptiles are able to offset this handicap by suitable exposure to available sources of heat. When they are cold, they bask in the sun on hot rocks; when too warm, they seek shelter in holes or under vegetation. Bogert (1959) has shown how lizards regulate their temperature by orienting their bodies for maximum or minimum exposure to the sun's rays. Bartholomew and Tucker (1963) regard lizards as an evolutionary stage in the transition between ectothermy and endothermy. No reptile can retain an independent body temperature, and all must hibernate in temperate climates. Some lizards are found as far north as the Arctic Circle.

As a result of their extensive adaptive radiation, modern lizards include terrestrial, burrowing, arboreal, and aquatic types. Most of them are highly carnivorous, but some have acquired herbivorous habits. The suborder Lacertilia, or Sauria, consists of the iguanids, agamids, geckos, chameleons, skinks, and the common lizard *Lacerta.* There are about 2200 species of modern lizards (Savage, 1960). The largest are the monitor lizards, such as the Komodo dragon that grows to be more than 10 feet long. The family Lacertidae, which includes *Lacerta* (Fig. 1-6), occurs in Europe, Asia, and Africa. New World lizards include the American "chameleon" *(Anolis carolinensis),* the collared lizard *(Crotaphytus collaris),* the blue-tailed skink *(Eumeces fasciatus),* and the

Fig. 1-6. *Lacerta,* the common lizard.

horned "toad" *(Phrynosoma).* The only known poisonous lizards belong to the genus *Heloderma* found in the American southwest.

Adaptive nature of reptilian tissue. The organization of a reptile is adapted for a land existence. In this respect they display many advances over amphibians. Their skin is dry and specialized for withstanding fairly high temperatures. Throughout their whole organization they have methods for economizing water. An active terrestrial life requires that the limbs lift the body off the ground to prevent friction and conserve energy. In contrast to the glandular skin of modern amphibians, the reptile body is covered with horny scales, or scutes.

Their hard, dry skin prevents exchange of gases through it. To offset this deficiency, reptiles have developed better lungs, which most of them expand by rib movements. No reptile has gills, and the shift to lung breathing has brought about significant changes in their circulatory system. The heart is partially developed (wholly developed in the crocodiles) into two halves, so that oxygenated blood is kept mostly separated from the venous blood with its heavy load of waste.

Most modern lizards have sturdy, well-developed limbs, although some are limbless. Many have five toes, usually with many joints. Skeletal modifications were necessary for limb adjustments and for a land existence. The lower jaw articulates at each side of the posterior margin of the skull; the quadrate bone is capable of movement relative to the squamosal and the brain case by a loose joint with the pterygoid (in the Squamata). This streptostylic condition (most developed in snakes) allows the animal to open its jaws with a wide gape.

It is generally agreed that reptiles were adapted in many ways to a land existence while they were still aquatic-like amphibians. Many apparently never left the water, although provided with many terrestrial adaptations. Many others are thought to have remained aquatic long after the development of the amniotic egg made possible a terrestrial existence.

Land existence also led to the development of a firmer type of tissue for lizards and other reptiles. Most of their tissues have a lower water content than the same tissues in amphibians. This is reflected in a more definite arrangement of cellular structures and a clearer concept in cell organelles.

AVES

The birds represent one of the most highly specialized groups in the animal kingdom. They are perhaps the most easily recognized of all animals, since they are the only ones that possess feathers. The class Aves has a worldwide distribution and an adaptive radiation that has resulted in more than 8500 species. Adapted as they are for an aerial life, they are found over all seas and terrestrial regions. Although they all have feathers, not all have the power of flight. Their highly adaptable nature is shown by their wide range of locomotion. Many are excellent swimmers, both flightless and flyers; some are perfectly at home running over the land; others spend most of their time in the air. Some flightless birds such as the penguin are awkward on land but are amazing swimmers, whereas others such as the ostrich travel over land with the speed of a horse.

There is plenty of evidence to show that birds came from reptilian ancestors. Nearly all their characters can be matched in some ancient reptilian type. Some of their modifications are in connection with flight. Because of their fragile bones, and for other reasons, birds do not easily form fossils. The fossil evidence most convincing of their reptilian ancestry is *Archaeopteryx,* known from four specimens found since 1861. The skeleton was distinctly reptilian. The jaws contained teeth implanted in sockets, a feature unknown among other birds with the possible exception of the fossil bird *Hesperornis,* found in Kansas. At present there are at least 28 orders of birds, representing 161 families; at least 6 orders and 39 families have become extinct.

In their adaptive radiation, birds have diverged in many ways, both morphologically and physiologically. Locomotor adaptations occur mainly in the wings and legs. It is generally believed by most specialists that the first birds were arboreal. Early birds, beside flying, hopped from branch to branch. But other forms of locomotion also evolved, such as walking, climbing, wading, swimming, and diving. Adaptations for feeding show equal divergence. The earliest birds probably ate insects, but modern forms are adapted for particular kinds of food, both plant and animal. Most of their modifications for food-getting involve the bill and the hind limbs. The tongue is specially adapted in hummingbirds and honey eaters for feeding on nectar and pollen.

Size ranges of birds vary enormously. The principle of surface-volume ratio plays an important part in size. Since the surface varies as the square and the volume as the cube, the minimum size of birds is chiefly determined by the balance between the amount of food taken in and the heat loss from the body. This balance must be a very delicate factor in the small size of the hummingbirds, which go into a torpid state at night when they cannot feed. The maximum size for flying birds is a function of the surface-volume ratio and the speed of flight. In general, the larger the bird, the swifter it must fly in order to stay in the air. The largest birds are flightless because of these factors.

Of the 28 to 30 orders of birds now recognized, the largest in number of species is the Passeriformes (literally, sparrowlike birds). These include the familiar song birds and many others. The pigeon, *Columba* (Fig. 1-7), belongs to the order Columbiformes, which includes the doves and the famous extinct dodo.

Adaptive nature of avian tissue. The histology of many avian bird structures is primarily a result

Fig. 1-7. *Columba livia,* the pigeon.

of arboreal adaptations. Their lightness of structure, the adaptations of their hind limbs for grasping and perching, the highly developed eyes, their dexterity of limbs, the development of feathers, and the enlarged size of the brain, especially the cerebellum and striatum, are all understood as highly effective adaptations to their mode of life from their earliest evolution. Although the type specimen for histologic study will be mainly *Columba,* the closely knit organization plan in all birds is so similar that almost any bird could be a typical representative.

In some respects birds have tissues that resemble the reptiles, their ancestors. One marked physiologic difference between birds and reptiles is the bird's warm-blooded nature. Among existing animals, only birds and mammals are homeotherms although the extinct *Pterosaurus* may have been warm-blooded or well on the way to such condition. Birds have carried their respiratory adaptations into their bones and other parts of the body—features that are mainly lacking in mammals. To maintain their body heat, birds have developed a feather pattern instead of scales. It is thought that this physiologic development preceded the development of true flight.

The presence of feathers may account for the nature of many avian tissues. A thinner and more delicate skin may be correlated with the protection afforded by feathers. A greater blood supply to the dermis in certain regions may be due to the brooding habits of birds, which causes the warmth of the blood to be next to the eggs. The

Fig. 1-8. *Lepus,* the rabbit.

skeleton is similar in many ways to that of reptiles, especially the subclass Archosauria (pterosaurs, dinosaurs, crocodiles). Archosaurs developed early the practice of running on their hind legs, and this resulted in a division of labor whereby the forelimbs became adapted for flying and hind limbs for walking, as in birds. The skeleton of many birds has been lightened by pneumatization; that is, in the hollow of their bones are air sacs that communicate with the nasal sacs of the bronchial system (Fig. 9-3). The avian bones also have more spongy material for the same reasons.

The general histologic structure of many bird tissues, however, tends to approach that of mammals, as is to be expected in the organization of warm-blooded forms.

MAMMALIA

The mammals (Fig. 1-8) represent the highest class of vertebrates. In many respects they are set apart from the other animals in their general structural features and physiologic adaptations. Their general organization is such that they are able to meet conditions of existence impossible for most other animals. They are distinguished from all other vertebrates by having hair on the body at some stage of their life cycle and having the young nourished by milk from special mammary glands of the female. All are viviparous except the duck-

billed platypus, *Ornithorhynchus anatinus,* belonging to the primitive monotremes. These aberrant mammals lay leathery, shelled eggs like reptiles.

Other characteristics of the mammals include the four-chambered heart, a left fourth aortic arch instead of a right arch as in birds, warm-bloodedness, nonnucleated red blood corpuscles, lungs for breathing, and a thoracic cavity that contains the heart and lungs and is separated from the abdomen by a muscular diaphragm. In mammals the placenta serves as a highly efficient organ of exchange for food and waste. The brain reaches its greatest size and organization in mammals. The brain has large cerebral hemispheres with great development of the cerebral cortex. The hemispheres are united by commissures, especially the corpus callosum that often roofs over the lateral ventricles. The skull has two occipital condyles (reptiles and birds have only one). The mandible, complex in reptilians, has been reduced to a single pair of bones, the dentaries, which articulate directly with the skull instead of through quadrate bones. The malleus and incus have become specialized auditory ossicles and are lodged with the stapes in the cavity of the middle ear.

Mammals arose from the therapsid reptiles in the late Triassic Era. These mammal-like reptiles probably came from several distinct therapsid stocks (polyphyletic origin). Mesozoic mammals were extremely small, and fossils show that they were intermediate between the reptilian and mammalian conditions. The fossil record gives no indication when hair and mammary glands first appeared. It is thought that the various therapsid stocks appeared at different times, and it is impossible to pinpoint a time when reptilian stocks became mammals. The full mammal-like condition was probably established by the middle Jurassic period, some 150 million years ago.

Mammals are divided into two subclasses: Prototheria (the monotremes, or egg-laying mammals) and Theria, which is made up of infraclass Metatheria (pouched marsupials) and infraclass Eutheria (true placental mammals). There are 18 existing orders. The Prototheria consists of 1 order, and the Theria contains the other 17. Rothschild (1961) lists the number of mammalian species at 4500. There are many subspecies.

Adaptive nature of mammalian tissue. In mammals, tissues have reached their highest development in the animal kingdom. Although histologists commonly assume that the building units (cells and tissues) are much alike in all animals, this narrow viewpoint has been due mainly to the lack of investigation of the basic constituents and morphologic organization of the tissues of different animals. It is well known to cytologists that there are many differences in cellular structures and that tissue cells are similar only in the broad general organization. There may be some degree of structural uniformity in a particular tissue among several kinds of animals, but this uniformity is greater among the tissues of the same kind of animal (Sumner, 1924). Specialization occurs in many types of animals, and differences are reflected not only in the organs but also to some extent in the tissues. Also, tissues as well as organs are adapted to the demands of the particular environment. In fact, morphologic differences can be recognized in the same tissue even in closely related animals (Yocom and Huestis, 1928). More is known about the structure and function of mammalian tissues than about those of any other group of animals. This has been due no doubt to the extensive investigations that have been made upon the best known of all mammals, man himself.

One unique characteristic of mammalian tissue is the high development of glands. Perhaps no other type of animal has made such extensive use of glandular structures. Thus to the many functions of epithelium have been added the development and elaborations of cells and groups of cells suitable for glandular functioning. Such glands vary all the way from simple unicellular ones to those of many thousands of cells. Along with the outpocketing of epithelial cells for the formation of glands, a large amount of accessory tissues has also been required, such as connective tissue in the partitions and framework of the larger glands. Special supplies of nerves, blood vessels, canals, and ducts are also involved. Both intercellular and extracellular secretory canals become necessary to convey secretions of glands to those surfaces where they are discharged.

References

Bartholomew, G. A., and V. A. Tucker. 1963. Control of changes in body temperature, metabolism and circulation by the agamid lizard, *Amphibolurus barbatus*. Physiol. Zool. **36**:199-218.

Bentley, P. J. A. 1966. A scientist enjoying science. Science **156**:1657-1658.

Bogert, C. M. 1959. How reptiles regulate their body temperature. Sci. Am. **200**:106-120.

Brodal, A., and R. Fänge. 1963. The biology of *Myxine*. Universitetsforlaget, Oslo, Norway.

Fowler, J. A. 1961. Anatomy and development of racial hybrids in *Rana pipiens*. J. Morphol. **109**:251-268.

Moore, J. A. 1967. Evolutionary divergence in the frog. Science **156**:541.

Norman, J. R. 1963. A history of fishes, 2nd. ed. by P. H. Greenwood. Hill & Wang, Inc., New York.

Rothschild, Baron N. M. V. 1961. A classification of living animals. John Wiley & Sons, Inc., New York.

Ruibal, R. 1959. The ecology of a brackish water population of *Rana pipiens*. Copeia, pp. 315-322.

Savage, J. M. 1960. Review of R. C. Stebbins, Reptiles and amphibians of the San Francisco Bay region. Copeia, p. 78.

Sumner, F. B. 1924. The stability of subspecific characters under changed conditions of environment. Am. Natur. **58**:481-505.

Yocum, H. B., and R. R. Huestis. 1928. Histological differences in the thyroid glands from two subspecies of *Peromyscus maniculatus*. Anat. Rec. **39**:57-62.

2 THE ANIMAL CELL

CELLS AND ORGANISMS

The concept of the cellular makeup of the bodies of plants and animals goes back to the early part of the nineteenth century. The clear statement of the "cell theory," however, is generally accepted as being the achievement of Schleiden and Schwann and is associated with the years 1838 and 1839. Schleiden, a botanist, and Schwann, a zoologist, stated that organisms are made up of units, the **cells,** and of their products. When the long history of magnifying instruments previous to this time is considered, it may seem rather surprising that the definitive statement of the idea came this late—during the first two years of the long reign of Queen Victoria!

The term "cell" itself dates far back to 1665 when the English microscopist, Robert Hooke, gave this designation to the innumerable little chambers that he observed in a thin section of cork. It was natural that the clear-cut division of tissues by means of cell walls, in plants, and cell membranes, in animals, should have impressed the early observers, and this was what led eventually, through accumulation of observations and synthesis of thought, to the cell theory.

Almost at the time, however, when the theory was being published, attention was beginning to focus on the interior of cells and on their substance. Brown in 1831 had discovered the nucleus in the cells of the orchid. He could not have guessed that in these smoothly rounded structures he saw in the "cell juice" were stored in intricate arrangement all the hereditary traits that make an orchid what it is and that transmit the characteristics from generation to generation.

Important studies were made on the protozoans. Dujardin (1835) called the substance of these animals "sarcode" and investigated its physical and chemical nature as far as he was able. The realization that the same, or a similar, substance constitutes the cells (or the contents of the cells if the earliest use of the term cell is considered) of many-celled organisms led to great interest in it. A new name for this "first substance"—**protoplasm**—now appeared in the literature. The term "protoplasm theory" was not used until 1892, when Oskar Hertwig applied it to the concept that this substance is essentially the same in one-celled organisms and in plants and animals.

Thus during the second half of the nineteenth century the cell had gradually come to be conceived of as a small mass of protoplasm surrounded by a membrane (or a membrane and a wall) and possessing a nucleus.

The remarkable behavior of the nucleus in both plant and animal cells attracted the attention of many investigators. Its division (generally into two equal parts) just prior to, or along with, the division of the whole cell was described. In addition to direct division (a simple constriction and separation), a far more complicated process was seen to be common. This indirect division involved a series of changes within the nucleus: (1) the formation of a set of bodies (called **chromosomes** because readily colored by various stains); (2) a breakdown of the nuclear membrane; (3) a mobilization and lining up of the chromosomes on a special framework, the spindle; (4) a neat longitudinal separation of each chromosome; and (5) the resulting halves travelling to opposite ends of the spindle and thus eventually becoming part of the nucleus of the new, "daughter" cells. This process was called karyokinesis because of the kinetic appearance of the karyon, or nucleus.

Another term, however, **mitosis,** derived from the thread formation seen in the nucleus in early stages of the process, has come to be used to a greater extent. The individual nuclei in various stages of the process, as viewed under the microscope, are called "mitotic figures."

Another notable activity of the nucleus was discovered in the reproductive process. It was learned that the head of the male gamete, or sperm cell, is essentially a nucleus and that when the sperm penetrates the egg, fusion of the nuclei of sperm and egg brings together the chromosomes, or hereditary material, of the parent organisms.

The earlier studies on the cell and the development of these basic concepts are well described in some excellent reviews (Conklin, 1940; Baker, 1948).

STRUCTURE OF THE CELL

For a general description of the structure of a cell, the name of the organism in which the cell is found is not necessary. This is not to deny that great differences exist among various kinds of cells and among cells derived from different species of plants and animals, but there are enough common features to permit us to speak of a "typical cell" (Fig. 2-1).

First, a cell will be described as it can be studied by means of the light microscope. With the onrush of advances in technology, there is actually some danger that earlier methods will be looked on with less regard than they deserve,

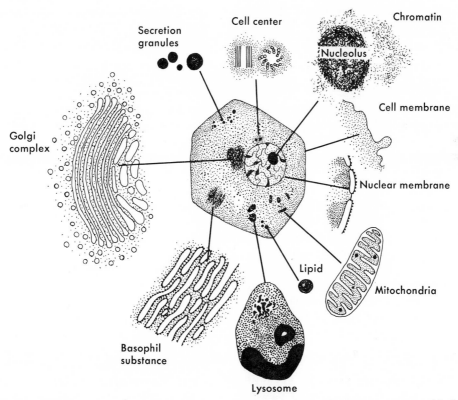

Fig. 2-1. Typical cell. The cell in the center of this figure shows the features as seen with the light microscope. At the periphery each organelle is indicated as it appears with the electron microscope. One centriole of the cell center is shown in longitudinal section, the other in cross section.

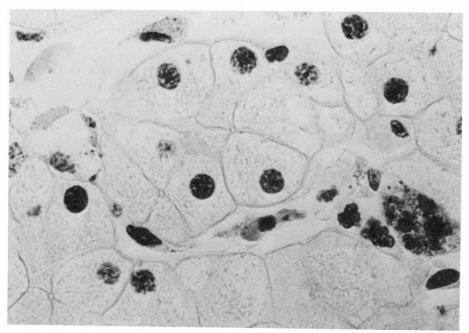

Fig. 2-2. Cells of the liver of *Ambystoma mexicanum*. The majority of cells seen here belong to the hepatic parenchyma. They are polyhedral in shape and show distinct nuclei and cell membranes. The clearer spaces are blood sinusoids in which a few lining or reticuloendothelial cells are observed. (Hematoxylin-eosin stain; ×400.)

although they have led to an imposing accumulation of knowledge and are still useful. The advent of the electron microscope has tended to place the light microscope in the category of such unappreciated tools. Yet this instrument, with its many modifications, had given scientists a great knowledge of tissues and of individual cells before the electron mircoscope was developed. With the light microscope we have been able to study living cells and to learn the many ways in which cells and tissues can be colored, or stained, due to their affinity for specific dyes. Even at a linear magnification of ×1000, the light microscope with the oil-immersion lens gives an increase in area of 1000^2, or 1 million times.

Light-microscope appearance of the cell

When stained with hematoxylin-eosin (the most common histologic, or tissue, stain) and viewed under the light microscope, the typical cell appears as a rounded or polyhedral body surrounded by a thin membrane (Fig. 2-2). It contains, generally near the center, a **nucleus.** The nucleus usually shows a somewhat detailed structure. Thin, weakly staining threads have been called linin. On the reticulum formed by these threads are granules of **chromatin,** which stain deeply with hematoxylin. The nucleus is basophilic (hematoxylin is a basic stain).

The nuclear membrane was recognized as a real structure when methods of microdissection became available, for it can be punctured or torn, allowing the karyolymph or "nuclear sap" to exude from the interior. It is in this fluid interior that the complex organization of the chromosomes must reside between divisions of the nucleus and cell. In a few cases in the animal kingdom, chromosomes remain identifiable between division, but generally they seem to have become lost in the obscurity of threads, granules, and karyolymph. However, there are good reasons for the belief that they are still present in an organized state.

In the nucleus of the typical cell there will be

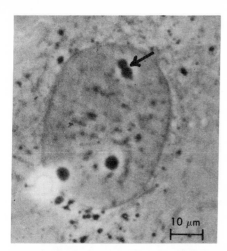

Fig. 2-3. Phase-contrast photomicrograph of nucleus of single interphase cell cultured from heart tissue of *Triturus viridescens.* Two single nucleoli are prominent, as well as two others in process of fusing (arrow). Brilliant extranuclear body in this field is fat droplet. (×800.) (From Amenta, P. S. 1961. Anat. Rec. **139:**155-165.)

also one or more **nucleoli** (Figs. 2-1 and 2-3). These are rounded, often deeply basophilic structures. At times, with the oil-immersion lens one can see a more eosinophilic central circle, with basophilic masses embracing the periphery. Nucleoli differ from chromatin granules in showing a much more specific shape and size.

The part of the cell between nucleus and cell membrane is the **cytoplasm**—or we may say that the cell is a mass of cytoplasm containing a nucleus. With hematoxylin-eosin stain, the cytoplasm generally shows a rather indefinite structure, even at high magnifications of the light microscope. In specific kinds of cells one may recognize minute crystals, clear vacuoles, or pigment material of different natural colors. In general, however, the cytoplasm is more likely to show a granular, alveolar (foamy), or fibrillar appearance.

In the majority of cells, but by no means all, the cytoplasm is stained more readily by the eosin

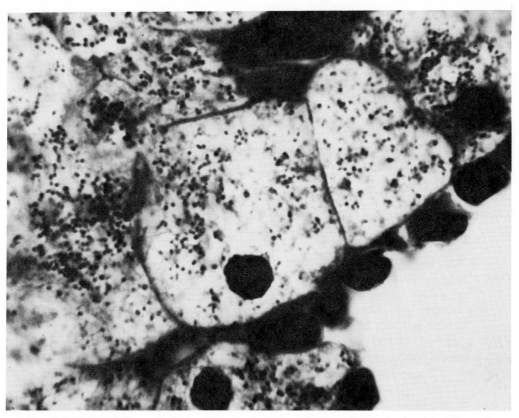

Fig. 2-4. Cells of liver of *Ambystoma mexicanum* stained to show mitochondria. (Regaud fixation, acid fuchsin stain; ×1200.)

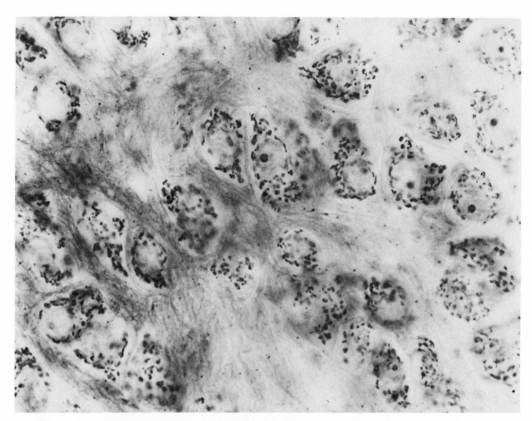

Fig. 2-5. Golgi apparatus in the cells of a ganglion of the cat. The apparatus in such preparations appears as a coarse network. (Silver impregnation; ×400.)

and is described as usually eosinophilic or acidophilic (eosin is one of the acid stains), in contrast to the basophilic nucleus. The student who wishes to go into this subject more deeply will soon find that it is a more complicated one than is indicated by this brief description. The terms here given, however, are useful and can be applied to many prepared specimens.

The structures just described are by no means all of those which were made known by the light microscope. With a small amount of added technical knowledge and effort, other features were shown in clear detail in many animal tissues.

With special methods of fixation and staining the cytoplasm shows an array of minute structures called **mitochondria** (Fig. 2-4). These vary in shape from spheroidal or ovoidal to rod-shaped and filamentous. There may be a very few or there may be many hundreds in a single cell. Time-lapse motion pictures have shown these bodies to move not only with streaming of the cytoplasm but also on their own. A strictly morphologic relation of mitochondria to other structures in the cell is seldom apparent, but they apparently play an important role in the chemical and metabolic life of the cell.

With other methods—especially by impregnation with salts of silver, osmium, or other metals —a relatively large netlike structure, the internal reticular apparatus, is seen (Fig. 2-5). Camillo Golgi who first demonstrated it, using certain large cells of the brain in the barn owl, gave it this name. Today it is more frequently called the **Golgi apparatus.** It lies usually in a specific location in the cell. In gland cells it is almost invariably between the nucleus and the apex or free end of the cell.

The presence of **secretion granules,** products of gland cells, in the vicinity of the Golgi apparatus, led to the opinion that this structure is

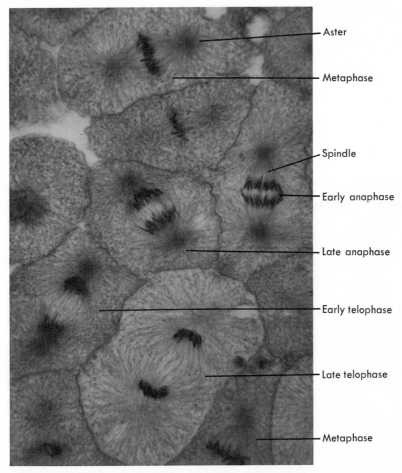

Fig. 2-6. Various stages of mitosis in the cells of the whitefish during cleavage. Asters and spindle usually are prominent in this material. (×500.) (From Hickman, C. P., C. P. Hickman, Jr., and F. M. Hickman. 1974. Integrated principles of zoology, 5th ed. The C. V. Mosby Co., St. Louis.)

involved in some way with the process of secretion. This subject and its clarification by use of the electron microscope will be discussed later.

Another structure that was clearly demonstrated with the light microscope is the central body or **centrosome.** It appears in the resting cell as a clear area containing one or two minute bodies called centrioles. Usually the centrosome lies close to the nuclear membrane. In many cells the beginning of mitosis brings the formation of a beautiful "aster," or star, at the site of the centriole, with rays passing out from it far into the cytoplasm (Fig. 2-6). Also with the light microscope it was shown that in many cells a close relationship exists between the centrosome and the

basal bodies of cilia and flagella. Thus the function of the centrosome is involved with cell division and with special motile structures of cells. The stain most frequently used to demonstrate the centrosome and its centrioles has been Heidenhain's iron hematoxylin, a favorite with the earlier cytologists and still useful.

Electron-microscope appearance of the cell

A description of the typical cell as revealed with the electron microscope will include all the structures that are made visible with the light microscope by various methods. Some of them, however, at the tremendous magnifications of this new instrument, hardly can be recognized as the same entities with which cytologists had for so long been dealing. Then, too, structures never before seen have been brought into view by the high resolving power made available through replacement of light waves by the electron beam with its far shorter wavelength. The appearances of the organs of the cell, or "organelles" as they are generally called, can be compared in Fig. 2-1, which shows them as seen with the two types of microscope.

The nucleus, probably the most highly organized of all parts of the cell, is somewhat disappointing at these high magnifications. A rather indistinct mass of fibrillar and granular material is seen, although certain areas, which were shown well by staining and study with the light microscope, are electron-dense and appear dark. They also stain well with the special stains that have been developed for use on electron microscope

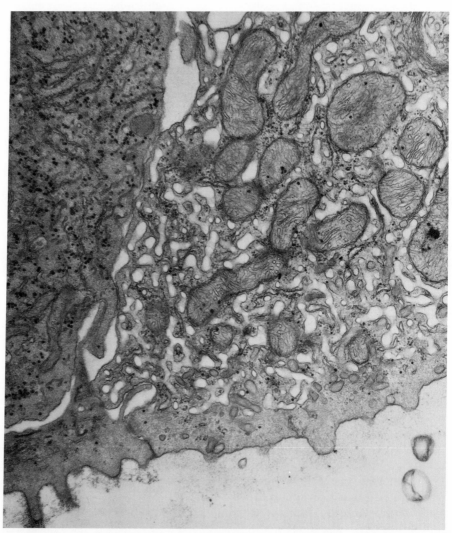

Fig. 2-7. Portion of a cell of the epithelium covering a gill filament of the rainbow trout, *Salmo gairdneri.* The conspicuous organelles here are the mitochondria. Each one shows a double outer membrane and, extending inward from it, the cristae mitochondriales that greatly increase the area of membrane surface. Clear channels of endoplasmic reticulum pervade the cytoplasm except near the apical portion of the cell. There blunt, irregular protrusions are seen. (Glutaraldehyde fixation, followed by Palade's fixative; uranyl acetate stain; ×36,000.)

specimens. The nucleoli show a considerable amount of complex structure, including multitudes of fine granules that chemically are ribonucleoprotein. The nuclear membrane is seen as a double structure, and in many, if not all, cells it has a complex system of pores in one or both of its layers (Watson, 1955, 1959). It has been suggested that small molecules and ions are exchanged through the membranes, large molecules through the pores (Watson, 1955).

Although one does not make out a great amount of detail in the interior of the nucleus even by use of the electron microscope, recent studies (Porter, 1969) are revealing interesting information about the nature of the DNA (deoxyribonucleic acid) molecule, which is the most important constituent of the chromatin or hereditary material. The molecule itself can be visualized. It is a long, slender structure with a thickness of only 20 angstroms (Å). Some idea of its smallness may be gained from the estimation that if the molecule were extended so far as to amount to a weight of 0.5 gram, it would reach from the earth to the sun (approximately 92 million miles)!

With the electron microscope the mitochondria are seen as bodies in which a constant structure

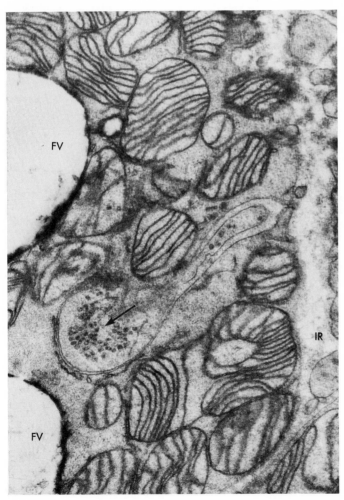

Fig. 2-8. Specialized type of mitochondrion as found in brown fat tissue. The cristae are narrow but very prominent and extend all the way across the mitochondrion. *FV,* Fat vacuole; *IR,* intercellular space; arrow, nerve ending. (From Ochi, J., M. Konishi, and H. Yoshikawa. 1969. Z. Anat. Entwicklungsgesch. **129:**259-267.)

occurs, whether they be of the spheroid, rodlike, or filamentous type. Each mitochondrion has a peripheral double membrane. From the inner portion of this membrane ridges or crests, the **cristae mitochondriales,** project into the interior (Figs. 2-7 and 2-8). The interior of the mitochondrion is occupied by the **matrix,** a material in which no organized structure may be seen or which may contain small electron-dense granules or occasionally delicate fibrils. The cristae are surrounded by matrix. The cristae usually are simple, but in some cells they branch to form complex networks. Villous, or fingerlike, cristae are not uncommon and seem to be found generally in cells that are secreting steroid products.

It might seem that the ultimate in minute structure had been reached with these details on the mitochondria, which with the highest powers of the light microscope appear still as tiny but intriguing shapes. Even finer detail, however, has been reported, for along the cristae are ranged multitudes of "elementary particles" and each of these particles in turn shows individual parts (headpiece, stalk, and base)!

What is the functional significance of the mitochondria and of their fine structure? The biochemists and cytochemists have gone far in helping to furnish answers. Mitochondria have been separated from other parts of the cell by means of high-speed differential centrifugation, and literally "liters of mitochondria" have been made available for chemical and physical studies. It has been shown that they are the chief sites of the respiratory enzymes succinoxidase and cytochrome

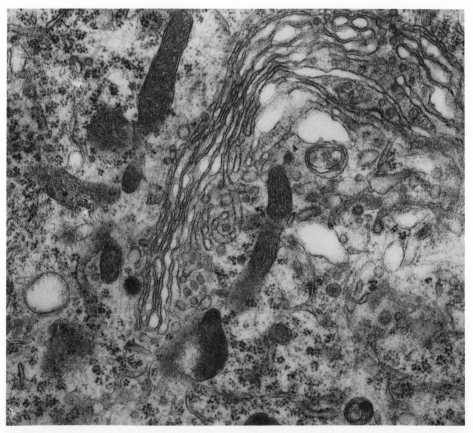

Fig. 2-9. Electron micrograph showing the ultrastructural aspects of the Golgi apparatus, as seen in a ventral horn nerve cell of mouse. It is a complex structure consisting usually of stacked lamellae, vacuoles, and vesicles. (×33,000.) (Courtesy Dr. S. S. Sekhon, Veterans Administration Hospital, Long Beach, Calif.)

oxidase in the cell. They are able, on their own, to oxidize a large number of substrates arising from the metabolism of proteins, fats, and carbohydrates in the series of chemical processes called the Krebs cycle (tricarboxylic cycle). Furthermore, the energy released by these oxidative processes is transformed within the mitochondria into high-energy bonds (adenosine triphosphate [ATP]) of much importance in the various activities of the cell. For further details of these processes recent articles and reviews on the mitochondrion (Andrew, 1956; Fernández-Morán, 1964; Green, 1964; Lehninger, 1964) are useful.

The Golgi apparatus has a characteristic appearance in electron micrographs that permits its differentiation from other cell organelles (Fig. 2-9). This appearance, however, is not similar to the reticular aspect seen with the light microscope in preparations impregnated by metallic salts or even in fresh and living cells by phase microscopy. Electron microscope study has supported the concept that the Golgi apparatus occurs in cells generally, both plant and animal, and that it is a highly organized structure. It comprises several morphologic elements: (1) cisternae, flattened sacs that appear in section as dense parallel membranes; (2) clusters of dense vesicles about 600 Å in diameter; and (3) large vacuoles, usually near the edge of the Golgi complex and probably derived from the cisternae. In various kinds of secreting cells the vacuoles may have granules or masses of material in their interior. The most constant of these components seems to be the cisternae, of which both vesicles and vacuoles probably are derivatives.

Experimental studies have to some extent confirmed an important role for the Golgi apparatus in secreting cells. It is concerned, however, not with the actual production of secretion granules, which are formed elsewhere in the cell, but rather with their "packaging," or enclosure within a membrane. Such packaging appears to be important for the eventual extrusion of the product from the cell, when its covering membrane fuses with the cell membrane at the apical surface and an opening in the common membrane thus formed permits release of the granules.

Our discussion of the Golgi apparatus has introduced us to the presence of membranes within the cytoplasm of the cell. The abundance and complex arrangement of such membranes, which in many cases are found throughout the cell body (Fig. 2-10), are outstanding features of ultrastructural organization.

At first rather vaguely seen with the electron microscope in cells prepared from tissue cultures, the system of membranes and channels (**endoplasmic reticulum**) has been described now in great detail for a variety of cells (Porter, 1957). With progress in sectioning tissues for electron microscopy, this system was seen with much greater clarity. It was found to consist of an anastomosing series of channels, dilated in some places, narrow in others, and evidently having their contents fairly well separated from the "true" ground substance of the cell, or cytoplasmic matrix, which lies outside of the membranes.

Another component of the cytoplasm, so small that it is completely invisible by any form of light microscopy, is the **ribosome.** The ribosomes are dense bodies consisting largely of ribonucleoprotein. They vary greatly in abundance in different kinds of cells and in different parts of one and the same cell. Often they are arranged in an orderly fashion, lined up along the membranes of the endoplasmic reticulum within the cytoplasmic matrix. Functionally, they are of great importance in synthesis of proteins.

There are two chief types of endoplasmic reticulum: rough endoplasmic reticulum, in which ribosomes are an important element, and smooth endoplasmic reticulum, in which they are few or lacking. The rough type, and indeed the presence of large numbers of ribosomes in general, is bound up with **basophilia** of the cytoplasm, as seen in stained preparations for light microscopy.

What can be the function of such a system of channels, often extending to almost every part of the cytoplasm? It must be confessed that our knowledge of its role at present, in spite of many excellent morphologic and biochemical studies, is an imperfect one. It has been suggested that (1) it may act as a kind of intracellular circulatory system, important in transport of materials; (2) it may serve as a segregation apparatus, collecting materials that are being synthesized and storing them temporarily before delivery to the Golgi apparatus or other parts of the cell; and

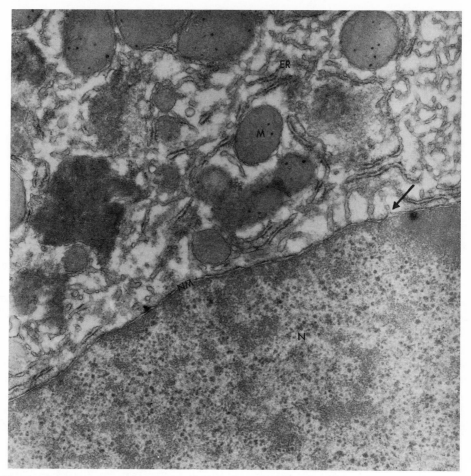

Fig. 2-10. Nucleus and cytoplasm of a hepatic parenchymal cell of *Siredon mexicanum*. The double nature of the nuclear membrane is seen. There are large amounts of smooth endoplasmic reticulum. At some places, as at arrow, continuity of nuclear membrane and channels of endoplasmic reticulum are seen. *ER,* endoplasmic reticulum; *M,* mitochondrion; *N,* nucleus; *NM,* nuclear membrane. (×28,400.)

(3) it may act as an intracellular conductor, with conic gradients across the membranes initiating differences of potential. It seems probable that some combination of these functions eventually will be able to be assigned more specifically to the endoplasmic reticulum.

With the electron microscope the centrioles are seen as cylindrical bodies with an electron-dense wall and an interior of low density. Smaller tubular structures are within the wall, oriented parallel to the axis. There are nine groups of these tubules, as will be found to be true in cilia and flagella (Chapter 3). In fact, the pairs of tubules,

or filaments, that are typical of cilia and flagella are formed in direct continuity with the corresponding structures of the centriole.

MITOSIS

Mitosis is a continuous process, but for the sake of convenience in description and discussion, a number of phases have been named. These are (1) interphase, (2) prophase, (3) metaphase, (4) anaphase, and (5) telophase.

Interphase is the phase of the nondividing or resting nucleus. The term "resting" refers only to the fact that no active phase of division or visible

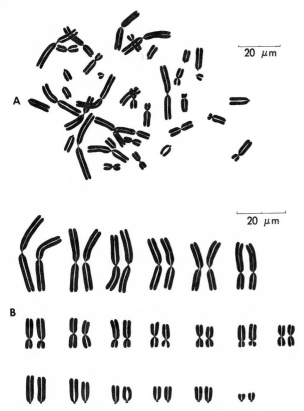

20 µm

20 µm

Fig. 2-11. A, Chromosomes of a female *Necturus;* smear made of a cell in metaphase. Camera lucida drawing. **B,** Idiogram of the chromosomes seen in **A.** Here the members of the nineteen pairs are placed together. (From Seto, T., C. M. Pomerat, and J. Kezer. 1964. Am. Natur. **98:**71-78. Copyright 1964 by the University of Chicago.)

animals have shown that each chromosome has a core, an inner thread (chromonema) that is spirally coiled, and an outer matrix, a jellylike material surrounding the core and often much thicker than it. The shortening and thickening of the spireme appears to be due to a tightening of the coils of the inner thread and an increase in the amount of matrix around it. As the thickening proceeds, the individual chromosomes can be made out. Their individuality is, indeed, one of their most remarkable features. They are distinctive, yet each does have a "mate" in one sense, for there are pairs of chromosomes in each cell, always a specific number of such pairs for any given species. They are not found in proximity to each other, but the investigator has to seek them out and put together the drawings or photographs in such a way as to bring out the paired condition (Fig. 2-11).

In the last part of prophase the nuclear membrane disappears. A new structure now makes its appearance—the achromatic spindle, a barrel-shaped body composed of slender fibrils. It lies between the two centrioles, which have migrated to positions at some distance from one another and serve now as the poles of the spindle. They may be surrounded by rays at this time (astral division) or they may lack such rays (anastral division).

The next stage of mitosis is **metaphase.** The chromosomes now line up on the equatorial plate of the spindle. Here they undergo a longitudinal division. Thus metaphase is a relatively short, but critical, stage in mitosis.

In **anaphase** the chromosomes are seen to migrate along the spindle fibers, which are thought to have a contractile activity as well as a guiding function.

The last stage of mitosis is **telophase.** The chromosomes, having reached the opposite poles of the spindle, go through a change that has been described as "prophase in reverse." There is an uncoiling of the chromonemata and an imbibition of nuclear sap. The matrix of the chromosomes seems to dissociate or at least to alter in appearance in such a manner that individual chromosomes no longer are visible. A new nuclear membrane is formed for each of the daughter nuclei. Usually division of the cell is occurring at this

preparation for it is present. However, the nucleus no doubt is very active in a metabolic sense during interphase, serving as the regulator of the activities in the cytoplasm as well as having to manage its own metabolism. As has been stated previously, some type of chromosomal organization is probably present in the nucleus even during interphase.

In **prophase** a threadlike aspect is seen in the interior of the nucleus. It is difficult to tell whether there are many threads or one thread. With the idea of one thread, the name "spireme" was given to this stage, which, in turn, is subdivided into a thin-spireme stage (early prophase) and a thick-spireme stage (late prophase). Studies by many investigators on various species of plants and time, initiated indeed by the events of nuclear division.

Mitotic division of living cells has been studied with phase microscopy, and excellent motion pictures are available. In viewing these, it is important to remember that they are speeded up. The events from beginning of prophase to end of telophase generally take, in actuality, from 1 to 2 hours.

References

Amenta, P. S. 1961. Fusion of nucleoli in cells cultured from the heart of *Triturus viridescens*. Anat. Rec. **139**:155-165.

Andrew, W. 1956. The mitochondria of the neuron. Int. Rev. Cytol. **5**:147-170.

Baker, J. R. 1948. The cell theory, a restatement, history and critique. Q. J. Microsc. Sci. **89**:103-125.

Conklin, E. G. 1940. Cell and protoplasm concepts: historical account. Pages 6-20 *in* Moulton, F. R., ed. The cell and protoplasm, Vol. 1, American Association for the Advancement of Science, Washington, D. C.

De Robertis, E., W. W. Nowinski, and F. A. Saez. 1965. Cell biology, 4th ed. W. B. Saunders Co., Philadelphia.

Fernández-Morán, H. 1964. New approaches in correlative studies of biologic ultrastructure by high-resolution electron microscopy. J. Microsc. **83**:183-195.

Green, D. E. 1964. The mitochondrion. Sci. Am. **210** (1):63-74.

Lehninger, A. L. 1964. The mitochondrion: molecular basis of structure and function. W. A. Benjamin Co., New York.

Porter, K. R. 1957. The submicroscopic morphology of protoplasm. Harvey Lectures, Series LI, pp. 175-228. Academic Press, Inc., New York.

Porter, K. R. 1969. High voltage electron microscopy. Address at Ninth Annual Meeting of American Society of Cell Biologists, Nov. 6, Detroit, Mich.

Schwann, T. 1847. Microscopical researches into the accordance in the structure and growth of animals and plants. (Transl. by H. Smith.) Sydenham Society, London.

Seto, T., C. M. Pomerat, and J. Kezer. 1964. The chromosomes of *Necturus maculosus* as revealed in cultures of leukocytes. Am. Natur. **98**(899):71-78.

Watson, M. L. 1955. The nuclear envelope. Its structure and relation to cytoplasmic membranes. J. Biophys. Biochem. Cytol. **1**:257-271.

Watson, M. L. 1959. Further observations on the nuclear envelope of the animal cell. J. Biophys. Biochem. Cytol. **6**:147-156.

3 THE TISSUES

SIGNIFICANCE OF TISSUES

Tissues are to bodily structures as trees are to forests. Important as cells are as basic units of the body, they do not, with few exceptions, occur as isolated entities in multicellular animals. They are bound together by intercellular substances to form a few fundamental units known as tissues. Aggregations of these tissues are integrated and combined to make up the functional structures, or organs, of the body. The term histology comes from the Greek words *histos* (tissue) and *logos* (science of). The actual significance of tissues and their role in the build up of the body dates from the nineteenth century. X. Bichat (about 1800) first clearly described tissues in the modern sense. It may be stated that the concept of tissue structure is a direct outcome of the development of the microscope. As microscopic studies were made of the organic body, it became apparent that the various body parts were all constructed from certain basic arrangements of a few building units by which the body was formed.

It must be remembered and understood by the student that microscopic study of tissues is as much a part of physiology as of anatomy. Histology does not merely emphasize a minute and precise examination of the details of each tissue, but it is also concerned with the particular function of that tissue. As such, histology bridges the gap between gross anatomy and physiology as revealed in the finer distinctions of living matter. This means that the rapid development in the study of molecular or finer structure of tissue composition is bringing about striking changes in many of our functional concepts.

In a sense, tissue study is based on a simplicity of design in nature, in which diversified patterns are formed by assembling over and over in varied and logical ways a few basic building materials. This type of organization aids the student in comprehending the structures and functions of integrated patterns as represented in organ systems. The greatest problem confronting the student of biologic phenomena is the relative roles of structure and function. An understanding of this relation can come only by histologic study, or at the microscopic level. The unity of all knowledge, both structural and functional, about the life process is strikingly shown by histology, since the working and integration of the subject matter of life cannot be understood without a comprehension of the details revealed by a microscopic study. Histology is therefore a broad and not a narrow subject; it serves as a key of understanding for all the disciplines concerned with a study of the organism. Its major aim is to give a perspective of the whole and not of the part.

CLASSIFICATION OF TISSUES

Four fundamental tissues are found in body construction. These are epithelial, connective muscular, and nervous tissues. Sometimes a fifth one (vascular or blood) may be considered, but it is actually a specialized portion of connective tissue. The four or five basic tissues are arbitrary divisions because no tissue occurs in a pure form. Other tissues are always involved in some way. The tissues form an integration that is manifest in all aspects of the life process. No tissue can perform its work wholly by itself, but each requires the coordination of others as well.

Epithelial tissue is specialized for protection, but it has other functions such as secretion and absorption. It is arranged in protective sheets of cells, covering surfaces and lining cavities both inside and outside the body. It develops from all three germ layers, but especially from ectoderm and endoderm. **Connective tissue** has for its primary function the general occupation of connecting, padding, binding, and supporting; it also includes defense mechanisms, blood and blood-forming tissues, and storage (fat). It contains many types of cells, which are separated by large amounts of nonliving intercellular substance laid down by certain cells, such as cartilage, bone, tendons, and ligaments. Connective tissue is thus adapted for providing strength and support for the body. It develops almost entirely from the mesodermal germ layer. **Muscular tissue** is highly specialized for contractility, and its elongated fibers (cells) are adapted for contraction in their long axis. Within them are the specialized myofibrils, which, in turn, contain the still smaller myofilaments involved in the actual contractile process. Muscular tissue is also derived from mesoderm. **Nervous tissue** is adapted for irritability and con-

ductivity. It is made up of cells with long processes that are specialized for reception of stimuli and conduction of impulses, together with other related elements. This tissue comes from ectoderm. **Blood,** often considered a fifth tissue, consists of a cell-containing fluid circulating within specialized vascular channels. All these tissues fulfill in an integrated way the basic functions required for the maintenance of the life process. Although they are studied separately, it must be understood that each is interdependent on the others.

Within each organ, a certain tissue may be responsible for the major function of that organ; it is called the **parenchyma.** However, other tissues play a part in the general coordination and integration of the major tissue in its functioning; they are called the **stroma.** All organs must contain these two constituents in order to function.

BASIC ORIGIN OF TISSUES

Tissues really originated as specialization of the basic functions of protoplasm, or the substance that represents life as we commonly think of it. All protoplasm and its intercellular substance, no matter how primitive or unspecialized, have certain vital properties that are intrinsic in

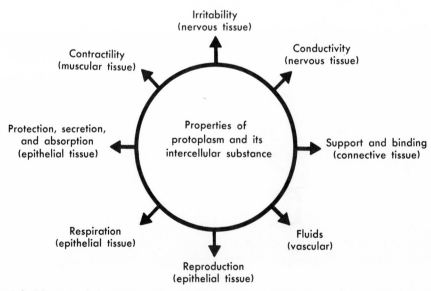

Fig. 3-1. Diagram of the basic origin of body tissues. All tissues have originated by specialization of properties found in all primitive protoplasms.

their nature. These properties have been built into the structure of protoplasm during its long formative period of many eons ago. Such properties of protoplasm are irritability, conductivity, absorption, contractility, assimilation, excretion, secretion, respiration, growth, reproduction, and support. These properties may have evolved one by one as life emerged. To perform these varied functions efficiently, specialized tissues have gradually evolved. Nervous tissue, for example, by division of labor and specialization has become highly efficient in taking care of the demands of irritability (from environmental stimuli both within and without the body) and the demands of conductivity (of impulses in coordination and integration). In a similar manner, the other tissues have been variously specialized to perform the other functions of life in the evolutionary process.

Fig. 3-1 indicates the final specialization of the properties shown in the diagram. The student will see an overlapping of functions in the scheme, for as previously pointed out, no tissue is entirely independent in its functioning but there must be integration of other tissues in the major functioning of each one. Many influences (e.g., endocrine or metabolic) are involved in the interrelationships of the tissues as they are woven together in organized patterns. Inherent specificities are nebulous in the varied patterns of animal life. When a tissue is moved experimentally from one potential growth region to another, it may assume an entirely different role in its new surroundings.

COMPOSITION OF TISSUES

The body is composed of cells, intercellular substance, and body fluids. Cells are the most common elements of most tissues. They are the building blocks, but they also produce and regulate the intercellular components.

The intercellular constituents are made up principally of fibers and amorphous ground substance. Fibers consist of protein macromolecules in long chains. Three kinds of fibers occur, and all of them are formed by connective tissue cells (Figs. 3-4, 5-2, and 5-3). They lie between cells and groups of cells. The amorphous ground substance is not easy to demonstrate and is composed of protein-polysaccharide complexes. They com-

prise the matrix of cartilage, bone, and a great deal of connective tissue. Ground substance ranges from semifluid, sol conditions to gel states. Intercellular substances are responsible for the form assumed by bodies. Some authorities regard the body as an edifice of intercellular substance in which are found billions of cells, many of which live for only a short time. As residents, these cells exchange their products with one another, die, and are replaced by a constant renewal of cells.

Fluids exist in a considerable variety. They occupy two hypothetical compartments. Intercellular fluid is found within the protoplasm of the cell, and extracellular fluid is contained in the vascular system, lymphatic spaces and vessels, and interstitial spaces between the vessels and cells. This interstitial fluid compartment is also subdivided into tissue spaces and other spaces in joints, body cavities, brain, and elsewhere. All body fluids originate from food and water in the digestive gut; there they are absorbed through the epithelial lining. But the body fluid components are dynamic, not static, and move in and out of the gut. The amount of fluids, especially water, varies with the fluctuating conditions of the environment and metabolic changes. The amount of fluid, however, remains very constant in the normal body. This is especially the case with blood plasma. Of the fluids within the body, plasma makes up about 5% of the average weight of an adult human, tissue fluid 15%, and intracellular fluid 45% to 50%.

The human body is composed of about 60% to 70% water, 15% protein, 10% to 15% lipids or fat, 5% inorganic material (ash), and 1% carbohydrates. Proteins represent one of the most important constituents of all living tissues. They also play important roles in blood plasma and other body fluids. Because proteins furnish enzymes that are responsible for the reactions of all life processes, the chief realm of energetics is this class of substances. Proteins are constantly changing in cells, and their building blocks of some amino acids are responsible for building the enormous macromolecules necessary for the life process. Protein must of necessity be considered the basis for the fixation and staining of the slides students employ for the study of tissues.

Both acidic and basic stains will color proteins because proteins are amphoteric, acting either as acids or bases. If body fluids are acid, proteins will act as bases; or the reverse. Their amino acids have side chains, some of which dissociate as acids and others as bases. Extra acid tends to suppress dissociation of acid side chains. The basic side chains dissociate to neutralize the excess acid; the opposite occurs if there is an excess of bases.

Other important macromolecules of importance in studying tissues are the nucleic acids, lipids, glycoproteins, and mucopolysaccharides. These and other molecules are encountered in histologic material and have specific influences on staining.

EMBRYONIC ORIGIN OF TISSUES

In an early transient stage of embryonic development there is a period when the embryo contains only three primitive germ layers: ectoderm, endoderm, and mesoderm. The germinal disk, or gastrula, first separates into an upper layer of large, polygonal cells (ectoderm) and a lower layer of flattened cells (endoderm). Later, invaginated cells migrate laterally between the ectoderm and the endoderm to form the third germ layer, mesoderm. In time, the mesoderm splits into two layers, the somatic layer closely applied to the ectoderm and the splanchnic layer closely applied to the endoderm. The cavity between the two layers thus formed is the body cavity, or coelom. Subsequently, this cavity is subdivided into three divisions: the peritoneal, pleural, and pericardial cavities. Such a description applies to mammals; other vertebrates deviate from this in certain ways.

From these three primary or primitive germ layers the four fundamental tissues mentioned before arise. The embryonic ectoderm gives rise to the outer portions of the skin (epidermis) and its derivatives (hair, glands, etc.) and also to the epithelium lining the body orifices. In addition, the ectoderm forms the central and peripheral nervous systems, the retina, the adrenal medulla, and a few other structures. From the endoderm comes the epithelium lining the alimentary canal, respiratory tract, and bladder, the parenchyma and ducts of digestive glands, and the epithelium of the thyroid, parathyroid,

and other glands. The mesoderm gives rise to connective tissue (of fibrous or homogeneous matrix), muscle, blood cells, endothelium of the vascular and lymphatic systems, mesothelium of cavities, and the epithelium of the kidneys, adrenal cortex, and reproductive gonads. Abnormal conditions may alter these developments from the basic germ layers, so that cells and tissues have a fluid potentiality in the early stages of their development.

Histogenesis is the development of tissues. With some tissues this process seems rather simple; with others, very complex. Take epithelium, for example. The early arrangement of cells in the embryonic germ layers is very much like that of epithelium. In its early stages, the whole embryo is epithelial in structure, and hence epithelium may be called the most primitive tissue. Some germ layers may actually keep their epithelial nature throughout life. Ectoderm and endoderm are germ layers that tend to develop directly into characteristic forms of epithelium. On the other hand, mesodermal structures tend to undergo radical transformation in the formation of tissues.

CHARACTERISTICS OF TISSUES

Each of the four basic tissues (epithelial, connective, muscular, and nervous) has been briefly described before, but a more detailed account will be given here. Each tissue has certain characteristics that stamp it more or less with an individuality. These characteristics, in general, are not difficult to identify; however, each tissue has subtypes, which complicates their study somewhat. The ability to recognize the type of tissue in histologic study cannot be overemphasized. It is of equal importance to understand their nature and the role each plays in organic structure.

The combination of tissues in an organized fashion results in an organ (e.g., the heart) with a specific function. When several organs are coordinated and integrated to perform a general body function (e.g., circulation or respiration), an organ system is formed. Although the terms histology and microscopic anatomy are often used as synonyms, there is a difference between the two. Histology is a general description of tissues, their nature, relations, and functions. Microscopic

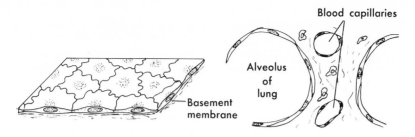

SIMPLE SQUAMOUS EPITHELIUM

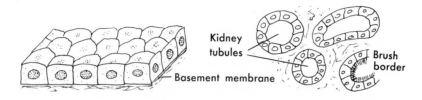

SIMPLE CUBOID EPITHELIUM

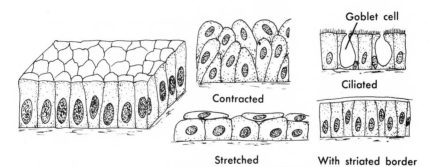

SIMPLE COLUMNAR EPITHELIUM

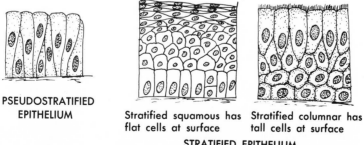

STRATIFIED EPITHELIUM

Fig. 3-2. Types of epithelium. The two basic types are simple and stratified, according to the number of layers. Subtypes are named according to the shape of their component cells. The basement membrane is a sheet of noncellular material supporting and binding down the epithelium. Evidence indicates that the basement membrane is synthesized partly by epithelium and partly by connective tissue.

anatomy shows the details of tissues and the arrangement of tissues to form organs, etc.

Epithelium. The derivation of epithelium is from the Greek *epi* (upon) and *thele* (nipple), referring to the skin covering the nipple of the female breast. The three general types are **simple** (one layer of cells); **pseudostratified** (actually one layer, but not all cells reach the surface, so it appears to have more than one layer); and **stratified** (more than one layer cells) (Fig. 3-2). There are subtypes of these types (e.g., squamous, cuboidal, and columnar). The cells are closely packed, usually in layers, and cellular membranes are formed at the outer surface of cells. Blood vessels are absent (with few exceptions), and there is usually no intercellular matter. Epithelium

is specially fitted for protection, for covering and lining membranes, and for secretory glands (Fig. 3-3). Many epithelial tissues rest on **basement membranes,** which serve to attach the tissue to underlying connective tissue. The basement membrane is composed of nonliving tissue and is synthesized by both the epithelium and connective tissue. Since epithelial tissue has no capillaries, nourishment is furnished by the tissue fluid from the underlying capillaries of connective tissue.

Connective tissue. Connective tissue is treated in greater detail in Chapter 5. It is derived chiefly from mesoderm in the form of mesenchymal cells, is internally located, and is separated from external and internal surfaces by epithelial tissue.

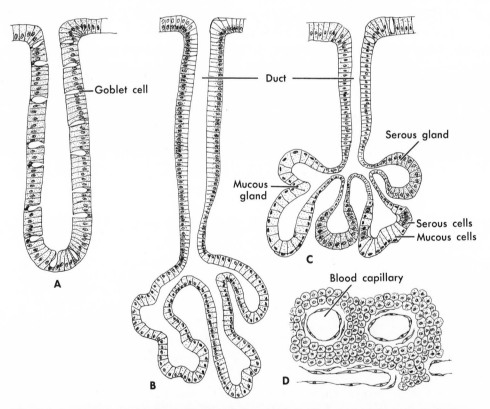

Fig. 3-3. Some types of glands. **A,** Simple tubular gland. **B,** Branched tubular gland. **C,** Compound tubuloacinous gland, combining tubular with two kinds of saccular or acinus type glands. **D,** Endocrine (ductless) gland, which empties its secretion directly into blood capillaries. Glands may be classified functionally or anatomically. In functional classification glands may be exocrine (having a duct for the secretion, **A, B, C**) or endocrine (ductless, **D**). The anatomic classification divides glands into two types: unicellular (goblet or other cell types) and multicellular (simple tubular, simple saccular, or compound).

The other three tissues—epithelial, muscular, and nervous—are mostly cellular, but connective tissue has nonliving intercellular tissue between its cells (Figs. 3-4 and 3-5). The two kinds of intercellular substance are the amorphous ground substance of soft to firm jellies and the fibrous substance of elastic and inelastic fibers. All kinds of connective tissue are made up of three components: cells, fibers, and amorphous ground substance. By variations in the amounts of these three components according to their functions,

the different types of connective tissue are formed. These are discussed in detail in Chapter 5.

Muscular tissue. Muscular tissue also originates from mesoderm and is enclosed by connective tissue. Muscular tissue is concerned with the outward manifestations of animal life and is involved in the various movements required by the living process. There are three classes of muscle: smooth or visceral, skeletal or somatic, and cardiac (Fig. 3-6). The parenchymal elements are elongated cells (the muscle fibers) that occur

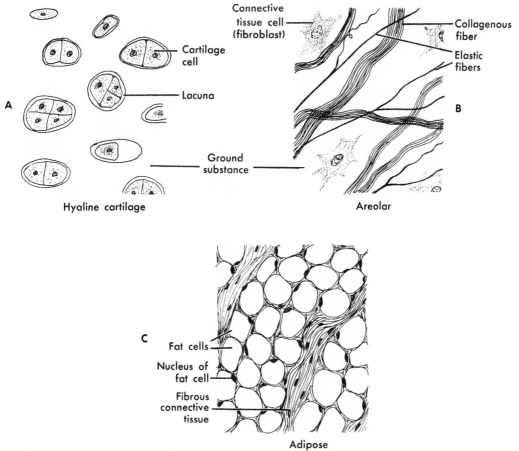

Fig. 3-4. Three types of connective tissue. **A,** Hyaline cartilage, a dense connective tissue adapted for withstanding weight because of its firm and flexible ground substance and collagenic fibers. **B,** Areolar connective tissue, a loose connective tissue containing a combination of cells, fibers, and ground substance that makes it a prototype of connective tissue in general. **C,** Adipose tissue. It emphasizes the cellular component, in which it stores fat, whereas its other components are inconspicuous. Note that all three types, especially **A** and **C,** are modifications of the same mother tissue, caused by stressing one or the other of the basic components of all connective tissues. (From Hickman, C. P., and F. M. Hickman. 1968. Laboratory studies in integrated zoology. The C. V. Mosby Co., St. Louis.)

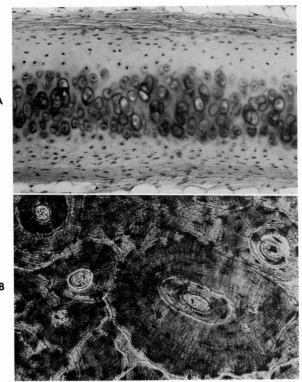

Fig. 3-5. Two types of dense supporting tissue. **A,** Cartilage of cat. **B,** Section of compact bone of mammal. Preparation ground thin, showing haversian canals. (**A,** ×200.)

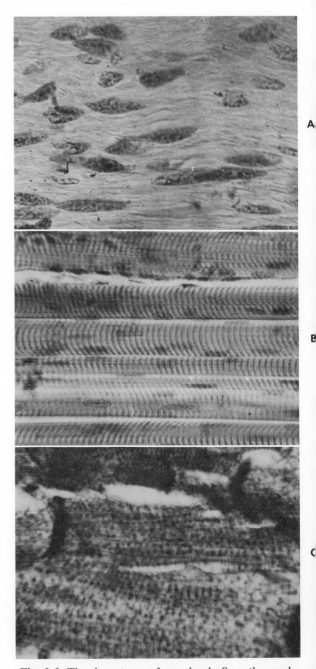

Fig. 3-6. The three types of muscle. **A,** Smooth muscle. **B,** Striated or skeletal muscle. **C,** Cardiac muscle.

in bundles or layers enclosed by connective tissue (the stroma). All types of muscles have myofibrils as the chief contratile elements (Fig. 3-7). The electron microscope reveals still smaller elements in the myofibrils, the myofilaments. The netlike myoepithelial cells around certain glands may be considered a fourth type of muscle. This tissue is fully described in Chapter 6.

Nervous tissue. Nervous tissue (Fig. 3-8) has exploited the basic properties of irritability and conduction. The vital unit of the tissue is the nerve cell, or neuron (Fig. 3-9); the functional unit is the reflex arc of varying numbers of cooperating neurons. Nervous tissue developing from a portion of the ectoderm gives rise to the parenchyma, and connective tissue forms the stroma. Associated with the neurons are the smaller, but more numerous, neuroglia cells, which are packed in among the nerve cells and fibers, and the neurolemma cells, which make up the cellular sheath

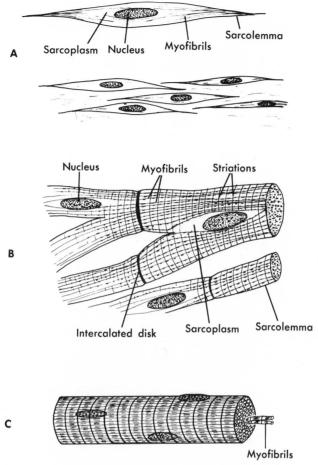

Fig. 3-7. Diagrams of three types of muscle fibers showing details of structure. **A,** Smooth muscle, showing a single fiber and several fibers as they would be arranged in muscle tissue. **B,** Cardiac muscle. Cardiac muscle is a branching network but is not a syncytium, for the intercalated disks are the sites of end-to-end attachments of individual muscle cells. Both smooth and cardiac muscles are involuntary. **C,** Skeletal muscle fiber, showing myofibrils. Skeletal or striated muscle is voluntary.

covering peripheral nerve fibers (Fig. 3-9). All these cells have a common embryonic origin from the ectoderm of the neural plate, although one type of neuroglia cells comes from connective tissue (mesoderm). The neurons in some form extend into nearly all tissues of the body.

STAINING IN RELATION TO TISSUE STUDY

It is not the purpose of this text to emphasize histologic techniques. Such methods are legion and are dealt with in many excellent manuals. However, a knowledge of certain basic principles of staining may prove useful to the student in his examination and study of the tissues he commonly finds on his slides.

The dyes that are used in microscopic technique may be general ones, staining the nucleus or the cytoplasm, but many are specific for certain components of the tissues. Most stains in general use are considered as either acids or bases, but they are actually neutral salts having both acidic and basic radicals. Basic dyes have the basic radical of the neutral salt as their coloring property. Structures that stain with it are termed basophilic. When the staining property is the

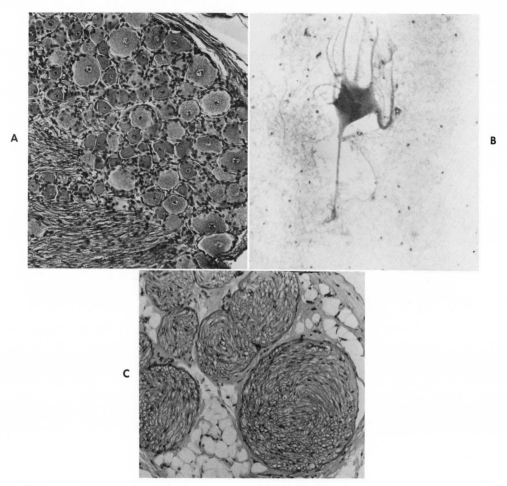

Fig. 3-8. Three examples of nervous tissue. **A,** Multipolar neuron from the anterior horn of the spinal cord. **B,** Section through a spinal ganglion showing nerve cell bodies and fibers. **C,** Cross section of a cluster of nerve fascicles.

acidic radical of the neutral salt, the dye is acidic, and the structures it stains are called acidophilic. Usually the basophilic substances, which attract basic dyes, are acids such as nucleic acids (DNA and RNA).

The student will find that many of the slides he will study are labelled H & E (hematoxylin and eosin). The staining property of hematoxylin depends on the presence in solution of its oxidative product, hematein. This is why a freshly prepared hematoxylin solution must be allowed to ripen (occurrence of oxidation) before it is ready for use. Hematoxylin acts very much like a basic dye and stains nuclei dark blue or black,

but some of its staining qualities cannot as yet be explained. Eosin, on the other hand, is an acidic dye and gives a pink-to-red color to whatever it stains, usually the general cytoplasm. Most histologic slides are stained with a basic and an acidic stain. Some constituents of a tissue will take on a color different than that of the dye employed (metachromasia).

The kind of fixation also plays a part in staining. After Zenker-formol fixation, smooth muscle stains lavender instead of blue-to-purple with hematoxylin-eosin.

Much attention is given at present to histochemistry, which refers to the use of specific stains for structural analysis of specific tissue constituents. It is really a form of chemical analysis based on the deposition of stains in tissues as a result of chemical and physical properties. Such biochemical analyses have been responsible for the localization of many substances that ordinary stains cannot demonstrate. Many enzymes have been detected in tissues by histochemical methods.

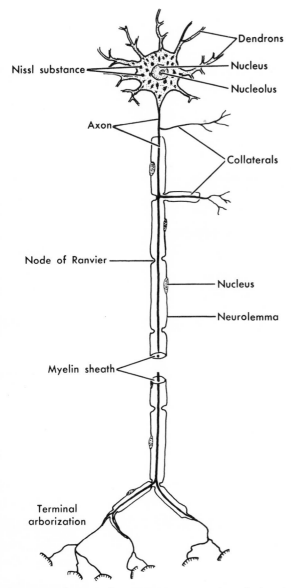

Fig. 3-9. Diagram of a motor neuron, such as is found in the anterior horn of the spinal cord.

SPECIFICITY OF THE SAME TISSUE DIFFERENTIATION IN DIFFERENT ANIMALS

In a general way it is known to many histologists that the same tissue may show detectable differences in widely different vertebrate animals. Amphibian liver cells, for example, usually show different characteristics from liver cells of other vertebrates. But little progress has been made along this line. Many years ago, H. B. Yocom and R. R. Huestis (1928) showed histologic differences in the thyroid gland of two closely related subspecies of *Peromyscus maniculatus,* white-footed mice. Investigations along this line have been extremely scanty, but much is to be expected in the future as more precise information comes to hand through histochemical methods, development of specific stains, genetic code interpretations, and electron microscopy.

The question is naturally posed, are there specific differences between the same tissue of one animal and that of another? When once the attention of histologists is focused on this problem, more definite results can be expected. Comparative anatomy has furnished many examples of gross structural differences among vertebrates, so it is logical that details of differences may exist down to the microscopic level.

The broader implication of the determination of species or subspecies classification may well hinge on this future line of investigation. Behavior criteria have been used in the classification of certain groups, but so far histologic differences of similar tissues have not been so used. The discovery of the Barr bodies by M. L. Barr and E. G. Bertram in 1949 indicated a difference between male and female cells. Such illustrations as those we have cited may be only mere examples, but they may point the way to future trends.

References

Baker, B. L., and G. D. Abrams. 1955. The physiology of connective tissue. Annu. Rev. Physiol. **17**:61-78.

Barr, M. L., and E. G. Bertram. 1949. A morphological distinction between neurons of the male and female, and the behavior of the nucleolar satellite during accelerated nucleoprotein synthesis. Nature **163**:676.

Barr, M. L., L. F. Bertram, and H. A. Lindsay. 1950. Morphology of nerve cell nucleus, according to sex. Anat. Rec. **107**:283-297.

Bell, L. G. E. 1952. The application of freezing and drying techniques in cytology. Int. Rev. Cytol. **1**:35-63.

Bodian, D. 1952. Introductory survey of neurons. Symp. Quant. Biol. **17**:1-13.

Bodian, D. 1962. The generalized vertebrate neuron. Science **137**:323-326.

Calvin, M. 1968. New keys to life processes. Perspect. Biol. Med. **11**:555-564.

Fawcett, D. W., and K. R. Porter. 1954. A study of the fine structure of ciliated epithelia. J. Morphol. **94**: 221-281.

Freeman, J. A. 1962. Fine structure of the goblet cell mucous secretory process. Anat. Rec. **144**:341-357.

Freeman, J. A. 1964. Cellular fine structure. McGraw-Hill Book Co., New York.

Gersh, I., and H. R. Catchpole. 1949. The organization of the ground substance and basement membrane and its significance in tissue injury, disease, and growth. Am. J. Anat. **85**:457-521.

Goss, C. M. 1944. The attachment of skeletal muscle fibers. Am. J. Anat. **74**:259-289.

Huxley, H. E. 1958. The contraction of muscle. Sci. Am. **199**:66-96 (Nov.).

Huxley, H. E. 1965. The mechanism of muscular contraction. Sci. Am. **213**:18-27 (Dec.).

Mark, J. S. T. 1956. An electron microscope study of uterine smooth muscle. Anat. Rec. **125**:473-493.

Oppenheimer, J. M. 1940. The non-specificity of the germ-layers. Q. Rev. Biol. **15**:1-27.

Porter, K. R. 1954. The myo-tendon junction in larval forms of *Amblystoma punctatum*. Anat. Rec. **118**: 342-356.

Porter, K. R., and M. A. Bonneville. 1963. An introduction to the fine structure of cells and tissues. Lea & Febiger, Philadelphia.

Singer, M. 1951. Factors which control the staining of tissue sections with acid and basic dyes. Int. Rev. Cytol. **1**:211-255.

Yocom, H. B., and R. R. Huestis. 1928. Histological differences in the thyroid glands from two subspecies of *Peromyscus maniculatus*. Anat. Rec. **39**:57-62.

4 INTEGUMENTS

CYCLOSTOMATA

The integument, or skin, of even the most primitive vertebrates, the cyclostomes, presents two layers: an outer epithelial one, the epidermis, and an inner connective tissue, the dermis (Fig. 4-1). Beneath the dermis is a subcutaneous layer, the hypodermis, which contains varying amounts of fatty tissue. The cyclostome epidermis is different from that found in man and other higher vertebrates. First, there is no superficial layer of dead and altered cells to form a **stratum corneum.** The epidermal cells actually correspond to the layer of living cells, the **stratum germinativum,** of our own epidermal covering. Second, the epidermis of the cyclostomes contains a number of different kinds of cells. The surface of the skin is smooth. There are no scales.

The epidermis of the lamprey is shown in Fig. 4-2. The basal layer consists of elongated cells with pointed apices. Above this layer follow

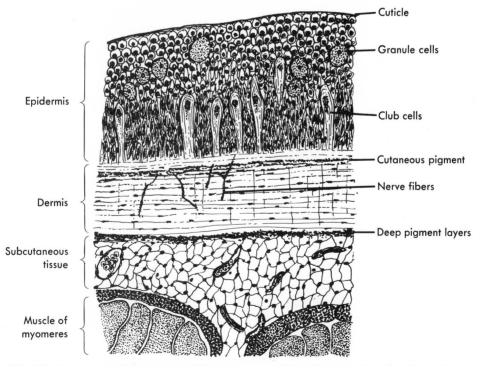

Fig. 4-1. Integument of the lamprey. Note numerous club cells and lack of keratinization in epidermis. (×400.) (Redrawn from Krause, 1923.)

35

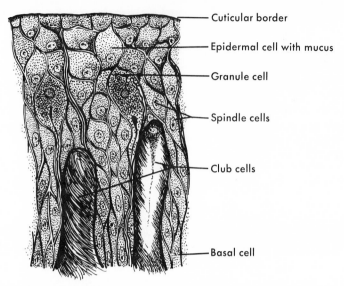

Cuticular border

Epidermal cell with mucus

Granule cell

Spindle cells

Club cells

Basal cell

Fig. 4-2. Epidermis of the lamprey. Two large club cells are prominent. Epidermal cell with mucus probably corresponds to small mucous cells of hagfish, and granule cells correspond to large mucous cells. (×1000.) (Redrawn from Krause, 1923.)

several strata of polyhedral cells. Above these, in turn, the cells are flattened. Mitotic figures can be seen in almost all the lower layers. The most superficial layer of cells bears a thin, noncellular cuticle, reminiscent of that of the invertebrates, which shows vertical striations. The regular epidermal cells have numerous fibrils, especially in the upper portions of the cell bodies. These are thought to have a supporting function and are called **tonofibrils.** Toward the surface, an accumulation of mucus is usually seen in the lower parts of many epidermal cells.

Scattered among the regular epidermal cells are special types: mucous cells, club cells, and granule cells. Mucous cells are always present. Club cells and granule cells are generally present but vary in quantity in different regions of the body. Thus there are few in the mouth area and none along the edge of the dorsal fin.

The upper half of the epidermis is occupied by small mucous cells. These secreting cells, like epidermal cells in general, are pushed gradually toward the surface by the growth and multiplication of cells in the lower levels and are finally lost from the body. Such cells often have basally located, crescent-shaped nuclei, as do many types of mucous cells, such as in the salivary glands of higher animals.

The club cells attain a length, or height, of up to 100 micrometers (μm). Their somewhat broadened base rests on the dermis, terminating above with a rounded clublike enlargement. The interior of these cells stains only lightly. Two or more nuclei can usually be seen in the apical cytoplasm. A thick layer of fibers seems to occupy the periphery of each cell (Fig. 4-3). In the hagfish epidermis, large "thread cells" are readily apparent. Each one usually shows a conspicuous mass of spiral threads located peripherally, dense granular cytoplasm, and a somewhat flattened nucleus. Although homology has not been definitely established, these thread cells probably correspond to the club cells of the lamprey skin.

The function of the club cells is obscure, but there is some evidence that like the cells of Leydig in the amphibians, they furnish a material that helps fill the intercellular spaces and protect the cells from invasion by bacteria or viruses.

The granule cells are always found in the middle or upper layers of the epidermis. They are rounded or pear-shaped, with a diameter of 20 to 50 μm. Their cytoplasm contains coarse acidophilic granules. The nucleus is small, deeply basophilic, and often obscured by the abundant granules. The peripheral portions of these cells also

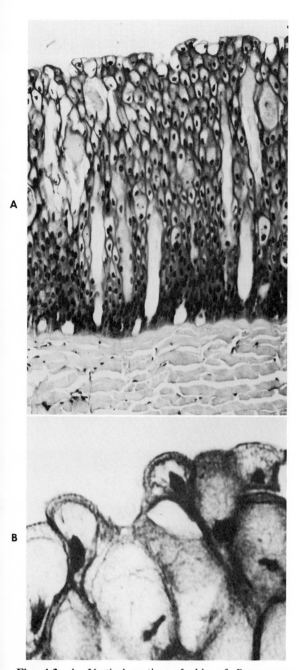

Fig. 4-3. A, Vertical section of skin of *Petromyzon*. Stratified epithelium shows many layers and includes the elongated club cells filled with secretion. Underlying dermis consists of thick collagenous fibers. **B,** Superficial layers of the epidermis of *Petromyzon*. The outermost cells show a well-defined cuticle with vertical striations. (**A,** ×100; **B,** ×400.)

show many tonofibrils. At places, the cytoplasm with its numerous fibrils is drawn out into processes that pass for long distances through the spaces between the other cells.

Special sensory cells also occur in the epidermis of the hagfish. They are found to some extent over the entire body, but are especially numerous on the tentacles. Groups of these cells bear a strong resemblance to taste buds. Fig. 4-4 shows the distribution of photoreceptor structures in the hagfish, as determined by the functional studies of Newth and Ross.

The dermis, like the epidermis, varies in thickness in different body regions. In some places it is about as thick as the epidermis; in others it is far thicker. The predominant elements are bundles of collagenous fibers, 5 to 10 μm in thickness. The great majority of them are arranged in a circular manner, around the body of the lamprey. Some bundles, however, also course longitudinally or obliquely.

Another type of fiber, morphologically resembling the elastic fibers of higher forms, is found in two thin layers, one at the inner (dermohypodermal) boundary and one at the outer (dermoepidermal) region, with a broad-meshed interconnecting network. Although these fibers are difficult to stain with stains that are specific for elastic fibers in other vertebrates, it seems probable that the two are homologous.

The dermis contains many stellate fibroblasts, as well as ameboid wandering cells. Numerous nerves are also found in this layer, from which individual fibers penetrate the epidermis. In addition, networks of blood vessels are seen, although no vessels penetrate the epidermis. Branched pigment cells occur in places in the dermis. In fact, two or three layers of pigment cells form a boundary between the dermis and hypodermis. The bodies and processes of these cells are filled with brown or black melanin.

Under the dermis is the hypodermis, which consists chiefly of bundles of collagenous connective tissue separated by large quantities of adipose tissue. This subcutaneous tissue thus serves as an immense depot for fat storage. The bundles of collagen pass without a sharp boundary line into the myosepta between the muscle segments, or myotomes.

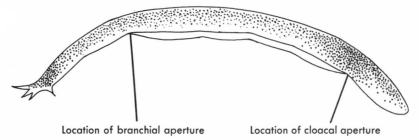

Location of branchial aperture   Location of cloacal aperture

Fig. 4-4. Distribution of dermal photoreceptors in *Myxine,* according to results of functional tests on sensitivity to light. (Redrawn from Ross, 1963; after Newth and Ross, 1955.)

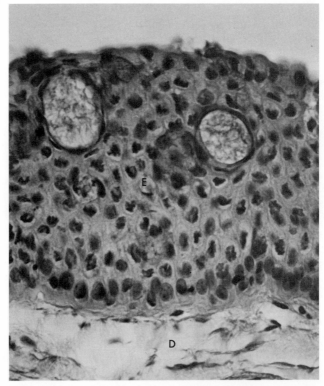

Fig. 4-5. Skin from clasper, a secondary sexual organ, of male elasmobranch *Raja clavata.* Intraepithelial glandular structures, each apparently composed of a number of cells, appear in epidermis of this region. *E,* Epidermis; *D,* dermis. (×250.)

CHONDRICHTHYES

The skin of the shark is tough. It is covered with several layers of epidermis. The surface is rough to the touch when stroked forward, due to the presence of placoid scales (dermal denticles). Many authorities believe that, phylogenetically, the earliest fishes had a continuous covering that later broke up into scales from which the denticles were derived. The epidermis is a thick, compact structure with a well-marked stratum germinativum resting on a basement membrane (Fig. 4-5). The striated cuticle is found on the superficial cells of elasmobranch epidermis only during embryonic development. Beneath the epidermis is a

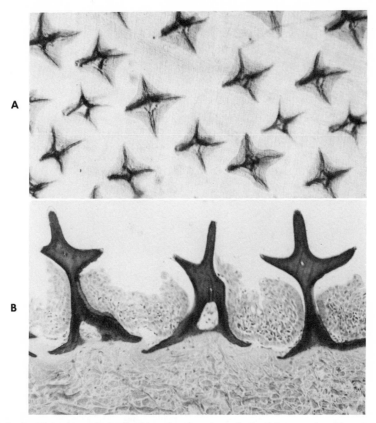

Fig. 4-6. A, Flat view of shark skin, showing dermal denticles or placoid scales. **B,** Placoid scales of shark. Side view showing spines upright and dental bases with dental tubercles. (**A,** ×50; **B,** ×120.)

thick dermis, or **corium,** of connective tissue with its fibers arranged similarly to the warp of a carpet, giving it strength and flexibility. The upper layer of connective tissue is somewhat looser than the deeper portions.

The denticles are different in size and form among the various shark genera. Some are spine-like, and others are flat or rounded like a knee. Indeed, they may differ in size in the same animal. In rays and skates, they are scattered unevenly over the body, but in most sharks the denticles are evenly distributed over the entire body. They do not increase in size as the shark grows. Rather, new scales are added between the existing ones. At any one time, some scales are newly erupted and others are disintegrating.

Each denticle has a broad base embedded in the skin and a superficial spine that projects back-

ward (Fig. 4-6). Structurally, each dermal denticle consists of a core of dentin, which surrounds a pulp cavity containing blood vessels and nerves, and is overlaid with a hard substance, durodentin or vitrodentin. Durodentin was formerly thought to be the same as tooth enamel but is now known to be a particularly hard kind of true dentin, and thus of dermal origin. There is no enamel on the denticles (Kerr, 1955). The calcareous dentin is secreted by a layer of odontoblasts around the outer edge of the pulp cavity.

An important difference between the scales of elasmobranchs and teleosts is that elasmobranch scales pierce the epidermis, whereas teleost scales do not. The reader will recall the general roughness of the skin of the shark as contrasted with the smooth skin of the majority of the bony fishes.

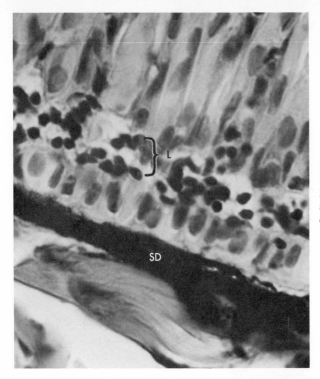

Fig. 4-7. Skin from dorsal surface of *Conger conger,* the conger eel, a teleost fish. Just above basal cells is layer where leukocytes, chiefly lymphocytes *(L),* are aggregated. Superficial layer of the dermis *(SD)* is deeply pigmented. (×400.)

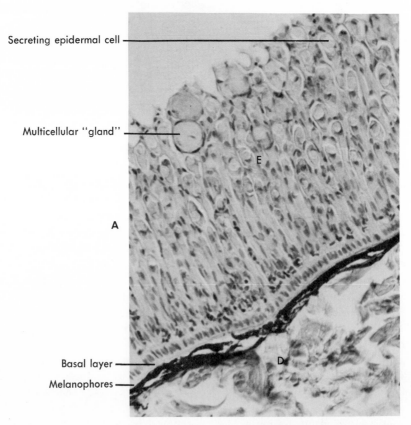

Secreting epidermal cell

Multicellular "gland"

Basal layer

Melanophores

Fig. 4-8. A, General view of skin on dorsal surface of *Conger conger. E,* Epidermis; *D,* dermis. Note regularity of basal cell layer. **B,** Detail of superficial layers of *Conger* dorsal epidermis. Complex nature of gland cells is seen. (**A,** ×125, **B,** ×400.)

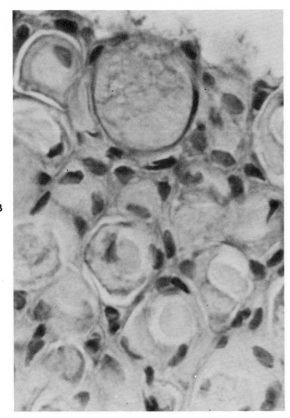

B

Fig. 4-8, cont'd. For legend see opposite page.

The epidermis of fishes frequently shows leukocytes of various kinds in the intercellular spaces (Fig. 4-7). In the elasmobranchs, eosinophils are often conspicuous in these locations. Gland cells, chiefly mucous cells, are numerous, as shown in Fig. 4-8. It seems probable that in some regions, as on the head of the ray, the function of mucus secretion is not limited to cells of special type but belongs to the regular epidermal cells as well.

The elasmobranchs were the first among the vertebrates to exhibit epidermal pigment cells, or chromatophores. Some of these occur in the lower layers of the epidermis, their processes ascending toward the surface. However, many more small chromatophores are found in the outermost layers of the dermis. They produce the gray color of the shark's body on its dorsal and lateral surfaces. They are not found on the ventral body surface or in the white, nonpigmented spots seen on the lateral surfaces. Shifting the position of chromatophores, so that more or fewer of them are exposed to view, can produce changes in the pattern of hues and shades. Changes in color may also be induced by quantitative changes in pigmentation.

Pigment cells typically form many branching protrusions. In the dogfish shark, the granular pigment is mostly brown or black melanin contained in melanophores and may be distributed through the branches, producing dark color, or displaced toward the centers of the cells, yielding a lighter effect. The branching protrusions do not extend or contract; the pigment simply shifts position during color changes. Color changes in the dogfish are known to be controlled both by blood-borne hormones and by direct innervation of the chromatophores (Parker, 1936). In all vertebrates the hormonal control of melanophores is thought to be by melanocyte-stimulating hormone (MSH) (Jorgensen, 1968).

OSTEICHTHYES

The epidermis of the bony fishes (Figs. 4-9 and 4-10) is actually soft and slimy. The scales usually lie so close beneath the epidermis that one has the impression of a hard skin. The epidermis shows a tremendous variation in thickness among different species and different body regions, ranging from a minimum of about 20 to 3000 μm or more.

The epidermis is a stratified epithelium consisting of many layers of cells. A large proportion of these cells may be considered as unicellular mucous glands in various stages of secretory activity. They give rise to the slimy mucus covering so characteristic of the body of many fishes, which makes them so elusive and hard to hold. In *Salvelinus* we have found long lines of basal cells with distal nuclei, an arrangement leaving a conspicuous expanse of cytoplasm just above the basement membrane (Fig. 4-11), but other areas show the nuclei centrally located.

Several layers above the basal layer consist chiefly of cylindrical cells perpendicular to the surface. Then the cells become polygonal, and the ones farther out are flattened so that the long axis is parallel to the surface. The surface layer is made up primarily of flattened cells, polygonal

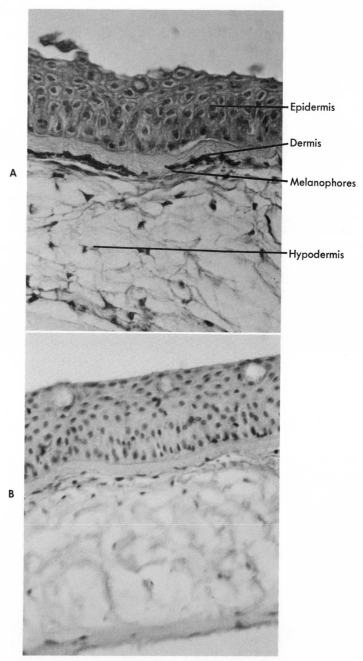

Fig. 4-9. A, Dorsal skin of flounder *Pleuronectes patessa*. Few intraepithelial glands, but many conspicuous pigment cells, occur in the skin of this surface. **B,** Ventral skin. Several intraepithelial gland cells present, but few melanophores. (**A,** ×250; **B,** ×125.)

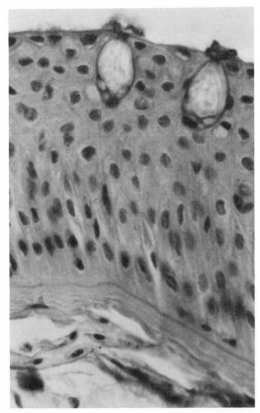

Fig. 4-10. Detail of ventral skin of flounder *Pleuronectes patessa*. Two intraepithelial gland cells are seen discharging secretion. (×400.)

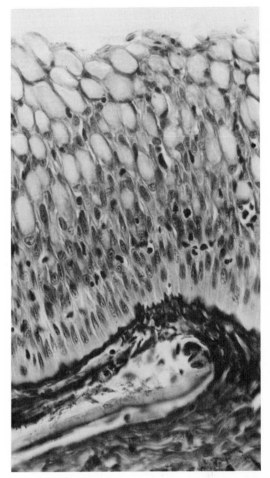

Fig. 4-11. Epidermis of *Salvelinus fontinalis*, brook trout. Columnar cells of the basal layer show considerable extension of cytoplasm below layer of nuclei. One scale seen in dermis. Numerous lymphocytes occur among epithelial cells. (×400.)

when seen from above, with flat, disk-shaped nuclei.

The epidermal cells thus undergo a kind of transformation as they move from the basal to the superficial position, but they do not become keratinized, and their nuclei continue to be visible even when the cells are cast off. For the most part, the striated cuticle had disappeared by the time the teleosts, or "higher fishes," evolved, although it remains in a few specific cases.

Many mitotic figures are visible in the basal layer of the epidermis. These cells, normally tall cylinders, become spindle-shaped when dividing, then somewhat rounded. Whereas mitosis may not be entirely confined to the basal layer, this apparently is the chief site of division.

The club cells, bearing the same name but of doubtful homology with those of the cyclostomes, tend to discharge their secretion while they are

in the middle layers of the epidermis. This secretion, found between the epidermal cells, may help the epidermis to resist pressure and may tend to ward off predators by its disagreeable or even poisonous nature. There seem to be few pigment cells in the epidermis, except covering the free portion of each scale.

The epidermis of teleosts frequently shows **leukocytes** of various kinds in the intercellular spaces. Some of these cells appear to be migrating or moving about between the epithelial cells. They can be seen all the way from the basal layer to a region close beneath the surface. But it is questionable whether they migrate so far as to be cast off with desquamating epithelial cells.

The dermis is generally several times as thick as the epidermis. It is a connective tissue layer subdivided into a more superficial layer of loose tissue, containing the scales, and a deeper, compact layer.

The scales are flat, almost circular plates of bone. The surface of each scale is marked by concentric lines. They do not show the architecture that we associate with bone in higher vertebrates. No haversian systems are present. In fact, bone cells are not seen: this is **acellular bone.**

The scales have an imbricating, or tilelike, arrangement; that is, they overlap one another from head to tail. Each scale slants backward and upward from the region of the deeper, dense layer of connective tissue through the loose layer to the epidermis. The manner of imbrication is such that a small three-sided part of each scale lies free just beneath the epidermis whereas all the rest of the scale is hidden by those immediately anterior to it.

The distal edge of each scale is finely serrated, the individual teeth representing the ends of the concentric lines that we see on the surface of an isolated scale. The epidermis lying just superficial to the distal edge shows a slight inward folding, so that, in section, the entire surface of the skin shows a regular series of gradual ascents, each succeeded by a sharp descent.

The cells that give color to the skin are large, branched pigment cells, or chromatophores. In the trout, these appear to lie rather deeply, at the lower border of the dense fibrous layer of the dermis and in the upper portion of the subcutaneous fascia.

AMPHIBIA

The integument of the amphibians is of special interest because these animals as a class of vertebrates, and even within the life history of many individuals, show the phenomenon of transition from life in the water to life on the land. During this transition, many basic changes in structure occurred, not the least of which involved the skin.

The integument of adults of the class Amphibia differs from that of the fishes in three significant ways: (1) the surface layer often is cornified;

(2) multicellular glands lie beneath the general level of the epidermis, allowing the epidermis as such to have a more uniform cellular population than in the lower forms; and (3) scales are practically absent in modern amphibians. According to the fossil record, early amphibians had scales. Today, a few in tropical countries have dermal plates fused to the neural spines, and Apoda have vestigial dermal scales.

The amphibians constitute a class of vertebrates in which there are often profound differences between immature forms (larvae) and adults. For example, the outer layers of most adults (such as frogs) are adapted for terrestrial life by being fully cornified, although many salamanders show cornification only on the feet, the surface cells over the rest of the body having a cuticular border. In contrast, the rather thin epidermis of the larval forms of many amphibians is completely nonkeratinized.

Furthermore, the epidermis of larval salamanders contains the large **cells of Leydig** (Fig. 4-12). These are unicellular glandular elements, reminiscent of unicellular glands found widely in the fishes. They resemble in particular the club cells of the cyclostomes. Their secretion, like that of the club cells, appears to be released largely within the epidermis and may well be useful in protection from viruses and bacteria. Other types of unicellular glands also occur in various larval amphibians. Their disappearance at metamorphosis is one of the many striking changes that occur during this process.

Epidermis. The epidermis of the frog has several layers, including an outer layer of horny, dead cells (the stratum corneum) that affords protection for the deeper living cells and prevents moisture loss. The epidermal cells are renewed at intervals by molting, under the combined control of the pituitary and thyroid glands. The cells of the innermost layer of the epidermis are columnar, but they become flattened toward the surface until those of the stratum corneum are broad and thin (Fig. 4-13, *B*). This gradual change in cell shape results from the continual production of new cells at the inner layer, which pushes the others out toward the surface, flattening them in the process.

In the outer part of the epidermis are scattered

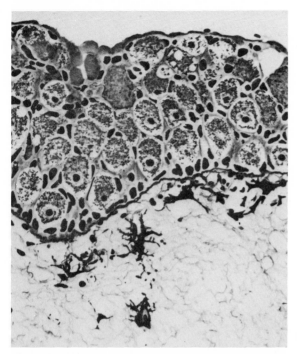

Fig. 4-12. Skin of dorsal surface of urodele amphibian, *Siredon mexicanum.* Much of the total volume of epidermis is taken up by unicellular cells of Leydig. Branched melanophores are seen in dermis. Well-developed basement membrane marks epidermodermal boundary. (Wilder's reticular stain; ×100.)

flask-shaped cells, which some biologists consider degenerate epithelial cells of no known function. Others think they may secrete a substance that aids the molting process. The necks of the many glands lying chiefly in the corium, or dermis, pass through the epidermis. The apertures of the glands may be raised slightly above the surface.

Dermis. The corium (dermis) is composed of two layers, as shown in Fig. 4-13: the stratum spongiosum, in which most of the glands lie, and the stratum compactum, a dense connective tissue. The stratum spongiosum is mostly loose connective tissue, containing many blood vessels and lymph spaces. Thick portions of the stratum spongiosum form small dermal papillae on the dorsal side of the body. These projections may have a tactile function.

The stratum compactum is made up chiefly of connective tissue fibers running in a more or less wavy course parallel to the surface of the skin. Vertical strands of connective tissue, smooth muscle, blood vessels, and nerves cross the stratum compactum at intervals, extending into the stratum spongiosum and epidermis. The elastic fibers and smooth muscles in these vertical strands may aid in squeezing out the glandular secretions (Fig. 4-13, *A*).

Beneath the stratum compactum is a loose layer of subcutaneous connective tissue, also shown in Fig. 4-13. This stratum is next to the muscle layers and contains large lymph spaces. The outermost part of the layer contains many blood vessels and stellate pigment cells with whitish granules. These cells are especially abundant on the belly side of the body, producing the white coloration seen in all frogs.

Glands. The glands of the frog integument, primarily mucous glands and poisonous glands, are far more developed than in fishes. They are derived from the epidermis but lie chiefly in the stratum spongiosum of the dermis. The mucous glands keep the skin moist, which facilitates "skin respiration" and serves as a lubricant in water. Mucous glands are less abundant in the more terrestrial amphibians such as the toad, which has a thick, dry skin. The poison glands (serous glands) (Fig. 4-13, *A*) apparently serve a protective function.

The mucous and serous glands are multicellular and numerous. Both types are richly supplied by blood vessels and nerves. The mucous glands (Figs. 4-13, *B*, and 4-14) are usually smaller and much more numerous than the poison glands. Their ducts are narrow and lined with flattened epithelial cells. The body of the gland is lined with cuboidal cells that may form two layers near the openings of the glands. This cuboidal epithelium produces the mucus that is poured over the surface of the skin. The nuclei of the cells are usually small.

The cells of the poison glands tend to be longer and more flattened than those of mucous glands. The mucous gland is mostly fluidlike, whereas the poison gland is granular and whitish in color. The granular poison glands are better developed in the toads, especially the parotid glands of the back of the neck.

Glands reach their greatest development in the dermal plicae. The dermal plicae are two pale,

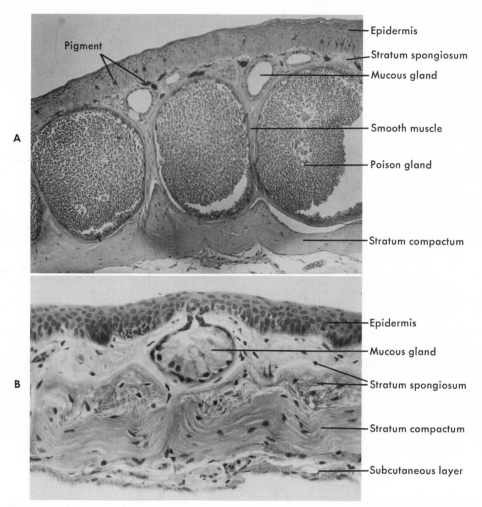

Fig. 4-13. A, Section through frog skin showing three poison glands (serous glands). Note smooth muscle around glands. **B,** Frog skin showing typical mucous gland. (**A,** ×145; **B,** ×380.)

thickened bands of dorsal skin, one on each side of the body, which extend from a region behind the eye to a region near the urostyle. They contain concentrations of glands of both types. Both mucous and serous glands have an outer coat of fibroblasts, and the serous glands, in addition, have a smooth muscle coat inside the fibroblast band. The ducts of the glands appear intracellular at the base but extracellular at the skin surface.

Pigment cells. Color is highly developed in most amphibians. Generally, there are three layers of pigment cells, or chromatophores. Melanophores (black) are the deepest, followed by the guanophores (blue-greenish) and then the yellow lipophores that overlie the others and filter out the blue rays. The guanophores form a network in the corium. The cells of this layer contain coarse granules about one half as large as a human red blood corpuscle. The chief function of the guanophores is to filter out ultraviolet rays.

Some scattered chromatophores may be found occasionally in the epidermis, but mostly they are aggregated in highly vascular regions of the corium (Figs. 4-13, *A,* and 4-14). Pigment cells are chiefly stellate, but as can be seen from Fig. 4-15, other shapes may occur. Changes in color are produced by expansion of the pigment into the fixed processes or its contraction toward the

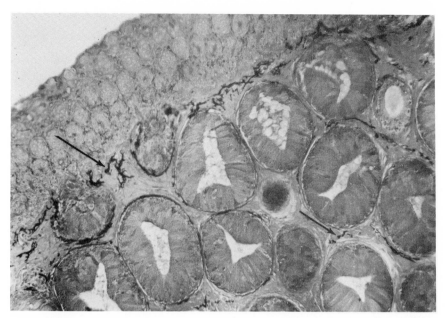

Fig. 4-14. Section through thumb pad of male frog. Numerous tubular glands are well developed during the breeding season because their secretions help the male to maintain a firm grip on the female at amplexus. Arrows indicate accumulations of pigment in dermis. (×190.)

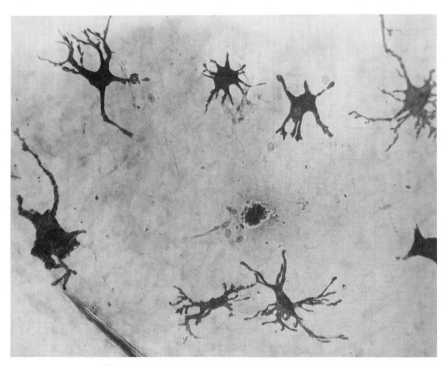

Fig. 4-15. Chromatophores in skin of frog. (×175.)

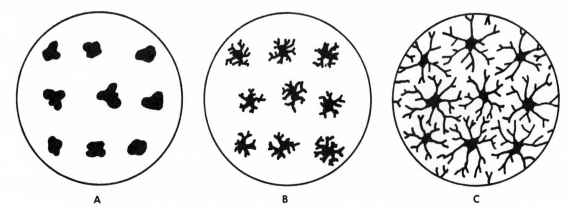

Fig. 4-16. Stages of extension and contraction of pigment in frog skin chromatophores. **A,** Mass of pigment concentrated in center of cells. **B,** Pigment partially extended. **C,** Pigment extended throughout branches of cells.

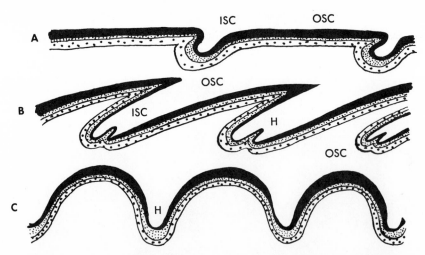

Fig. 4-17. Diagrams of longitudinal sections through different kinds of reptilian scales. **A,** Head scales of snakes and lizards. There is a reduced degree of overlap and a corresponding limitation of inner scale surface and hinge region. **B,** Overlapping scales of the body region as seen in snakes and in some lizards. **C,** "Tubercular" scales, characteristic of many families of lizards, such as Gekkonidae and of some of the Lacertidae. Flexibility of skin as a whole is due in part to deformation of tuberclia as well as to the flexibility of the hinge region. *ISC,* Inner scale surface; *OSC,* outer scale surface; *H,* hinge region. Black = B-keratin; light stipple = A-keratin; dark stipple = living epidermal cells. (Redrawn from Maderson, 1965.)

center of the cell (Fig. 4-16). There is no direct nervous control of the melanophores, as in the bony fishes; rather, the control is hormonal. It is apparently mediated by the effect of light on the eyes of the amphibian, with the secondary release of hormone from the pituitary, or by direct action of light on the pigment-bearing cell.

REPTILIA

A high degree of keratinization and a paucity of glands are two distinguishing characteristics of the skin of reptiles. In contrast to the amphibians, the reptiles have horny scales throughout the integument (Figs. 4-17 and 4-18). These structures, however, are not homologous with the scales of fishes. They are primarily **epidermal** rather than dermal in character, and the surface, exposed to the relatively dry air, is itself dry and hard. The scales do involve the dermis in that each one has its own connective tissue papilla.

Beneath the horny scales some lizards develop **osteoderms,** or bony plates, in the dermis over the body. Osteoderms may be confined to the

head where they lie superficial to the skull bones. In addition, the horny scales may be modified to form crests, spines, or other similar structures conspicuous in both living and extinct reptiles. An interesting specific modification is the digital lamellae found in many lizards (Fig. 4-19). The lamellae have many hairlike projections (setae), which are provided at their free ends with flattened spatulas for frictional traction.

The epidermis generally shows three layers: the stratum germinativum, the stratum granulosum with its basophilic granules of keratohyalin, and the stratum corneum. The stratum corneum is continuous over the surface, but it is thick and inflexible over the scales, and relatively thin and flexible between them.

Ecdysis, molting of the entire skin, occurs in snakes and lizards at intervals. Some lizards molt as often as twice a month. In turtles and crocodiles, on the other hand, desquamation is a gradual process, similar to that which occurs in birds and mammals. Reptiles exhibit the phenomenon of **reduplication** of epidermal layers

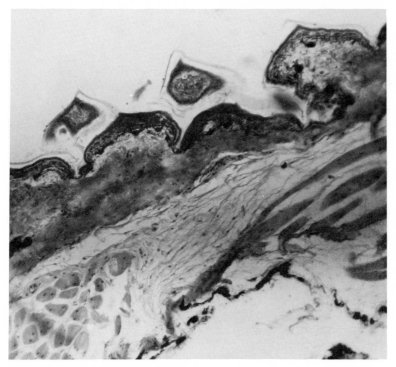

Fig. 4-18. Section of skin of lizard *Anolis*. Scales overlap only to slight extent. Note cornified layer on dorsal side of scale and the thin stratum corneum in regions between scales. Epidermis is well pigmented. (×150.)

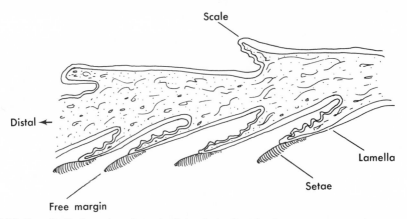

Fig. 4-19. Longitudinal section through digit of *Anolis*. The longest setae are about 30 μm in length. (Redrawn from Ernst and Ruibal, 1966.)

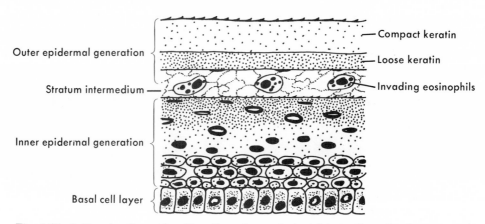

Fig. 4-20. Epidermis of a snake. New tissue layer, stratum intermedium, has become established between outer and inner epidermal generations. Invading eosinophilic granular leukocytes will soon bring about dissolution of new layer causing outer layers to slough off. (Redrawn from Maderson, 1965.)

(Fig. 4-20). After the stratum germinativum has given rise to the granulosum and corneum, it repeats the process, pushing the new layers up underneath the old. Experimental work on the Hydrophiidae (sea snakes) indicates that the more superficial cells are released from the lower ones by the activities of proteolytic enzymes secreted by eosinophil granulocytes that have migrated from the dermis. This phenomenon probably occurs in snakes generally, and perhaps in lizards.

Most reptiles exhibit elaborate color patterns. The pigment is often found in superficial layers of the stratum corneum. However, there also may

be a layer of pigment cells in the uppermost part of the dermis, with branches extending outward between epidermal cells.

The physiology of the control of skin color varies in different reptiles. In the chameleons, the autonomic nervous system controls expansion and contraction of chromatophores. In other lizards there seem to be no nerves to the chromatophores. They are controlled by the secretion of the pituitary gland, in turn dependent on visual impulses. The most active pigment cells are the brown-black melanophores, but there are also xanthophores (yellow-red) and leucophores (white).

AVES

The skin of birds generally is loose, dry, and thin. The only cutaneous gland is the so-called "preen gland," or uropygial gland, at the base of the tail. The capacity of the skin to produce keratin has been directed, or perhaps "elevated," to the production of the varied, beautiful, and highly useful structures known as feathers. There remain, however, reptilian-like scales on the legs and feet, and the claws and beak are specialized keratinous structures, which in some birds undergo periodic molting.

The skin is composed of the typical outer epithelial layer, the epidermis, and the deeper connective tissue layer, the dermis, or corium.

Epidermis. The epidermis is in two layers: a superficial nonliving stratum corneum of flattened, cornified cells and a living stratum germinativum next to the dermis. The thickness of the epidermis varies in different regions of the body, being thinnest in those regions protected by feathers.

The basal cells of the stratum germinativum form a single layer of cylindrical cells with their axes perpendicular to the surface of the dermis. The epidermis is without vascular supply and receives its nutrients from the dermis. The mitotically active stratum germinativum continually furnishes cells to the stratum corneum, replacing those cells that are shed at the surface. These cells become keratinized as they leave the stratum germinativum and migrate toward the surface.

Dermis. The thicker dermal layer is made up of fibrous connective tissue, which is netlike in nature where it merges with the subcutaneous tissue. Conelike projections, or papillae of the dermis, are found on the undersurface of the feet and around the beak. In the feather-covered regions, the surface of the dermis under the epidermis is smooth. The dermis contains many blood vessels, smooth muscle, and sensory nerve endings. Feather follicles extend deeply into the dermal and subdermal tissues and are abundantly supplied with blood vessels.

Feathers. The most characteristic derivatives of the avian skin are the feathers. Feathers are keratinized, epidermal outgrowths of extremely light weight and provide the bird with many adaptive features. They afford a very efficient insulation for maintaining the higher body temperature; they are water-repellent; and as a body covering, they are admirably suited for flight. In addition, feathers provide the color and color patterns so important in competitive courtship and protective coloration.

Except in the case of penguins, feathers are not uniformly distributed over the body but are restricted to certain areas or tracts (pterylae) separated by regions of naked skin (apteria). The arrangement and size of these feather tracts vary greatly among different birds.

Feathers are epidermal, nonvascular, and nonnervous appendages. It is generally considered that they are highly modified forms of the reptilian scales of the Mesozoic Era. They develop in a pouchlike follicle, formed by an inversion of the integument on a papilla, which causes the feather to assume its peculiar shape. Only the core comes from the dermis; the rest of the feather is an epidermal modification. When the feather is fully formed, the blood supply is discontinued, and the central shaft or quill becomes hollow and narrows into the **rachis,** which bears the barbs and barbules (Figs. 4-21 and 4-22).

Three general types of feathers are recognized: contour feathers, downy feathers, and filoplumes. The first of these make up most of the definitive plumage. In mature birds molting of the plumage normally occurs just after the breeding season, but many modifications of the molting process are found among the various species.

Pigments. The plumage of birds is often brightly colored both by various pigments and by structural coloration produced by physical surfaces that disintegrate the spectrum. The most important of the integumentary pigments are the melanins, which range from black through brown to yellow. They occur as discrete particles or granules produced by highly specialized, branched cells known as melanocytes or melanophores. It has been shown experimentally that melanocytes originate from the embryologic neural crest and produce the darker shades of different tints.

Other pigments found in bird coloration are the lipochromes or carotenoids, which are responsible for bright feather, feet, and bill pigments such as oranges, yellows, and reds. Ingested foods often provide a source of various

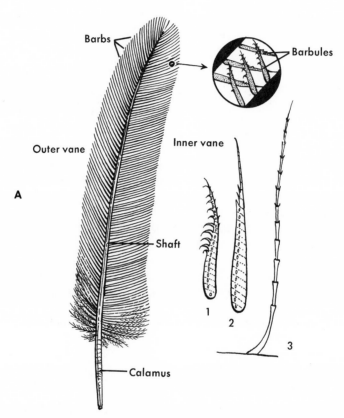

Fig. 4-21. **A,** Some features of mature contour feather from wing of domestic fowl. Circular inset (enlarged portion of feather vane) shows arrangement of minute barbules to form interlocking mechanism between barbs. *1,* Distal barbule borne on side of barb nearest tip of feather; *2,* proximal barbule borne on side of barb nearest base of feather; *3,* downy barbule from barb of fluffy basal portion of feather has no hooklets (barbicels) for interlocking. **B,** Photo of barbs and barbules, comparable to inset in **A.** (**A,** Redrawn from Rawles, 1960; **B,** ×160.)

B

Fig. 4-21, cont'd. For legend see opposite page.

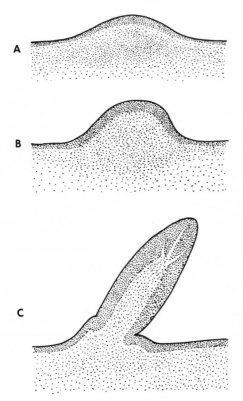

Fig. 4-22. Successive stages in the development of down feather. **A,** Beginning of feather-germ formation. Aggregation of mesodermal cells (constituting primordium of dermal papilla and pulp) and thickening of overlying ectoderm. **B,** Feather germ protruding from surface of skin. Both tissue layers have increased in thickness. **C,** Elongation of feather germ into tapering epidermal cylinder enclosing core of mesoderm. Epidermal cells of cylinder wall beginning to undergo rearrangement into barb ridges. (Redrawn from Rawles, 1960.)

pigments, which are chemically altered and deposited in the feathers.

MAMMALIA

The skin of mammals is a relatively elaborate structure with features not found in that of other vertebrates. It is generally thicker and has greater tensile strength. It has features connected with temperature control and with the formation of an impermeable but sensitive surface. Mammalian skin gives rise to many specialized derivatives such as hair, nails, claws, many kinds of glands, epidermal scales, and certain types of horns.

Although the skin has much in common in all

mammals, there is nonetheless considerable diversity among the different types. This difference is usually expressed in the derivatives of the skin. Thickness is another variable. The outer skin (epidermis) can be hundreds of layers thick in the pachyderms (e.g., the elephant), but exceptionally thin, only a few cells thick, in certain small rodents.

Epidermis. Rabbit integument is composed of the epidermis and the dermis, or corium. The epidermis is a stratified squamous epithelium. The mammalian epidermis consists generally of two to four layers. When the epidermis is thin, there are usually only a stratum germinativum and a stratum corneum (Fig. 4-23). In thicker epidermis a stratum granulosum and sometimes a stratum lucidum are found. The cells of the stratum granulosum contain coarse, deeply basophilic granules. These large, highly refractive granules are composed of masses of keratohyalin, which precedes the keratin. The stratum lucidum is a region of light-staining keratinized cells between the stratum granulosum and the outer stratum corneum.

Most of the rabbit's epidermis has three layers. The innermost stratum germinativum is often called the stratum malpighii. Its active cells undergo cell division and pass outward to the surface, the daughter cells being pushed up by still younger ones. The cells become flatter in each succeeding layer; when they reach the surface, they are shed as dead cells. Outside the stratum germinativum, as the cells proceed toward the surface, they eventually become keratinized and nonnucleated in the outer epidermal layer, the stratum corneum.

The second epidermal layer, called the stratum granulosum, is found next to the stratum germinativum. The horny outer layer of keratin is worn away at the surface, but its thickness is maintained by the actively dividing cells of the stratum germinativum. On the rabbit's footpads, as on the palm of the hands and soles of the feet of man, is a fourth layer, the stratum lucidum, found between the stratum granulosum and the stratum corneum.

The epidermis is nonvascular and derives its nourishment by fluid seepage from the dermis. The deeper layers of the epidermis may have

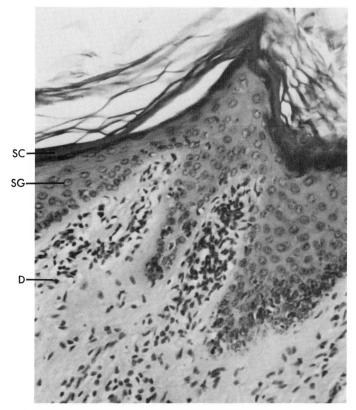

SC

SG

D

Fig. 4-23. Skin of the scrotum of the dromedary camel, *Camelus dromedarius.* Well-developed stratification is seen in this relatively thin epidermis. (×250.)

naked nerve endings, although the nerves usually end in the dermis. The basal layers of human epidermis show downward projections, which form a net, or **rete,** the individual projections into the dermis being known as the **rete pegs.** A rete is characteristic of larger mammals but is usually lacking in the small ones. In the rabbit the dermoepidermal junction is mostly smooth; there is no rete. Scalelike structures, reminiscent of reptilian ancestry, are seen on certain parts of the bodies of various mammals, such as on the tail of the rat. There are few evidences or none of reptilian epidermal scales in the adult rabbit.

Dermis. The second chief layer of the skin is the dermis, a mixed tissue of mesodermal origin. It consists largely of fibers and the scattered fibroblasts that secrete them. Most of the fibers are white and collagenous, but there are some yellow, elastic fibers. Blood vessels and nerves occur in the dermis; there are also scattered muscle fibers and adipose cells.

In some mammals, but not so evidently in rabbits, the dermis is further divided into two layers. The first has ridges and papillae protruding into the epidermis and containing tactile sensory corpuscles and vascular papillae. The second is a deeper reticular layer of densely interlacing fibers, with numerous chromatophores in regions where the epidermis is heavily pigmented (e.g., the areola around the nipple and the circumanal region).

Below the dermis is the subcutaneous layer, which is not a part of the skin; yet it may blend somewhat into the dermis. It consists of a network of connective tissue bands and septa, intermeshed with adipose tissue. This layer constitutes the superficial fascia. Where fat is abundant, the subcutaneous layer is called a panniculus adiposus.

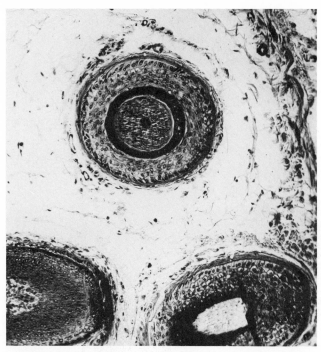

Fig. 4-24. Cross sections of hair roots in human scalp. Vacuolated medulla, or core, is surrounded by dark pigment–containing cortex, outside of which is root sheath made up of a number of layers of cells. (Iron hematoxylin and orange G; ×218.)

The skin of the rabbit is easily separated from this underlying layer. Some mammals (e.g., the horse) have extensive subcutaneous muscles (panniculus carnosus) for moving the skin. In man, these muscles are represented by the platysma and other muscles of expression.

Pigment. At the base of the epidermis in some mammals is a layer of melanocytes, peculiar branched cells derived from the neural crest. These cells produce melanin, the dark pigment mainly responsible for mammalian coloration. Melanocytes are found in the dermis in many mammals.

Epidermal derivatives. Hair is the most characteristic derivative of the mammalian epidermis (Fig. 4-24). The hair of rabbits is of the fur variety, consisting of fine, relatively short hair. In addition, vibrissae are found on the snout and above and below the eyes. The vibrissae, shown in Fig. 4-25, serve as sensitive tactile organs. They are long, stiff hairs with the root of each surrounded by a basketlike network of sensory nerve endings.

The part of the hair that projects above the surface of the skin is called the hair shaft; the nonprojecting portion is the root. The latter is enclosed in the hair follicle, a saclike structure consisting of a portion of the epidermis surrounded by dermal tissue. At the base of the hair the dermal part of the follicle forms the hair papilla. The papilla projects into the basal part of the hair root, carrying blood vessels to supply the growing and multiplying epithelial cells of the hair matrix in this region. As the lower epithelial cells multiply, they push those above them upward, and the hair elongates, or grows.

The hair itself is a horny thread formed by metamorphosis of the upward-moving cells of the matrix. It consists of a central core, or **medulla,** which often contains masses of bubbles of air; a **cortex,** which is pigment-containing and may appear yellow or brown even in unstained sections; and a covering **cuticle** of scalelike structures. An interesting feature of the cuticle is the fact that the edges of its "scales" face upward, interlocking with the downward-facing scales of the

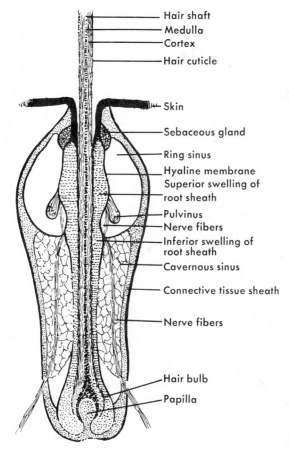

Hair shaft
Medulla
Cortex
Hair cuticle

Skin

Sebaceous gland

Ring sinus
Hyaline membrane
Superior swelling of root sheath
Pulvinus
Nerve fibers
Inferior swelling of root sheath
Cavernous sinus

Connective tissue sheath

Nerve fibers

Hair bulb
Papilla

Fig. 4-25. Diagram of a vibrissa, or tactile hair, of the mouse. (×125.)

innermost layer of the hair follicle. Thus the hair is held securely in place.

On the wall of each hair follicle is inserted a small muscle, the arrector pili. When these muscles contract, the follicles and hair shafts are drawn toward a vertical position, thus causing the hair to "stand on end." This arrangement is apparently an adaptation for increasing the insulating capacity of hair and also, of course, is involved in displaying emotional reactions.

Hairs are not derived from reptilian scales as formerly thought. Their origin is more ancient than that of epidermal scales. They appear to be modifications of the prototriches of bristle-containing sensory pits found in fishes, tailed amphibians, and some reptiles. In fact, these tactile sensory pits may have originated to counteract the hard, sensory-impervious scales.

Claws are also epidermal derivatives, modifications of the stratum corneum. They are primitive structures from which nails and hoofs have originated in some mammals. The claws of mammals are similar to those found in reptiles and birds. They are made up of a dorsal horny unguis and a ventral softer subunguis. By elongation, the unguis folds down over the edges of the subunguis to form a claw. The nails of primates are mostly the flattened horny unguis, covering the dorsal part of the finger or toe. In the case of the hoof, the subunguis is greatly thickened, and encloses or encases the end of the digit.

Glands. No other group of vertebrates has made wider use of glandular functions in the skin than have the mammals. The exocrine glands (i.e., glands with ducts) are all of epidermal origin and may be classified as eccrine, apocrine, and holocrine.

The eccrine glands are the regular sweat glands (Fig. 4-26). They are small tortuous tubes that sink into the dermis from the surface and end blindly after coiling. All sweat glands show contractile, or myoepithelial, cells in the peripheral portions of the coiled tubules. These presumably are active in helping to squeeze out the secretion. In the rabbit, they are found in the foot pads and a few other places. In man, sweat glands are distributed generally over the body but are concentrated in some areas such as the axillary regions. The water excreted from these glands is converted into vapor, thereby absorbing the excess heat of the body.

The apocrine glands are larger than eccrine glands, and their ducts are longer and more winding. Their secretory coils may extend down into the subdermis regions. These glands usually open into a hair follicle or where a hair has been. Phylogenetically, they are older than the eccrine glands with which they are often included, but they have nothing to do with regulating the heat of the body. Apocrine glands secrete a milkish yellow substance from the apical cytoplasm of the cells. They are active in certain aspects of the sex cycle and may have other functions as well.

This type of gland is modified to form the mammary glands, perineal glands, inguinal glands, and some others. The various scent glands wherever found among mammals are all of the

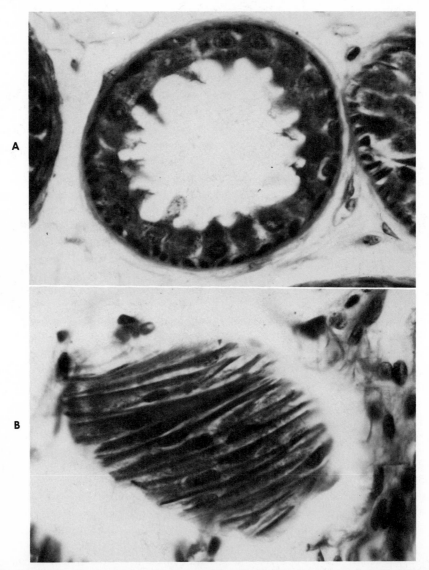

Fig. 4-26. A, Axillary sweat gland of human female. These cells, at least by light microscopy, seem to carry out "apocrine" secretion, giving off portions of cytoplasm along with other secretory products. At bases of epithelial cells are seen deeply staining nuclei of contractile myoepithelial cells. **B,** Tangential section through the peripheral portion of sweat gland of same subject. Elongated bodies of the myoepithelial cells are demonstrated. (Iron hematoxylin and orange G; ×440.)

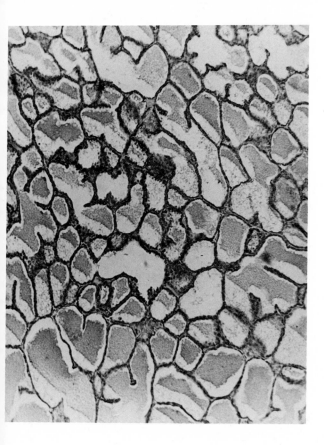

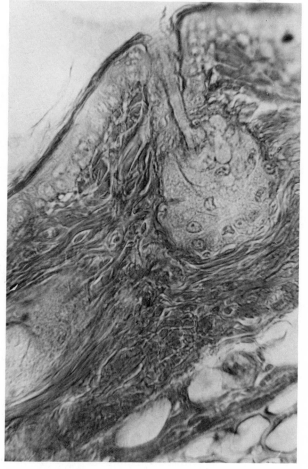

Fig. 4-27. Active mammary gland of a rabbit. Note degree of development of alveoli during period of lactation and their low epithelial wall when distended with milk. (×330.)

Fig. 4-28. Skin and multiple sebaceous gland of rabbit external ear. These glands vary in size, position, and number. Epidermis, dermis, and subcutaneous layers are evident in this photograph. (×840.)

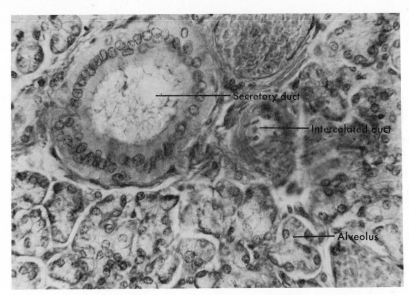

Fig. 4-29. Parotid salivary gland of rabbit. A pure serous gland with the basophilic part chiefly at the end of the cell. Large duct with many nuclei is secretory duct, and smaller one is intercalated duct terminating in an alveolus. (×820.)

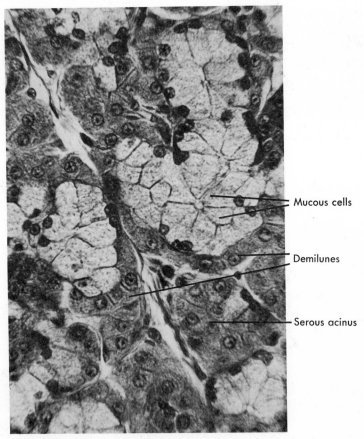

Fig. 4-30. Submaxillary gland of a cat. Mixed gland with serous cells forming darkish demilunes over the mucous cells of an acinus. (×348.)

apocrine type. Mammary glands occur in all female mammals (Fig. 4-27) and, in an undeveloped form, in males. In the rabbit, they consist of branched tubes between the skin and the underlying muscles on the ventral surface of the body, opening onto the surface by four pairs of nipples.

Holocrine glands are represented by the sebaceous glands (Fig. 4-28). They usually lie in the dermis and open into hair follicles. Their secretions keep the hair and skin soft and glossy. In addition, the ducts of sebaceous glands open onto the free surface of the lips, the mammary papillae, the forehead, the male glans penis, and female vulva.

The secretory portions of sebaceous glands are clusters of alveolar sacs opening into a common short duct, which is lined by stratified squamous epithelium. The central part of each alveolus shows large cells filled with masses of sebum and degenerate nuclei. The cells themselves are in the process of breakdown. The secretion involves whole cells, thus the term holocrine. Mitosis occurs in cells close to the duct walls; new cells then move into the secretory portions.

Another group of exocrine glands well developed in most mammals are the salivary glands. Three large pairs include those of the **submaxillary** between the angles of the jaws; **sublingual,** inside the lower jaw; and the **parotid,** behind the angles of the jaws. The parotid (Fig. 4-29) is a pure serous gland (cells that produce a serous fluid or watery secretion). The other two pairs are mixed glands of both serous and mucous cells (mucus is a slightly more viscid secretion). The submaxillary glands display in striking fashion the demilunes that are the serous cells forming caplike structures over the mucous cells (Fig. 4-30).

Histologically, in serous cells the nuclei are rounded and located near the base of the cells. Mucous cells, on the other hand, have flattened nuclei that are crowded against the bases of their cells. The cytoplasm of mucous cells is also less basophilic than that of the serous cells and shows up less darkly with the ordinary hematoxylin-eosin stain.

References

Ernst, V., and R. Ruibal. 1966. The structure and development of the digital lamellae of lizards. J. Morphol. 120:233-266.

Jorgensen, C. B. 1968. Chemical correlation and control. *In* M. S. Gordon, Animal function; principles and adaptations. The Macmillan Co., New York.

Kerr, T. 1955. The scales of modern lungfish. Proc. Zool. Soc. London 125:335-345.

Krause, R. 1923. Mikroskopische Anatomie der Wirbeltiere in Einzeldarstellungen IV. Walter de Gruyter & Co., New York.

Maderson, P. F. A. 1965. The structure and development of the squamate epidermis. Chapter 8 *in* A. G. Lyne and B. F. Short, eds. Biology of the skin and hair growth. Angus & Robertson (Publishers) PTY, Ltd., Sydney, Australia.

Newth, D. R., and D. M. Ross. 1955. On the reaction to light of *Myxine glutinosa* L. J. Exp. Biol. 32:4-21.

Parker, G. H. 1936. Integumentary color changes in the newly-born dogfish *Mustelus canis*. Biol. Bull. **70:**1-7.

Rawles, M. E. 1960. The integumentary system. Chapter 6 *in* A. J. Marshall, ed. The biology and comparative physiology of birds. Academic Press, Inc., New York.

Ross, D. M. 1963. The sense organs of *Myxine glutinosa* L. Pages 150-160 *in* A. Brodal and R. Fänge, eds. The biology of *Myxine*. Universitetsforlaget, Oslo, Norway.

5 CONNECTIVE AND SUPPORTING TISSUES

Connective tissue is one of several basic tissues of the animal body. It arises from mesenchyme, a loose spongy tissue that forms the early embryonic tissue of mesoderm (the third primary germ layer). All connective tissue including connective tissue proper, cartilage, and bone (Windle, 1969) is directly or indirectly formed from mesenchyme. Connective tissue performs the important function of mechanical integration of bodily structures by binding and anchoring parts and giving support to the body and organs. Its extensive variety of cells and structural features cannot be grouped under one functional heading, since it not only binds and supports tissues and organs but has many other functions of a more or less mechanical nature, such as cushioning organs, providing elasticity of blood vessel structures, separating organs by forming capsules around them, affording pathways for nerves and blood vessels, giving yielding but firm support where needed to the rigid framework of body form as in cartilage and bone, and forming separation and connection of joints (Arey, 1968).

Connective tissue also provides for fluid lubrication and reduction or friction in strategic places, as in joints and in the movement of viscera. Loose connective tissue invests bundles of nerves and blood vessels and fills in spaces as padding. Another type of connective tissue forms tendons (for transmitting the force of muscles onto the skeleton) and strong ligaments (for holding bones together). Certain of its cells store food in the form of fat. Connective tissue is responsible for dividing the body into the compartments so necessary for specialized functioning. Not the least of the functions of this highly adaptable tissue is the role it performs in protecting the body by immune responses and otherwise against invasion by foreign substances (Mancini, 1963). It is also potentially reparative, as in wound healing and regeneration of lost parts.

BASIC STRUCTURE OF CONNECTIVE TISSUE

Connective tissue differs from all other tissues in having a relatively large proportion of intracellular components. The intercellular components, or matrix, consist of the amorphous ground substance, which has a semifluid, jellylike consistency, and a fibrous structure of three kinds of fibers. Thus by varying the proportions of ground substance, fibers, and cells, many subtypes of connective tissue can be formed. Some types have more cells, and others have more matrix. The matrix may consist mainly of amorphous ground substance, or it may contain many fibers. The matrix is really nonliving and makes up so much of the animal body that the cells may be said to reside in this vast framework of nonliving substance.

There are two major types of connective tissue: connective tissue proper and specialized connective tissue such as the supporting tissue of bone, cartilage, and joints. Connective tissue proper occurs in two forms, loose and dense. Loose connective tissue has a loose and irregular arrangement of its fibers, with several cells scattered through its substance. On the other hand, dense connective tissue has its fibers compactly arranged, with one type of fiber predominant. This type of connective tissue also has fewer cells, usually arranged in rows between the fiber bundles. Loose connective tissue may be considered the prototype of all connective tissue, and the modification of the unspecialized type has made possible the variety of other types of connective tissue. It is also the most common of all connective tissue and has chiefly a binding function. In contrast, the dense form plays mainly a supporting role in bodily structures.

Connective tissue cells

First will be considered the many different kinds of cellular components found in loose connective tissue (Fig. 5-1). The cells of the specialized connective tissues will be discussed with the various types of those tissues.

Undifferentiated mesenchymal cells. Undifferentiated mesenchymal cells are characterized by an oval or round nucleus with one or two nucleoli. The cells are stellate or fusiform, making a network of cells whose processes meet but do not

form a syncytium. Their cytoplasm is agranular and lightly acidophilic. They have multiple developmental potentialities and can produce many kinds of connective tissue cells. Mesenchymal cells are found mainly in embryonic tissue, but some may also be found in adult tissues alongside blood vessels. They resemble fibroblasts but are smaller.

Fibroblasts. Fibroblasts are among the most numerous of the cells of loose connective tissue. They are large, flat, branching cells that appear spindle-shaped in side view (Figs. 5-1 and 5-2). The oval or ovoid nucleus has fine chromatin granules and one or two conspicuous nucleoli. Their cytoplasm stains poorly except in young, active fibroblasts where it is granular and basophilic. Fibroblasts make, or secrete, both fibers and ground substance. They are present in all connective tissues and are the only kind of cell found in tendons. **Tropocollagen** molecules are secreted by fibroblasts and are polymerized out-

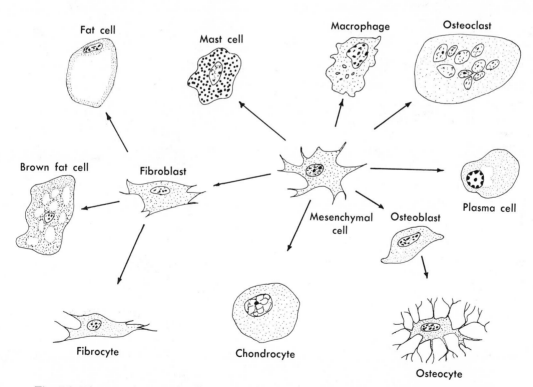

Fig. 5-1. Diagram of connective tissue cells, showing common origin and possible relationships. Note central position of mesenchymal cell. This type of cell is the source of all connective tissue cells and is the cellular tissue filling spaces between epithelial layers in embryos. As development proceeds, mesenchyme gives rise to the cells of mature forms of connective tissue.

side the cell. Mature fibroblasts no longer engaged in synthesizing protein are known as **fibrocytes.** They have the capicity for growth and regeneration and play a part in wound healing.

Macrophages (histiocytes). Macrophages are found in all loose connective tissue (Figs. 5-1 and 5-2) and exhibit phagocytic properties (Maximow, 1932). They are most abundant in richly vascularized regions. Macrophages are irregularly shaped with short, blunt processes. The nucleus is oval and smaller than that of a fibroblast and stains darkly. The cells are capable of ingesting foreign matter, such as particulate matter and dyes. They greatly increase in number when stimulated by inflammation and may coalesce into multinucleated giant cells to attack large particles. Macrophages may originate from monocytes, lymphocytes, or original mesenchymal cells. They form one important component of the reticuloendothelial system.

Plasma cells. Plasma cells are oval and resemble lymphocytes, with which they are often associated. The nucleus is located eccentrically and has coarse clumps of chromatin arranged like the spokes of a cartwheel. The cytoplasm is basophilic and agranular except in the region of the cell center. The electron microscope shows plasma cells to be rich in endoplasmic reticulum, with many ribosomes. These cells produce most of the antibodies in the defense mechanism of the tissue. They are plentiful in inflammations but are rarely seen in normal connective tissue. They also occur in serous membranes and lymphoid tissues. Their cytoplasm often contains acidophilic **Russell bodies,** which may be products of degeneration. In their immunologic reactions, plasma cells de-

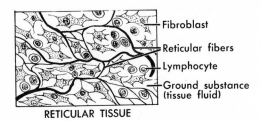

RETICULAR TISSUE

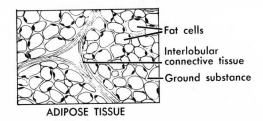

ADIPOSE TISSUE

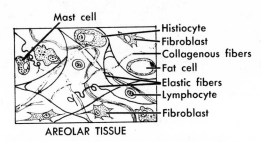

AREOLAR TISSUE

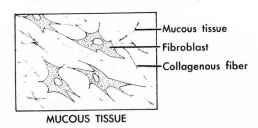

MUCOUS TISSUE

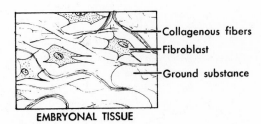

EMBRYONAL TISSUE

Fig. 5-2. Common types of loose connective tissue. The reticular, adipose and areolar tissues make up the loose connective tissue of the adult animal. The areolar may be considered the prototype of all connective tissue, since it has more cells and fibers than the others and is found in almost every part of the body. Mucous and embryonal tissues are characteristic of the early embryo; they are transient elements and may differentiate later. Their cells are mostly mesenchyme and fibroblasts, with collagenous fibers scanty or absent in the early stages. They are soft, jellylike tissues.

fend against specific diseases by forming and secreting gamma globulin.

Mast cells. Mast cells are widely distributed in connective tissue but are especially common near blood vessels. They are large plump cells with a cytoplasm containing many basophilic granules (Smith, 1963). The granules are soluble in water and, for this reason, can be easily lost in tissue preparation. Mast cells are round or oval and may have short pseudopodia (Fig. 5-1). The nucleus is small and stains only faintly. These cells produce heparin and histamine and, in some vertebrates, serotonin as well (Chayen and coworkers, 1966). They are also supposed to aid in the synthesis of hyaluronic acid and collagen. They arise from mesenchyme and microcytes and sometimes from fibroblasts. The easiest way to see mast cells is to inject methylene blue into the loose connective tissue of a rat or other rodent and then make a whole mount of the teased tissue. Mast cells must be distinguished from the basophilic cells of the blood that may occasionally wander into connective tissue.

Fat cells. Fat cells are derived directly from mesenchymal cells that collect fatty droplets in their cytoplasm (Figs. 5-1 and 5-2). In the typical adult fat cell, these droplets in the cytoplasm coalesce to form a single large vacuole that forces the nucleus and the remaining cytoplasm to the periphery of the cell. Alcohol may extract the fat, producing a signet-ring appearance. Sudan IV stains fat an orange red; osmic acid stains it black. Lower vertebrates concentrate their fat into **fat bodies;** higher vertebrates disperse fat in clumps in loose connective tissue under the skin and around organs, as well as in mesenteries and other places. Individual cells are often surrounded by a fine network of reticular fibers. New fat cells never arise by mitosis but may arise by the differentiation of cells within connective tissue. Many different kinds of cells (fibrocytes, macrophages, endothelial cells, etc.) are now known to become fat cells. When the fat is used, the cell reverts to its original condition. There are a few places in which fat is never found, for example, in the nervous system, eyelids, scrotum, and penis.

Blood elements. Blood elements are cells that emigrate from the blood into loose connective tissue. They include leukocytes, monocytes, lymphocytes, and eosinophils. Some may have originated in connective tissue and remained there. Neutrophilic leukocytes are known to escape from capillaries in inflamed regions. Actually, leukocytes are largely inactive in the bloodstream and perform their functions in connective tissue. Lymphocytes are recognized by their spherical, darkly staining nucleus and their thin rim of basophilic cytoplasm. The large monocytes will develop into macrophages in regions of inflammation. Plasma cells are thought to represent a special differentiation of the lymphocytes, which are the smallest of the free cells of connective tissue.

Chromatophores. Pigment cells (chromatophores) occur mostly in the dense connective tissue of the skin, pia mater, and choroid coat of the eye. Melanocytes are chromatophores that synthesize melanin, the dark pigment common in skin. Chromatophores are irregular in shape, with cytoplasmic processes.

Fibers

The intercellular components of connective tissue are fibers (Fig. 5-3) and ground substance. The three types of fibers are collagenous fibers, elastic fibers, and reticular fibers.

Collagenous fibers. Collagenous fibers are often called white fibers, but they are actually colorless. They are the most common fibers in connective tissue, consisting of the albuminoid, collagen, which makes up about 40% of all the protein found in the animal body. Collagen is synthesized chiefly by fibroblasts. A collagenous fiber is a bundle of finer fibrils, the whole being encased by a cement substance (Figs. 5-2 and 5-3). The electron microscope shows the fibrils to be made up of still smaller microfibrils (0.04 μm thick), which are invisible with the light microscope. Collagenous fibers vary in size, depending on the number of fibrils in them. Most fibers are 1 to 12 μm in diameter, although some are known to be larger. The fibrils are usually 0.3 to 0.5 μm in diameter (Freeman, 1964). Fibers frequently branch and recombine by the interchange of clusters of unbranching fibrils between one fiber and another.

Collagen owes its properties to its chemical and physical structure. The collagenous fiber is of low refractile index and may be straight or

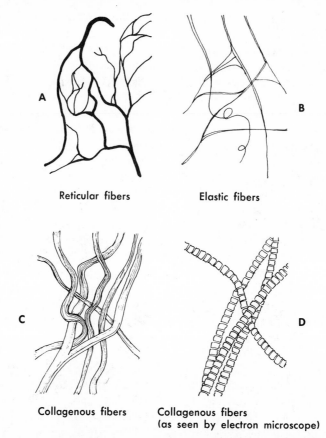

Reticular fibers

Elastic fibers

Collagenous fibers

Collagenous fibers
(as seen by electron microscope)

Fig. 5-3. Three kinds of fibers in adult connective tissue. White (collagenous) fibers, **C,** are different physically and chemically from elastic fibers, **B.** The collagenous fiber is made up of a bundle of smaller fibrils, each of which in turn consists of finer microfibrils as revealed by the electron microscope, **D.** Each microfibril is composed of macromolecules (tropocollagen) that lie side by side and produce, by staggered overlaps of one fourth the length of a molecule, the periodic cross-banding. Reticular fibers, **A,** are inelastic and have other features similar to the white, or collagenous, fibers; they are the first fibers to develop in wound healing.

wavy with longitudinal striations. Wavy threads may be straightened by pulling on them, but there is little or no stretching. They have great tensile strength. There is no specific stain for collagenous fibers, but hematoxylin-eosin stains them red. Mallory's connective tissue stain will reveal them clearly as a blue color. When broken, the fibers fray like the end of a rope. When boiled, fibers form a colloidal solution of animal glue.

The basic collagen molecule consists of a group of three polypeptide chains, each of which is made up of several hundred amino acids linked together end to end. These three chains intertwine in a helix. Two amino acids, hydroxyproline and pro-

line, form about 25% of the links in the collagen molecule and impart stability and rigidity to the molecule. Every fourth position in the collagen chain is occupied by the amino acid, glycine, followed in an adjacent site by proline or hydroxyproline. It is thought that the finer microfibrils are composed of macromolecules known as tropocollagen, which is about 2800 Å long. These macromolecules lie in parallel alignment with a staggered overlap by one fourth of their length. This overlapping causes the cross-bands, each with a periodicity of about 640 Å (Freeman, 1964), characteristic of collagenous fibers (Figs. 5-2 and 5-3).

Elastic fibers. Elastic fibers are thin, refractile threads that branch and anastomose (Figs. 5-2 and 5-3). They are normally straight and stretched but appear wavy or curled in teased preparations. In large masses they are yellowish; they can be stained selectively by orcein or resorcin-fuchsin. They are composed of another albuminoid protein, elastin. They are solitary and never occur in bundles. Their diameter ranges from a fraction of a micrometer to a few micrometers. When stretched and released, these fibers snap back like rubber bands, although they tend to lose their resilience with age. Elastic fibers as a whole are very resistant to chemical change, resisting boiling water, acids, and alkalies. Pancreatic juice, however, has an enzyme, *elastase,* that digests the fiber. Elastic fibers are optically homogeneous, not fibrillar, and have no periodicity of structure as have the collagenous fibers. Some histologists believe elastic fibers are formed by a type of fibroblast, but nothing is known conclusively about their deposition (Fahrenbach and co-workers, 1966).

Reticular fibers. Reticular fibers are similar to collagenous fibers (with which they often blend) but are usually much smaller and have a different arrangement (Figs. 5-2 and 5-3). They unite and branch to form a network. They are inelastic, and their banding has the same periodic spacing as that of collagenous fibers. The molecular arrangement in both is the same. Reticular fibers are also made up of bundles of microfibrils. Silver impregnation stains distinguish them from the white fibers because they absorb metallic silver. They also stain well with the PAS-Schiff technique. In development, reticular fibers are the first to appear, but in time many assume collagenous fiber characteristics. Reticular fibers supply support for the cells of many structures such as lymph glands, epithelial organs, and blood-forming organs. Reticular fibers are formed by fibroblasts.

Amorphous ground substance

The ground substance serves as a viscid medium in which connective tissue fibers and cells are embedded (Fig. 5-2). Its composition and nature vary from organ to organ. Chemically it belongs to a group of substances known as **mucopolysaccharides,** which contain equal numbers of molecules of hexosamine and glucuronic acid combined with protein (Ham, 1969). This complex compound (a nonsulfated type without sulfuric acid) is hyaluronic acid, which can attract and hold fluid in its interstices. In the eye and umbilical cord, the ground substance contains mostly hyaluronic acid. When the complex mucopolysaccharide combines with sulfuric acid, two principal types of compounds are formed—**mucoitin sulfuric acid** and **chondroitin sulfuric acid.** The former is found in the ground substance of connective tissue, and the latter in the ground substance of cartilage. These two compounds are known collectively as high–molecular ester sulfuric acids. The connective tissues of young animals contain a much higher concentration of these esters than do these of older animals (Sobel, 1967). An enzyme, **hyaluronidase,** found in certain bacteria, depolymerizes the large polymer of hyaluronic acid, allowing the invading bacteria to spread (Meyer, 1947). In the mammal the same enzyme may aid in the penetration of the egg by the spermatozoon.

The ground substance is usually not preserved in histologic preparations; it is dissolved out by the fluids used in fixation. It is optically homogeneous and transparent and can be seen in fresh tissue only when the latter is put in a medium of different refractive index. The freeze-drying method is probably the best way to demonstrate it.

CONNECTIVE TISSUE PROPER

Connective tissue varies in different parts of the body. As already mentioned, the relative proportions and arrangements of cells, fibers, and amorphous ground substance determine the nature of the connective tissue. The distinction between loose and dense connective tissue is not always clearcut because the two types may grade into each other. Therefore arbitrary distinctions between the two are often made.

Loose connective tissue

Loose connective tissue is characterized by a semifluid and gelatinous matrix in which there is a loose meshwork of the various fibers (Fig. 5-2). Collagenous fibers are far more common than the other two types, reticular and elastic. Loose connective tissue is soft and pliable and contains a great variety of cells and many blood vessels and nerves. It has a widespread distribution and provides support for tissue cells. It penetrates into crevices and fills up spaces between structures.

Embryonal connective tissue. All connective tissues are derived from mesenchyme, a mixture of cells originating from the middle germ layer. Embryonal connective tissue (Fig. 5-2, *E*) is this unspecialized connective tissue characteristic of the early period of embryonic life. Mesenchymal cells (p. 63) have large vesicular nuclei and scanty cytoplasm. The ground substance of mesenchymal tissue is mostly a coagulable fluid. Although mesenchymal tissue as such disappears when its cells undergo tissue differentiation, some cells of this type do persist in adult tissues, apparently as reserve units.

As the embryo develops, certain mesenchymal regions become differentiated toward the adult condition. This is a transient state in the normal development of all connective tissue and is known as mucous connective tissue. In the case of Wharton's jelly of the umbilical cord, there is no further differentiation. At this stage, the cells are large, stellate fibroblasts with cytoplasmic processes. The ground substance is gelatinous and gives a mucin reaction. A few other cells (macrophages or lymphocytes) may be present. There are a few collagenous fibers, but others types are usually lacking. The nearest counterpart to this ground substance in the adult is the vitreous body of the eye.

Areolar connective tissue. The term "areolar" refers to the small spaces occurring within this type of connective tissue. Areolar connective tissue is formed by the direct differentiation of mesenchyme. Areolar tissue is often called the prototype of all connective tissue because it displays a greater variety of cells and fibers than any other type. It is a loosely arranged, fibroelastic connective tissue, the most widespread of all (Fig. 5-2, *C*). Some histologists consider all loose connective tissue as areolar connective tissue. It is found in nearly every microscopic section; it forms the packing and anchoring substance as well as the embedding medium of blood vessels and nerves. Fibroblasts and macrophages are dominant in areolar connective tissue. All the different

types of fibers also occur within it, although collagenous fibers are most common. Its ground substance is fluidlike and often found in small areas where there is no other structure. In whole preparations such areas may pick up air, which gives a bubbly appearance to the tissue. The spongy nature of areolar connective tissue serves as a storage place for water, glucose, and salt. When tissue fluid is increased by capillary dilation or when evacuation of the fluid is interfered with by insufficient lymphatic drainage, the tissue becomes edematous and swells.

Study of areolar tissue is best done from spread preparations that have been stained by Mallory's or Masson's trichrome stain. No single stain will show all structures equally well.

Reticular connective tissue. Reticular connective tissue is characterized by a network of reticular fibers that are intimately associated with the primitive reticular cells (Fig. 5-2, *A*), which closely resemble mesenchymal cells. Their nuclei are large and ovoid and stain lightly. The cells are stellate, with long cytoplasmic extensions with basophilic plasma. Some reticular cells are phagocytic. Those reticular cells which produce reticular fibers are similar to fibroblasts, differing only in their developmental potentialities; that is, they may give rise to free macrophages, precursors of red blood corpuscles, leukocytes, or other elements.

Reticular fibers are arranged in a fine branching network containing lymphocytes and other cells in its meshwork. Some reticular fibers are associated with fibroblasts, but the regular reticular tissue forms the framework of lymphoid organs, spleen, bone marrow, and liver. It is also found in the linings of the alimentary canal, trachea, and bronchi. The relation of the fibers to the reticular cells is an intimate one; the fibers are closely invested by the protoplasmic processes of the cells. The phagocytic reticular cells form a definite part of the reticuloendothelial system that protects the body through phagocytosis.

Adipose connective tissue. Scattered fat cells are found in areolar connective tissue, but aggregations of these fat cells are called adipose connective tissue (Fig. 5-5). Each fat cell is surrounded by a network of reticular fibers, and in the spaces between them are such cells as lymphocytes, fibroblasts, eosinophils, and mast cells. The fat cells may be so closely packed that they form lobules separated by fibrous septa. They usually have a rich vascular supply. Adipose tissue may develop wherever areolar connective tissue is found, but certain regions are common sites. Collections of fat may be present in subcutaneous tissue, around the kidneys and adrenals, in mesenteries, around the heart region, in bone marrow, and in some other regions. Marine mammals have a subcutaneous sheet, the **panniculus adiposus,** which serves as an important fat reserve and as an insulator against heat loss. Adipose tissue is also an important storage place of energy-yielding material. It is not yet clear whether the cells that accumulate fat are specific or indifferent. Some experiments indicate well-established re-

gional differences in the capacity of cells to become fatty cells. It is also known that fat cells are rich in enzymes and are not merely static deposits, since they undergo a continuous turnover. When fat is used for body metabolism, the cell reverts to a type similar to a fibroblast.

In some adipose tissue the fat droplets do not run together in the cytoplasm to form typical fat cells. Because it has a high degree of vascularity, this type of cellular adipose tissue has a brownish tint and is called **brown fat** (Menschik, 1953) (Fig. 5-1). Deposits of brown fat are often called hibernating glands because they are especially prominent in hibernating animals. They may arouse animals from hibernation (Smith and Hock, 1963) by the stimulation of a syndrome of reactions involving the autonomic nervous system,

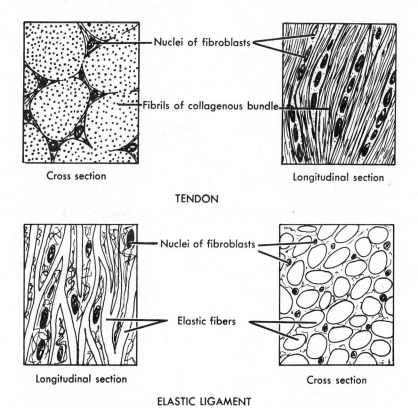

Fig. 5-4. Two types of dense connective tissues (elastic ligament and tendon) of cat, each shown in cross and longitudinal sections. One kind of compactly arranged fibers is found in each, but ground substance is scanty. Most fibers are collagenic, but elastic fibers may also be present as in elastic ligament. Dense connective tissue fibers may be arranged regularly, as in ligaments and tendons, in the plane in which the tensile strength is needed, or they may be irregularly organized as in sheaths and reticular layers of the skin where the collagenic fibers run in different directions and in different planes. (×300.)

the release of norepinephrine at the nerve endings, and the formation of fatty acids from the stored triglycerides (Hayward and associates, 1963). The metabolism of fatty acids within the fat cells and elsewhere provides more heat and increased metabolic activity. Brown fat is found in the human fetus but not in the adult. The real significance of its function is still controversial.

Dense connective tissue

In dense connective tissue, the fibers are closely packed together, the cells are fewer, and there is less amorphous ground substance (Fig. 5-4). There are two main types of fiber arrangement in dense connective tissue: the irregular, or interwoven, and the regular, or parallel.

In the irregular, or interwoven, type the fibers are so arranged that they withstand exertion from different directions. It is usually found in sheets with the fibers interlacing to form a coarse feltwork. This type has a subdivision, the primarily elastic forms. The collagenous irregularly arranged tissues include the fasciae, the capsules of various organs (spleen, testis), the sheaths (periosteum of bone, perichondrium of cartilage, epimysium of muscle, dura mater of nervous system), the septa and trabeculae (partitions in organs), and the dermis of the skin. The dominant cells are fibroblasts. Elastic irregularly arranged tissues exist as tubular sheets of elastic fiber networks. This type includes the elastic membrane of large arteries, which are concentric cylindrical tubes that expand and relax in response to the blood volume delivered from the heart.

In the regular, or parallel, arrangement of dense connective tissue, the fibers are closely packed and lie parallel to each other, forming very strong structures with the tension in one direction. This group includes tendons, ligaments, and aponeuroses. Ligaments and aponeuroses may be less regularly arranged than tendons, but in general all have a similar organization. In tendons, which attach muscles to bones, the collagenous fibers are organized as **fascicles** of **primary tendon bundles** and run parallel courses. Each fiber, or bundle, consists of a large number of fibrils. Fibroblasts are crowded in rows in between, so that they appear flattened in longitudinal sections and stellate in transverse view. Around each primary bundle there is some loose areolar connective tissue (the **endotendineum**). Secondary bundles (groups of primary bundles) are bound by a coarser type of connective tissue, the **peritendineum.** The whole tendon, containing a variable number of fascicles, is surrounded by a still denser connective tissue sheath, the **epitendineum.** Nerves and blood vessels do not enter the tendon fascicles but are restricted to the major connective tissue septa.

Aponeuroses are broad and flat, but they have the same composition as tendons. Ligaments usually have the same organization as tendons, but some are composed mostly of elastic fibers. In elastic ligaments, parallel elastic fibers are bound together by connective tissue. The fibers branch and fuse with one another. Individual fibers are surrounded by a network of reticular fibers. Such ligaments have many oval or elongated nuclei of fibroblasts between the parallel elastic fibers. Examples of yellow elastic ligaments are the ligamentum nuchae of quadrupeds, the ligamenta flava of vertebrae, suspensory ligament of the penis, and the true vocal cords.

SPECIALIZED CONNECTIVE TISSUE (SKELETAL TISSUE)

Specialized connective tissues (e.g., cartilage and bone), like all connective tissues, are composed of cells, fibers, and ground substance. They differ from the types of connective tissues already discussed by possessing a rigid matrix (ground substance and fibers). Cartilage and bone are often classified together as supporting tissue, but bone especially goes far beyond structural support and has metabolic implications of the utmost importance.

Cartilage tissue

Cartilage, or gristle, is specialized toward rigidity but also has flexibility and resiliency. Cartilage forms the skeleton of the lower vertebrate animals (cyclostomes and elasmobranchs) as well as that of mammalian fetuses. The ground substance of cartilage is mostly chondromucin, a glycoprotein rich in chondroitin sulfate. The cells that form cartilage are derived from mesenchymal cells as chondroblasts, but during the process of cartilage formation they become entrapped within the matrix and are then called chondrocytes (Figs. 5-1 and 5-5). Collagenous or elastic fibers are embedded in the ground substance of cartilage but are usually invisible because they have the same refractive index as the matrix. They can be revealed by special staining methods. These fibers increase the tensile strength and elasticity and adapt the cartilage to the mechanical requirements of the body. The ultrastructure of cartilage has been recently described by Anderson (1964).

Surrounding cartilage is a dense fibrous connective tissue known as the **perichondrium** (Fig. 5-5). It covers the cartilage except at articular surfaces. Blood vessels and nerves enter the perichondrium from the surrounding connective tissue. The outer layer of the perichondrium consists of compact bundles of fibers with fibrocytes; the inner layer is a transition region where fibers merge with the matrix and chondrocytes are being formed. In general, cartilage lacks intrinsic blood vessels, nerves, and lymphatics. In large cartilages such as occur in elasmobranchs there are channels in which blood vessels run. Cartilage gets its nutrition by a process of seepage into the matrix from the perichondrium, since there are no canaliculi to carry the nutrients and waste. Articular cartilage gets its nutrition from the synovial fluid. Nutritional requirements are modest; cartilage has a low metabolism.

The composition of the cartilage matrix is complex. It has a high content of chondromucin, a glycoprotein that yields chondroitin sulfate on hydrolysis. The matrix also has a high content of water and other chemical substances not well understood. Chondromucin is especially concentrated around cells and cell groups forming the **cartilage capsules.** When stained with toluidin

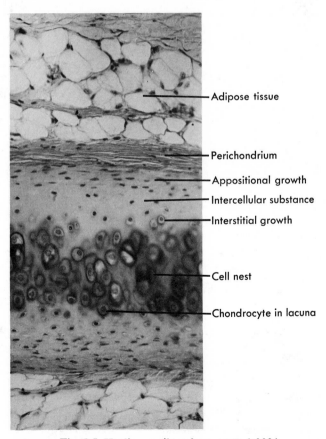

Adipose tissue

Perichondrium

Appositional growth

Intercellular substance

Interstitial growth

Cell nest

Chondrocyte in lacuna

Fig. 5-5. Hyaline cartilage from a cat. (×300.)

blue (metachromatically), the matrix gives a positive Schiff reaction (PAS), which indicates a carbohydrate component because chondroitin sulfate is not positive to this test. When the matrix is boiled in water, it forms chondrin, or cartilage glue.

Cartilage develops from mesenchymal cells, which round up, become closely packed, and form the perichondrium membrane. The cells (chondroblasts) secrete ground substance, become separated from one another by the intercellular substance, and are embedded in the matrix as chondrocytes. The continued growth of cartilage is by two methods: interstitial growth and appositional growth (Fig. 5-5). Interstitial growth is characterized by the mitotic division of young chondrocytes, which lay down new matrix around them. The formation of new cells and the intercellular substance around them causes the cartilage to expand from within. As cartilage becomes older, the intercellular substance increases in amount and becomes stiffer, so that cartilage cannot expand from within, and thus interstitial growth is limited to young cartilage. In appositional growth the inner part of the perichondrium (where cells divide, increase in number, and differentiate into chondroblasts and finally chondrocytes) lays down a new layer of cartilage. Additional layers can be laid down as needed.

The basis for classifying the subtypes of cartilage is the difference in the kind and abundance of fibers within the matrix. There are three subtypes: **hyaline cartilage, elastic cartilage,** and **fibrocartilage.** Hyaline cartilage is by far the most common and also the most characteristic of all the cartilages. The adult skeletons of the elasmobranch and cytostome, for instance, are almost entirely of this variety. It is the embryonic endoskeleton of all vertebrates above the cyclostomes and persists as part of the endoskeleton of all higher vertebrates.

Hyaline cartilage. The name of hyaline cartilage comes from its white, translucent, glassy appearance, which is due to the character of its intercellular substance. It is widespread but varies with the developmental stages of the animal as well as with the fundamental evolutionary level of the vertebrate animal. In the adults of higher vertebrates, it forms the articular surfaces of diarthroses, the costal cartilages of the ribs, parts of the nose and larynx, the trachea, the bronchi, and the external acoustic meatus. In the embryonic stage, nearly all the skeleton is laid down as hyaline cartilage and replaced later by bone. However, cartilage undergoes retrogressive changes with age. In old age, many cartilages become calcified and are opaque, hard, and brittle. Cartilage also becomes less cellular, and the matrix less basophilic. In some cases coarse, silky fibers unlike collagenous fibers may appear. These are called **asbestos fibers.**

Hyaline cartilage is made up of cells and matrix (Figs. 5-5 to 5-7). The cartilage cells, or **chondrocytes,** are found in small cavities called lacunae within the matrix. In some lacunae there may be a single chondrocyte; in others there may be many cells, each group representing the offspring of a single chondrocyte. A group of cells in a single lacuna is called a cell nest. The chondro-

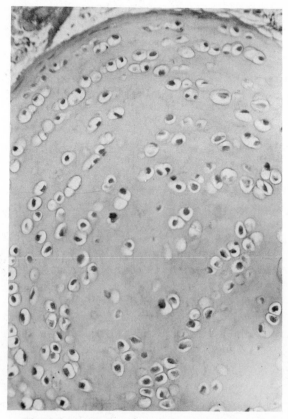

Fig. 5-6. Hyaline cartilage from lamprey. (×145.)

cyte is usually spherical and contains a centrally placed nucleus with one or more nucleoli. Cytoplasm is finely granular and somewhat basophilic and contains mitochondria, vacuoles, fat droplets, and glycogen. The chondrocytes have differentiated from mesenchymal cells. In living cartilage the chondrocytes completely fill their lacunae, but in fixed preparations they often shrink and seldom conform to the shape of the lacunae. When many cells are present in a single lacuna, there may be very fine partitions of intercellular substance between the individual cells, so that the area is broken up into a number of smaller lacunae.

The intercellular substance of hyaline cartilage is a firm gel and appears homogeneous in the fresh condition. After fixation, it may display fibrous and amorphous kinds of intercellular substance. The fibrous type is represented by collagenous fibers, which can be detected in thin sections with the polarizing microscope and the electron microscope. They can also be observed after digestion with trypsin. The fibers mostly appear as a fine feltwork.

Hyaline cartilage has some regenerative capacity but does not easily recover from an injury. New cartilage cells can develop at the junction of scar tissue with the cartilage. Fragments of articular cartilage must be removed when there is no chance for them to be reunited.

Elastic cartilage. Elastic cartilage occurs where flexibility is required, as in the external ear, internal auditory canal, epiglottis (Fig. 5-8), certain cartilages of the larynx, and the eustachian tube in higher vertebrates. In addition to collagenous fibers, its intercellular substance contains scattered elastic fibers that give it a yellow color. Elastic cartilage is more opaque than hyaline cartilage, and its cells have less fat and glycogen. Its branching elastic fibers may be visible, but the collagenous fibers are masked. The cartilage is surrounded by a perichondrium and growth occurs by both the interstitial and appositional methods. Elastic cartilage may be considered a modification of hyaline cartilage.

Fibrocartilage. Fibrocartilage has general stiffness and great tensile strength. It is found in the intervertebral and articular disks, the pubic symphysis (Fig. 5-9), rims of some sockets, lining of tendon grooves, and insertions of some tendons and ligaments. It is not a modification of hyaline cartilage and lacks a perichondrium. Wherever it is found, it always merges into the adjacent hyaline cartilage or dense fibrous tissue. Fibrocartilage consists of dense collagenous fibers, between the bundles of which are rows of chondrocytes in their lacunae. The collagenous fibers are arranged in a plane parallel with the tension pull. The intercellular substance is between the cells and the fiber bundles; its basophilic nature may indicate the presence of a great deal of chondroitin sulfuric acid.

Bone tissue

Bone, or osseous tissue, is a hard, specialized connective tissue that makes up most of the skeleton of higher vertebrates. It consists of cells and an intercellular matrix of both organic (mostly collagenous fibers) and inorganic components

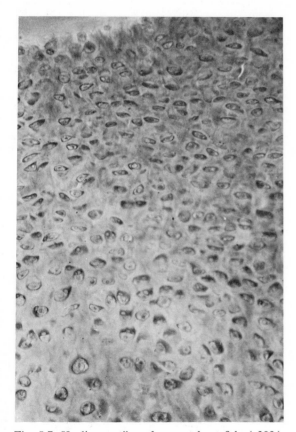

Fig. 5-7. Hyaline cartilage from a teleost fish. (×300.)

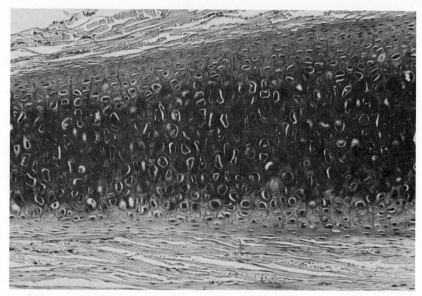

Fig. 5-8. Elastic cartilage in the epiglottis. Note scattered elastic fibers; collagenous fibers are hidden. Epithelial layers (not shown) vary from stratified squamous to pseudostratified columnar. Glands are usually found toward the base of the epiglottis. (×50.)

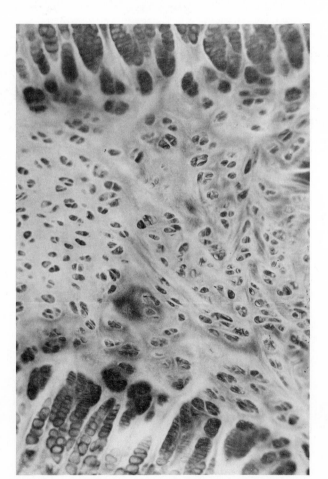

Fig. 5-9. Fibrocartilage in the pubic symphysis of a rabbit. (×140.)

(Ruth, 1947). The presence of mineral salts in its ground substance is one of the most distinguishing features of bony tissue. Bone not only forms the skeletal support of the body and protection for vital organs in cranial and thoracic cavities, but it also has metabolic functions of great significance to the body. Although it has a low metabolic rate, bone is not an inert tissue. It stores most of the calcium, which can be turned over and shifted to other parts of the body as needed. It houses the bone marrow, which is the chief hemopoietic tissue of the adult. Radioisotope studies show that bone is no more stable than any other organ but has a constant turnover of mineral salts, so that the whole skeleton is replaced every few months (LeGros Clark, 1965).

Bone is beautifully adapted for its functions, for it can perform its supporting function with an economical use of materials and with the least weight. A highly specialized adaptation related to this function is seen in the bones of birds (Fig. 5-10), which are filled with air spaces for maximum lightness. Bone is very sensitive to alterations in its mechanical functioning. When not used, it will atrophy; when used, it becomes hypertrophied. It undergoes internal reconstruction in response to external stimuli and can accommodate to changing stress. Bone can also be

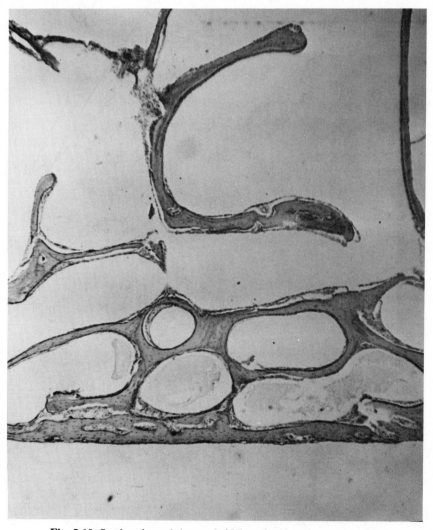

Fig. 5-10. Section through bone of chicken showing air spaces. (×35.)

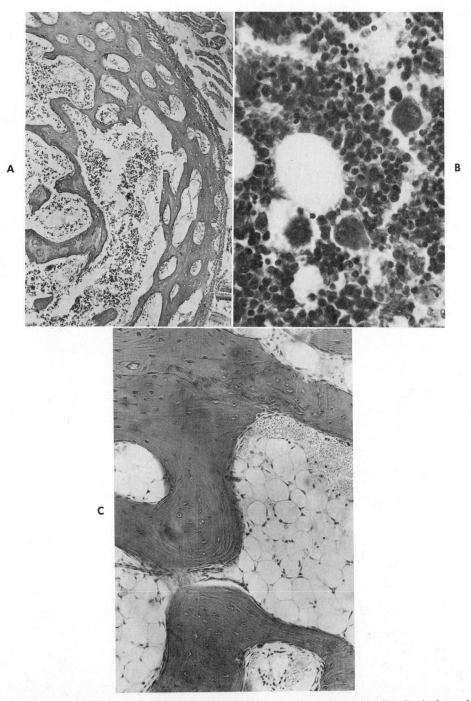

Fig. 5-11. A, Section of cancellous (spongy) bone showing irregular trabeculae. **B,** Active red bone marrow. The two large cells are giant marrow cells (megakaryocytes) that fragment to form blood platelets. The smaller nucleated cells are forerunners of the erythrocytes and leukocytes. Several fat spaces are present. **C,** Section of cancellous bone with yellow marrow. (**A,** ×58; **B,** ×600; **C,** ×300.)

modified by experimental procedures (Bourne, 1956).

The inorganic part of bone makes up about two thirds of its weight. This part consists of calcium phosphate (85%), calcium carbonate (10%), and small amounts of calcium fluoride and magnesium fluoride (5%). The inorganic components give hardness and rigidity to bone. The organic part of bone consists mostly of collagenous fibers (Bevelander and Johnson, 1951). This kind of collagen is often called **ossein** and will produce gelatin when boiled. The organic component provides tenacity, elasticity, and resilience to bone.

Bone requires special techniques for histologic study. One method involves removal of the inorganic components by decalcification in an acid medium. The bone can then be treated like other tissues. Many artefacts creep in by this method, since the cells are shrunken and the collagenous fibers tend to swell. Another method is to grind down a thin section of bone before it is mounted on a slide. This method destroys the bone cells, but the details of the matrix are well preserved.

Macroscopic structure of bone. Two types of bone can be distinguished—spongy (cancellous) and compact (dense). **Spongy bone** consists of intercrossing and connecting, slender, irregular trabeculae, or bars, of varying thickness and shape. These branch and unite with one another to form a meshwork within which are intercommunicating spaces filled with vascular bone marrow (Fig. 5-11, *A* and *B*). In early development this is red marrow (the myeloid tissue) of active hemopoiesis. Myeloid tissue contains the giant cells, megakaryocytes (Fig. 5-11, *B*), which fragment into blood platelets in mammalian bone marrow. These giant cells may also be found in the hemopoietic tissue of liver and spleen during embryonic development. Later the red marrow is more or less replaced by fat cells and called yellow marrow (Fig. 5-11, *C*). **Compact bone** appears as a solid, hard mass except for small microscopic spaces (Figs. 5-12 and 5-13).

However, no sharp boundary can be drawn between the two types of bone. They both have the same histologic elements, which are arranged differently in the two types. Nearly all bones have spongy and compact bone in their makeup, but the amount and distribution of each varies. In

typical long bones such as the femur and tibia the shaft **(diaphysis)** is mostly compact bone surrounding a medullary, or marrow, cavity of spongy bone. At the end of the shaft, the **epiphysis** consists of spongy bone with a thin, peripheral shell of compact bone. The cavities of the spongy bone are continuous with the marrow cavity of the diaphysis. In the growing animal the epiphysis and diaphysis are separated by the **epiphysial cartilage** plate, which is united with the diaphysis by columns of spongy bone **(metaphysis)** (Fig. 5-14). The epiphysial cartilage at each end of the long bone is the growing point in length of the shaft. Remodeling of the shaft occurs simultaneously with growth in length and thickness, so that the shape of the bone is retained. When the region of the epiphysial cartilage becomes ossified in the adult, the shaft can no longer grow in length. In

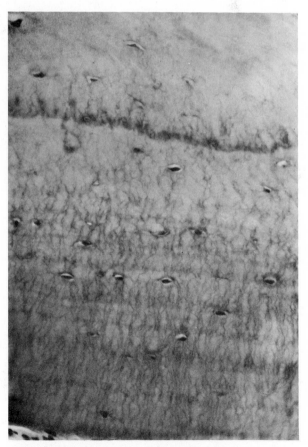

Fig. 5-12. Compact bone of frog, decalcified. Cross section shows osteocytes and canaliculi. (×650.)

Fig. 5-13. Thin ground section of compact bone showing haversian systems. Note remnants of old lamellae between the complete systems. (×400.)

flat bones (e.g., the skull) the compact bone forms a thick layer on both surfaces, enclosing a middle layer of spongy bone **(diploe)** of varying thickness. The short and irregular bones are made up mostly of spongy bone covered by a thin layer of compact bone.

Bone cells. Three types of cells are peculiar to bone: osteoblasts, osteocytes, and osteoclasts. They are closely interrelated, and transformation from one to another may occur. **Osteoblasts** differentiate directly from mesenchyme and do not divide when mature. They are associated with bone formation and come to lie on the surface

of bone where osseous matrix is being deposited. They frequently form a continuous layer, similar to cuboidal epithelium. Their nuclei are large, usually with a single nucleolus. The osteoblast cytoplasm is basophilic and rich in endoplasmic reticulum and ribosomes for the synthesis of protein components of the matrix. The cells contain the enzyme alkaline phosphatase, which is involved with calcification.

Osteocytes are formed from osteoblasts when the latter become imprisoned in the matrix. The osteocyte has a darkly staining nucleus and fine cytoplasmic processes that extend for some dis-

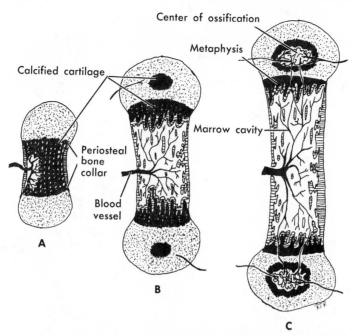

Fig. 5-14. Stages in the ossification of a long bone. **A,** Periosteal bud of embryonic connective tissue and blood vessels has eroded an opening through the periosteal bone collar. **B,** The cells (chondroclasts, osteoblasts, and fibroblasts) carried in by the periosteal bud dissolve away the calcified cartilage and replace it with bone and bone marrow. Early centers of calcification in epiphyseal centers shown. **C,** Further stage at time of birth shows the ossification of the epiphyseal centers and vascularization of the marrow cavity. Cartilage in black. (Modified from Windle, 1969.)

tance into the canaliculi (Fig. 5-15). These cytoplasmic processes are mostly withdrawn in mature osteocytes, but the canaliculi remain for the exchange of metabolites between the bloodstream and the osteocytes. Some of the cells are faintly basophilic and may have two nuclei.

Osteoclasts are multinucleated giant cells that vary in size. Their cytoplasm appears foamy and somewhat basophilic when young, but it becomes acidophilic in older cells. Osteoclasts are commonly found in regions where bone is being resorbed, but evidence for their capacity to dissolve bone matrix is not convincing. The cell is the product of cell fusions and not of repeated nuclear division (Mathews and collaborators, 1967). Some may be derived from the stromal cells of the bone marrow, and some by the fusion of osteoblasts. Some osteoclasts are found in shallow pits (Howship's lacunae).

Bone matrix (microscopic structure). Although the bone matrix appears to be homogeneous, it is really double. The organic part is composed of osteocollagenous fibers, collected in bundles 3 to 5 μm thick. These fibers are united by a mucoalbuminoid cement substance and can be demonstrated by silver impregnation and other methods. The inorganic component is located in the cement between fibers. Calcium phosphate is the most common inorganic deposit and has an **apatite** pattern. The basic structure of the bone material is hydroxyapatite [$Ca_{10}(PO_4)_6(OH)_2$]. The minerals are deposited as dense particles in a regular, orderly manner with respect to the fibers.

The most characteristic feature of bone is the lamellar arrangement of the calicified interstitial substance, or **bone matrix.** The arrangement of the lamellae is different in spongy and in compact bone (Pritchard, 1952). The lamellae are fibrillar in structure. Small cavities (lacunae) occur in the interstitial substance and are completely filled with the osteocytes, or bone cells. Radiating from each lacuna are numerous thin canals, the bone **canali-**

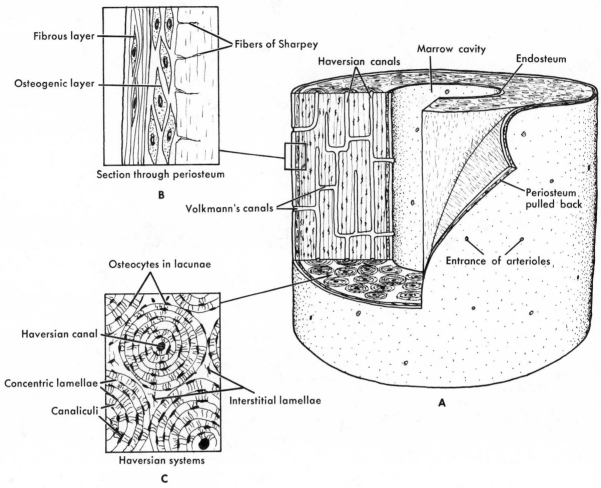

Fig. 5-15. Diagram of section through long bone, **A,** showing arrangement of structures. Box inserts, **B,** and **C,** show periosteum and haversian systems in greater detail. Only compact bone is indicated in the diagram. (Modified from many sources.)

culi, which pass into the hard interstitial substance in all directions (Fig. 5-12). These canaliculi branch and anastomose in a network that connects all the lacunae into a system of small channels (Fig. 5-16).

The lacunae and canaliculi have a border of special organic cement that differs from the rest of the intercellular substance in that it lacks fibers. This capsule often appears as a shining ring in unstained sections, but silver methods blacken it. Postnatal bony matrix is arranged in lamellar layers 3 to 7 μm thick, resulting from the rhythmical manner in which matrix is deposited. The fibers of each lamella are parallel to each other,

in a spiral pattern. Fibers in one lamella often make an angle with those in adjacent lamellae.

Architecture of bone. The cortex, or outer part, of a bone is usually compact, whereas spongy bone is found in the medulla. The structure of spongy bone is rather simple, although varied in form. It is made up of a network of trabeculae, the pattern of which is varied by the mechanical functions of individual bones. The parts are usually so arranged as to take care of the lines of maximum pressure or tension. The trabeculae consist of varying numbers of lamellae, in the interstitial substance of which are the lacunae containing osteocytes and an intercommunicat-

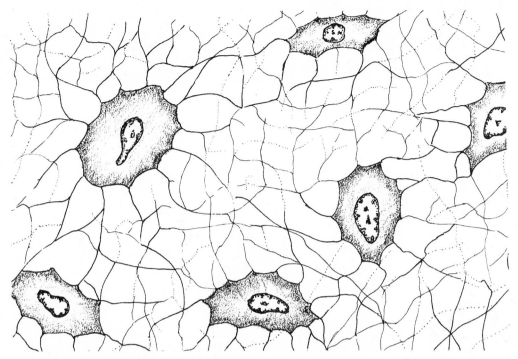

Fig. 5-16. Group of bone cells (osteocytes) with anastomosing processes that lie within canaliculi. Interstitial substance is found among the processes. (×900.)

ing system of canaliculi. Haversian canals are lacking, although there are some cylindrical cavities surrounded by concentric lamellae, which simulate haversian systems.

In compact bone the lamellae are regularly arranged in a pattern determined by the distribution of blood vessels carrying nourishment to the bone. The compact part of the diaphysis of any long bone is traversed by numerous cylindrical, branching, and anastomosing channels, the **haversian canals** (Fig. 5-13). These canals carry blood vessels, connective tissue, and nerves. They communicate with the canals of Volkmann, which come from the periosteal and endosteal surfaces at right angles to the long axis of the bone (Fig. 5-15).

The haversian system (0.05 to 0.1 mm in diameter) is the unit of structure of compact bone. Each haversian canal is surrounded by six to twenty concentric lamellae, each of which is 3 to 7 μm thick. Thus each haversian system consists of the central canal, the cells in the lacunae between the lamellae, and the lamellae of bone

matrix. Haversian systems are directed chiefly in the long axis of the bone. In cross section the canals appear as round openings with the ring-shaped lamellae around them; in longitudinal sections the canals are long slits, bordered by columns of lamellae. These canaliculi that border a haversian canal communicate directly with the canal. Those on the periphery loop back. The areas between the haversian systems are filled with the interstitial lamellae, which are highly irregular in size and shape. Some are parallel to the surface of the bone and represent the surviving remnants of earlier periosteal lamellae. Other lamellae are the curved remnants of haversian systems partly destroyed in the internal reconstruction of the bone. Some of these remnants can be seen in Fig. 5-13.

Coarse collagenous bundles known as penetrating fibers pass in different directions from the periosteum through the systems of lamellae (Fig. 5-15, *A*). They are not found in the haversian system nor in the endosteal lamellae. They are inserted at right angles to the surface and are surrounded by uncalcified, or partially calcified, matrix. They serve to anchor the periosteum firmly to the bone. Some **elastic fibers** from the periosteum also penetrate the bone.

A special, dense connective tissue membrane, the **periosteum** (Fig. 5-15) covers bone except at articular cartilage surfaces. It forms a close attachment at the epiphyses of long bones and at the places where tendons and muscles are attached. The membrane consists of two layers not sharply demarcated from each other. The outer layer is a network of dense connective tissue and is highly vascular. The deeper layer next to the bone consists of loosely arranged bundles of collagenous and elastic fibers and contains many spindle-shaped connective tissue cells. On stimulation these cells become activated into osteoblasts (Taylor and Yeager, 1966). Osteoblasts are not found in the normal adult bone, but a fracture or injury will cause them to reappear. From the deeper layer, blood vessels pass through **Volkmann's canals** to the haversian canals. A delicate layer, the **endosteum,** lines the marrow cavities, covers the irregular surfaces of spongy bone, and passes as a lining into the canal system of compact bone. The endosteum consists of condensed connective tissue; during active development it contains active osteoblasts. It may also have some hemopoietic potencies.

Development of bone (osteogenesis). Bone tissue has a number of unique features. Merely stiffening cartilage by calcification would not be satisfactory because the deposit of lime salts would prevent metabolic exchange. Bone has overcome the problem of rigid support in four distinctive ways:

1. Bone has a canalicular system so arranged that its fluid is continuous with the fluid in tissue spaces. Metabolites can thus be exchanged for the nourishment and well-being of the bone cells, even though they are surrounded by a dense, calcified intercellular substance.

2. Bone has an extensive vascular supply within its matrix. Bone cells can live only within a maximum range of about 0.5 mm from a capillary. Hence bone has many capillaries carried in special canals (haversian and Volkmann's).

3. Bone can grow only by an appositional mechanism; interstitial growth, as seen in carti-

lage, is impossible because lime deposits prevent expansion. Growth in bone thickness is accomplished by the addition of matrix directly under a fibrous and vascular periosteum; in the elongation of bone, the **epiphysial disk** grows near the bone's outer surface, while the bone's inner surface is destroyed and replaced by bony tissue.

4. Bone is a plastic substance that is continually undergoing change. It is destroyed locally and reformed by ossification time after time. Osteogenesis, or ossification, is not the same as calcification. Osteogenesis involves the formation of all the components of bone and not merely its mineral deposits. Osteogenesis can occur only when osteoblasts can secrete the special organic intercellular substance characteristic of bone.

In its formation, there are only two kinds of bone, immature bone and mature bone. **Immature bone** occurs in embryonic development or in the repair of bone fracture. This type of bone is characterized by the presence of many cells and by thick bundles of collagenic fibers without any regular arrangement. It also has less cement substance and mineral content than mature bone. Almost all immature bone is replaced by mature bone, but where immature bone does persist, it is usually mixed with mature bone. In contrast, **mature bone** has new layers added to bony surfaces in an orderly fashion. The osteoblasts that produce the successive layers of lamellated bone are found within or between the layers of bone matrix that they form. The collagenous fibers in a given lamella are usually at an angle to those fibers of adjacent layers. Mature bone has more cement and mineral substance than immature bone but has relatively fewer cells and flatter lacunae. Mature bone stains evenly, compared to the uneven staining of immature bone.

No definite tissue such as muscle or epithelium can be transformed into bone. Wherever bone forms, the osteoblasts must first appear, for they alone can produce the substance of bone. Little pieces of bone are known to occur in places where bone normally is not found (scars, skin, tonsils, etc.), but this is because an undifferentiated mesenchymal cell in that region is stimulated to divide and give rise to osteoblasts and osteocytes. This type of ossification is called **ectopic** or heterotopic. The sesamoid bones found in the tendons and ligaments of certain reptiles represent a kind of heterotopic ossification although their presence is normal.

Bone can be formed by two different processes, intramembranous ossification and intracartilaginous (endochondral) ossification. The end product is the same—immature bone that is eventually replaced by mature bone.

Intramembranous ossification. Intramembranous ossification is so called because the bone forms in primitive connective tissue and there is no cartilaginous precursor. It may also be called **direct** bone formation, since it is initiated by the differentiation of mesenchyme into loose connective tissue containing collagenous and reticular fibers, producing a loosely woven feltwork, or membrane. The dermal scales of fish and the dermal bones of the skull are laid down by this direct method. Such bone, however, has no membranous character, nor does membrane turn into bone. Intramembranous ossification begins when a group of mesenchyme cells differentiate into osteoblasts. The sites where the osteoblasts first appear are called centers of ossification. Two centers of ossification commonly occur for each of the bones of the skull vault. The osteoblasts next form thin bars of dense intercellular substance, which mask the connective tissue fibers present within the matrix (Fig. 5-17). At this

stage the matrix is not calcified and is termed the **osteoid.** Later, spicules of bone arise in a center and spread to form a wheellike pattern.

Primitive connective tissue cells at the periphery continue to transform into osteoblasts and to lay down new spicules. During the process some of the osteoblasts become embedded in the calcified matrix and remain there as osteocytes. As calcified matrix is formed around the osteoblasts and their processes, lacunae and canaliculi appear. The canaliculi connect with one another and with those of adjacent lacunae because of the contact between cell processes. In this way, a network of bone is built up, containing in its meshes blood vessels and connective tissue with osteoblasts. Through the continued activity of the osteoblasts, new layers of bone are deposited by apposition on the existing surfaces, and the trabeculae become thicker, so that the network increases and encroaches on the vascular spaces. The osteoblasts on the surface increase by mitosis and by the formation of osteoblasts from undifferentiated cells in the surrounding connective tissue.

The bone first formed is called **primary spongy bone.** It is nonlamellar, and its fibers are haphazardly arranged in the matrix. It consists of spicules, plates, and trabeculae. The spaces between the bony plates, **primary marrow cavities,**

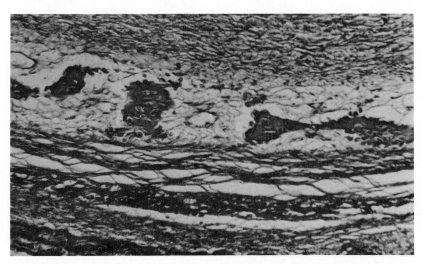

Fig. 5-17. Developing membrane bone in fetal pig. Small islands, or trabeculae, of new bone are shown across section. Around edges of islands are numerous osteoblasts. (×180.)

are filled with vascular connective tissue that is gradually transformed into hemopoietic, or blood-forming, tissue (Fig. 5-11, *B*). The original membrane is transformed into periosteum as the bone develops inside it. This cancellous, or spongy, bone is converted into compact bone as additional layers are added to the surface of the trabeculae, so that regions of little bone and large spaces are converted into those of much bone and narrow spaces. Compact bone occurs whenever bone substance becomes the dominant feature of the tissue. In intramembranous bone, compact bone is laid down into poorly defined concentric systems.

Membrane bones are represented by the flat bones of the cranium, irregular bones of the face, and to a certain extent, the temporal, occipital, and sphenoid bones.

Intracartilaginous ossification. In intracartilaginous, or endochondral, ossification the bones are preformed in cartilage. This cartilage model is replaced by bone roughly retaining the shape of the cartilage-bone in the process. The cartilage represents an intermediate stage not present in developing membrane bones; the cartilage is eroded, and ossification takes place in the eroded regions. During development the cartilage is replaced by bone except at the articular surfaces of the joints. The total process of ossification goes on until the bone has reached its full size and growth has ceased.

The process varies in complexity with the different types of bones. It is simple in short bones, because of the lack of epiphyses; it is most complex among irregular bones because they may have multiple centers of ossification (primary and secondary). It is customary to use a long bone of the leg or arm as an example of the essential features of intracartilaginous bone development. Such a bone has one primary ossification center for the shaft (diaphysis) and one secondary center for each end (epiphysis). Primary centers usually appear before secondary ones. The earliest primary center may appear in early fetal life: in man the latest occurs in tarsal and carpal bones sometime during childhood. Only a few secondary or epiphysial centers of ossification are present in cartilages before birth, although this is variable among vertebrates.

The extent of ossification is not the same in all classes of vertebrates. Of course, no ossification occurs among the Cyclostomata and Chondrichthyes because bone is not present in those groups. The Osteichthyes have an enormous range of variability because some of them are far more primitive than others, and their skeletons range all the way from mostly cartilaginous to almost complete ossification. The Teleostei are the most bony of all fish, yet many of them show a progressive reduction of ossification and a partial return to a cartilaginous condition (Goodrich, 1930). Only the higher vertebrates (reptiles, birds, and mammals) hold to a general scheme of ossification, and there are variations among them.

A cartilaginous model of a future bone may be taken as the first step in the intracartilaginous development of bone. Where a bone is to be laid down, a condensed mass of mesenchyme collects and begins to differentiate into cartilage. The mesenchyme adjacent to the sides of the cartilage becomes a surrounding membrane, the perichondrium; its outer (fibrous) part becomes a sort of connective tissue sheath. The mesenchymal cells of the inner part of the perichondrium remain more or less undifferentiated and become the chondrogenic layer of the membrane. The cartilage model increases in length and width by both interstitial and appositional mechanisms. A cartilage so formed becomes a provisional model for the bone (Fig. 5-14). The site of the first ossification is indicated by the hypertrophy of the chondrocytes in the midsection of the shaft of a long bone (the diaphysial center). As these cells enlarge, the intercellular material about them becomes thinned. When these chondrocytes become sufficiently large, they form **phosphatase,** an enzyme involved in cartilage and bone formation. Lime salts are deposited in the cartilage matrix, and it thus becomes calcified. This closes off the hypertrophied chondrocytes from their source of nutrition, and they soon die. With the death of the cells, the intercellular substance begins to break up and dissolve away, leaving cavities in the future bone model.

Bone formation actually starts within a ring-shaped area of the perichondrium. A differentiation pattern begins, which is associated with the invasion of the perichondrium by capillaries, and the transformation of the perichondrium into the periosteum. The former undifferentiated cells of the perichondrium now differentiate into osteoblasts and osteocytes, resulting in formation of a thin, ring-shaped shell of bone around the center of the diaphysis (Fig. 5-14). This collar encloses about one third of the cartilage model. This bone, formed in the manner of intramembranous bone, is called **periosteal bone.** Vascular connective tissue from the periosteum, called a **periosteal bud,** pushes through gaps in the bony collar into the cavities produced by the disintegration of the calcified cartilage previously described. As this bud advances, the thin partitions between the cartilage cells dissolve away, and the cartilage

cells die. In this region the osteoblasts gather around the remnants of calcified cartilage and deposit bone. Thus the first bone formed is cancellous, with its trabeculae having cores of calcified cartilage. This deposition of bone in the center of the diaphysis is called the **primary ossification center.** The zone of intracartilaginous ossification extends toward the ends of the cartilage in a sequence of changes similar to those establishing the primary center.

In the meantime, the periosteum continues to add further bone to the sides of the model bone, so that the periosteal bone collar becomes thicker and widens toward the epiphyses. This strengthening of the shaft makes support by the cancel-

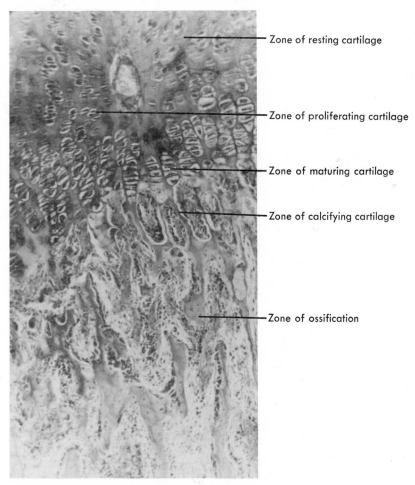

Zone of resting cartilage

Zone of proliferating cartilage

Zone of maturing cartilage

Zone of calcifying cartilage

Zone of ossification

Fig. 5-18. Developing bone. This type of osteogenesis is endochrondral bone formation and occurs in certain successive zones. (×200.)

lous bone unnecessary, and as it dissolves away, it leaves a **marrow cavity.** In this way the periosteal bone collar acts as a buttress to support the central zone of reabsorbing cartilage prior to its replacement by bone. The marrow cavity is always separated from each cartilaginous end (epiphysis) by a region of spongy cancellous bone.

As the cartilage in the epiphyses grows, cartilage of the entire model bone increases in size. Epiphysial centers of ossification appear later in the growing cartilaginous ends. The development of these centers and the extension of the primary ossification center produces a series of gradual changes or zones within the cartilage. Each zone changes character as ossification advances toward it. These zones are described as follows, listing from the ends of the cartilage toward the primary center of ossification (Fig. 5-18):

1. *Zone of resting cartilage.* This is composed of primitive hyaline cartilage adjacent to the bone. The zone shortens progressively as advancing ossification encroaches on it. Chondrocytes are scattered through its intercellular substance.

2. *Zone of proliferating cartilage.* This zone shows extensive mitosis among young cartilage cells (Fig. 5-18). The cells are somewhat wedge-shaped and piled on top of one another, so that they form columns whose axes are parallel to that of the bone. Cells are added near the resting zone. Cells are produced to replace those that die at the diaphysial surface of the disk.

3. *Zone of maturing cartilage.* Cartilage cells are in various stages of maturation. Those cells nearest the proliferation zone are the least mature. Cells and lacunae are cuboidal.

4. *Zone of calcifying cartilage.* This zone is thin and abuts directly on the bone of the diaphysis (Fig. 5-18). Most of the cells are dead, and others have reached the peak in their life cycles. Spaces resulting from destruction of cells and matrix give the appearance of a honeycomb. Vascular primary marrow extends into the spaces. The matrix surrounding the enlarged lacunae may be deeply basophilic.

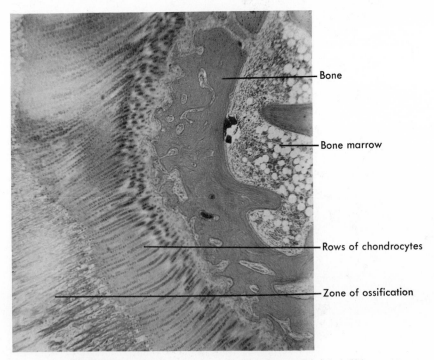

Fig. 5-19. Longitudinal section of rabbit femur at joint. (×60.)

Bone

Bone marrow

Rows of chondrocytes

Zone of ossification

5. *Zone of ossification.* Osteoblasts differentiated from mesenchymal cells lay down bone on the exposed plates of calcified cartilage. This results in bony trabeculae with cartilaginous cores (Fig. 5-19) that are continuous with the cartilaginous substance of the disk. Osteogenesis is very active at the diaphysial side of the epiphysial disk, so that bone is formed close to the cartilage of the disk. This process greatly extends the spongy bone already present.

6. *Zone of resorption.* As ossification advances toward the ends of the cartilage, there is a compensatory resorption of bone nearer the midpoint of the diaphysis. This occurs chiefly at the oldest, or central, ends of the bony mass. In this way the total spongy bone remains mostly constant, and the marrow cavity increases in length, now called the **secondary marrow cavity.** The growth of the periosteal collar compensates for the loss that occurs in resorption of the central endochondral bone.

Growth in length of a bone, as already mentioned, is made possible by the formation of new cartilage at the proximal surface facing the diaphysis in the epiphysial disks at the ends of the bone. When growth ceases, the epiphysial disk is replaced by bone. The epiphyses and diaphysis are united at the **epiphysial line.** Increase in length of bone is no longer possible. The two epiphyses of a long bone may not each contribute the same length to a bone. In the femur, for instance, growth in length takes place chiefly at the distal epiphysis; in the tibia, at the proximal one. Ossification at the epiphyses (secondary centers of ossification) is thus delayed until after birth and is not completed until the bone attains its final length. The sequence of changes of bone formation in these secondary centers is identical to that in the diaphysis. Cartilage cells in the epiphysis increase in number and size and vascular osteogenic, or bone-forming, buds enter from the periphery or through makeshift tunnels from the diaphysis. This results in the removal of cartilage and the deposition of bone. This ossification spreads in all directions until the cartilage is replaced except in two places. Cartilage remains over the free end as the **articular cartilage,** and between the epiphysis and diaphysis as the epiphysial plate.

Although bone cannot increase in length after the full ossification of the epiphyses, it can grow in diameter by the deposition of new periosteal bone, which forms intramembranously by appositional growth. As bone is added to the periosteal

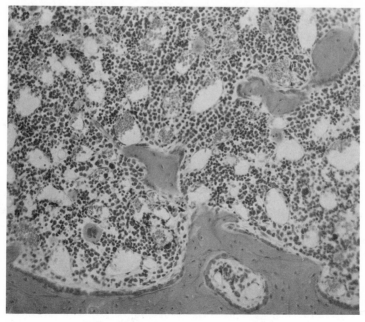

Fig. 5-20. Red marrow of rib. (×400.)

surface, bone is resorbed (although in lesser amounts) from the endosteal surface, so that the secondary marrow cavity increases. This cavity and the irregular spaces of spongy bone are filled with soft, vascular marrow tissue (the myeloid tissue) (Figs. 5-11, *B,* and 5-20). In the young animal (embryo and newborn) only red marrow is found; with age, the red marrow is partially converted into fat-containing yellow marrow (Fig. 5-11, *C*). The formation of blood cells (hemopoiesis) occurs chiefly in red marrow because in yellow marrow most of the tissue has been replaced by fat.

Bone is not static, but is constantly undergoing internal resorption in certain areas and deposition of new bone in other regions. Resorption and dissolution of bone are often associated with osteoclasts, whose role in the process is not fully understood. As spongy bone in many areas is replaced by layers of compact bone, postnatal osteogenesis makes use of the haversian systems coursing lengthwise in the shafts of long bones. Primitive haversian systems are formed in longitudinal tunnels produced in the peripheral regions of the epiphysial disks (Fig. 5-21). Others involve cylindrical cavities formed from the erosion

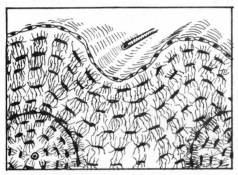

1. On shaft surface is longitudinal groove containing blood vessel.

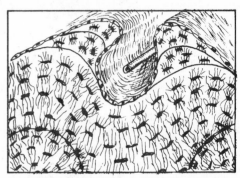

2. Sides of groove become higher by bone formation and tend to merge.

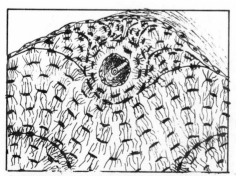

3. Ridges fuse to form tunnel, enclosing periosteum (now endosteum) with its osteoblasts and blood vessel.

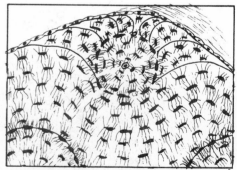

4. Layers of bone are laid down around haversian vessel by endosteal cells to form haversian system.

Fig. 5-21. Diagram to show how a haversian system is formed on the surface shaft of a growing bone. The surface of a growing bone is not smooth but has longitudinal ridges and grooves. The series of diagrams indicate how these grooves become tunnels and finally haversian systems. The numbers indicate the sequence of development. (Modified from Ham, 1969.)

of bone substance by vascular buds (Fig. 5-22). The tunnel becomes lined with osteoblasts, which differentiate from the primitive cells of the marrow. Successive lamellae of bone are deposited inward until the tunnel is reduced to a narrow canal containing blood vessels. These haversian systems are constantly destroyed and rebuilt. Those parts of the haversian systems not destroyed become interstitial lamellae and fill in between new canals. As bone growth nears completion, periosteal and endosteal lamellae are laid down as complete concentric layers. Rebuilding of bone continues much more slowly as aging occurs.

Short bones have no epiphysial disks and develop in a manner similar to the epiphyses of long bones. Ossification starts in the center of the cartilage and spreads in all directions. The peripheral cartilage proliferates until growth ceases, when it is replaced by bone. The surrounding layer of connective tissue becomes the periosteum, which adds compact bone by apposition.

When a fracture occurs, there is hemorrhage, tissue destruction, and clotting. Fibroblasts and capillaries invade the clot and convert it into granulation tissue known as the **procallus.** Cartilage develops within the granulation tissue and forms the **temporary callus** to unite the fractured bone. Osteoblasts of the periosteum and endosteum form new spongy bone, which replaces the cartilage of the temporary callus. Intramembranous bone formation with calcification and

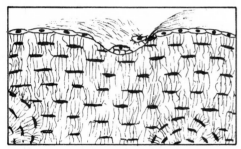

1. Osteoclasts erode groove in outer circumferential lamellae on surface of shaft.

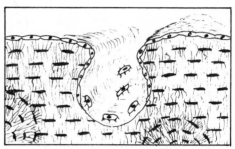

2. As groove deepens, periosteum with blood vessel enters groove.

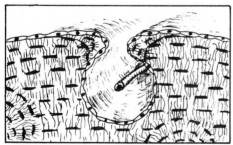

3. Groove is converted into tunnel by deposit of bone by osteoblasts and osteogenic cells on ridges.

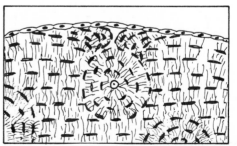

4. Successive lamellae of bone are now laid down by osteoblasts and osteogenic cells to form haversian system.

Fig. 5-22. Diagram of haversian system formation in older bones (cross section). When the surface of the shaft becomes smoother, appositional growth of bone under the periosteum forms circumferential lamellae, which must be replaced by haversian systems as the shaft becomes thicker. The numbers indicate the sequence of development. (Modified from Ham, 1969.)

erosion then gives rise to the **bony callus** of compact bone. Bony excesses are resorbed, and the healed bone assumes its original contour.

Many factors influence ossification. The growth hormone of the anterior pituitary is known to produce a growth pattern characteristic of the specific animal, but imbalance of this hormone may produce **dwarfism** or **gigantism.** Rickets may be produced in the animal by vitamin D deficiency. Lack of minerals (calcium, phosphorus, iron, etc.) in the diet may give rise to body deficiencies. Other vitamins and hormones also exert an influence on rate and extent of bone growth.

Joint structures

Most bones are connected by articulations, or joints. A joint may permit movement between the bones it connects, or it may be as solid as the bones themselves. Thus there are two principal types of joints: synarthroses, which are mainly immovable, and diarthroses, which are freely movable and possess a joint cavity. **Synarthroses** are of three common subtypes: **syndesmosis** in the sutures of the skull, where the connection is formed chiefly by dense collagenous tissue; **synchondrosis,** where the union is fibrocartilage, such as the pubic symphysis and intervertebral disks; and **synostosis,** where there is a gradual replacement of connecting cartilage or fibrous tissue by bone, as in the maturation of the skeleton (e.g., the union of an epiphysis and diaphysis).

In the freely movable **diarthroses,** hyaline cartilage covers the ends of the bones. This articular cartilage is left over from the general ossification. A **synovial joint cavity** is present between the bones, with a **joint capsule** surrounding the articulation. The outer part of this capsule is dense fibrous tissue, continuous with the periosteum over the bones at the margin of the articular cartilages. This tissue is sometimes thickened to form the ligaments of joints. The inner layer of the capsule is the **synovial membrane,** which lines the joint cavity except over the articular cartilage. The synovial membrane contains many elastic fibers and sometimes intra-articular fat pads. The membrane may form folds or projections and is lined with spindle-shaped fibroblasts. It secretes **synovial fluid** that lubricates the joint. Nerves occur chiefly in the outer fibrous layer of the capsule.

References

Anderson, D. R. 1964. The ultrastructure of elastic and hyaline cartilage of the rat. Am. J. Anat. **114**:403-439.

Arey, L. B. 1968. Human histology; a textbook in outline form, 3rd ed. W. B. Saunders Co., Philadelphia.

Bevelander, G., and P. L. Johnson. 1951. A histochemical study of membrane bone. Anat. Rec. **108**:1-22.

Bourne, G. H. 1956. The biochemistry and physiology of bone. Academic Press, Inc., New York.

Chayen, J., S. Darracott, and W. W. Kirby. 1966. A re-interpretation of the role of the mast cell. Nature **209**:387-388.

Edds, M. V., Jr. 1958. Origin and structure of intracellular matrix. In W. D. McElroy and G. Glass, eds. The chemical basis of development. The Johns Hopkins University Press, Baltimore.

Fahrenbach, W. H., L. B. Sandberg, and E. G. Eleary. 1966. Ultrastructural studies on early elastogenesis. Anat. Rec. **155**:563-576.

Freeman, J. A. 1964. Cellular fine structure. McGraw-Hill Book Co., New York.

Goodrich, E. S. 1930. Studies on the structure and development of vertebrates. The Macmillan Co., London.

Ham, A. W. 1969. Histology. J. B. Lippincott Co., Philadelphia.

Hayward, J. S., C. P. Lyman, and C. R. Taylor. 1963. The possible role of brown fat as a source of heat during arousal from hibernation. Ann. N. Y. Acad. Sci. **131**:441-448.

Jackson, S. F. 1964. Connective tissue cells. In J. Brachet and A. E. Mirsky, eds. The cell. Academic Press, Inc., New York.

LeGros Clark, W. E. 1965. Bone. In Tissues of the body, 5th ed. Oxford University Press, Fair Lawn, N. J.

Mancini, R. E. 1963. Connective tissue and serum proteins. Int. Rev. Cytol. **14**:193-205.

Mathews, J. L., J. H. Martin, G. J. Race, and E. J. Collins. 1967. Giant-cell centrioles. Science **155**:1423-1424.

Maximow, A. 1932. The macrophages or histiocytes. In Cowdry, E. V., ed. Special cytology, 2nd ed. Paul B. Hoeber, Inc., New York.

Menschik, Z. 1953. Histochemical comparison of brown and white adipose tissue in guinea pigs. Anat. Rec. **116**:439-455.

Meyer, K. 1947. The biological significance of hyaluronic acid and hyaluronidase. Physiol. Rev. **27**:335-359.

Pritchard, J. J. 1952. A cytological and histochemical study of bone and cartilage formation in the rat. J. Anat. **86**:259-277.

Ruth, E. B. 1947. Fibrillar structures of adult human bone. Am. J. Anat. **80:**35-53.

Smith, D. E. 1963. The tissue mast cell. Int. Rev. Cytol. **14:**327-338.

Smith, R. E., and R. J. Hock. 1963. Brown fat: thermogenic effector of arousal in hibernators. Science **140:** 199-200.

Sobel, H. 1967. Ageing of the ground substance in connective tissue. Adv. Gerontol. Res. **2:**205-283.

Taylor, J. J., and V. L. Yeager. 1966. The fine structure of elastin fibers in the fibrous periosteum of the rat femur. Anat. Rec. **156:**129-141.

Windle, W. F. 1969. Textbook of histology, 4th ed. McGraw-Hill Book Co., New York.

6 THE MUSCLES

In the lower invertebrates, muscle is a relatively simple tissue with smooth contractile fibrils in the cytoplasm of its cells. But even before the evolution of vertebrates as such, the muscle tissue of some invertebrate forms had differentiated into the striated type in which the fibrils have a complicated periodicity and run for great distances within columns, or muscle fibers. In further evolution of individual groups of invertebrates, striated muscle reaches a very high degree of morphologic differentiation and functional efficiency. These features are seen among the arthropods, especially in the insects.

The contractile material of striated muscle fibers in vertebrates shows a lamellar architecture in the early development of the tissue (Fig. 6-1). This lamellar structure resembles that found in the muscle of adult annelids, nematodes, and insects. It probably represents a prevertebrate, even a prechordate, condition. As development proceeds, the lamellae break up to form threads, or fibrils, rather than plates. For a while, the radiate arrangement of the fibrils is retained, as often seen in the cardiac muscle of vertebrates, even in humans.

GENERAL STRUCTURE AND FUNCTION

Muscle is the tissue of motion and is widely distributed in various organs of the body. Indeed, each skeletal muscle can be considered as an organ, although it must not be forgotten that the muscles generally act in groups and are dependent on each other in various ways. For example, the heart is an organ composed largely of a special type of striated muscle. Also, smooth mus-

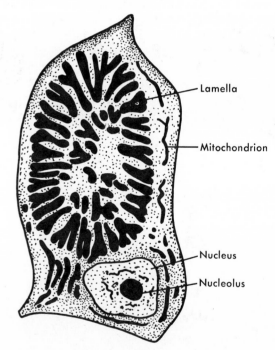

Fig. 6-1. Cross section of skeletal muscle fiber of newly hatched rainbow trout, *Salmo gairdneri*. Note lamellar structure of contractile material. A filamentous mitochondrion is seen near the sarcolemma. In further development contractile masses increase in number and decrease in size by a process of concentric and radial splitting. Early condition, as pictured here, is similar to that found in fibers of nematode and annelid worms and in arthropods. (Redrawn from Heidenhain, 1913.)

cle may be said to form tiny organs such as the arrectores pilorum that bring about erection of the individual hairs.

Muscle is composed of elongated muscle fibers, each an individual muscle cell, held together by connective tissue. There are three varieties of muscle: **striated** (voluntary), **smooth** (involuntary), and **cardiac** (striated, but involuntary). The fine structure of muscle cells, or fibers, has been revealed by the electron microscope to be basically similar among the vertebrates, and it will be described here before the discussions of the comparative aspects of this tissue.

Smooth muscle. Smooth muscle is relatively simple, with spindle-shaped fibers and usually a single, centrally located nucleus (Fig. 3-6). The ends of the fibers are generally unbranched. Smooth muscle fibers vary in length and thickness, being long and thin in the walls of the bladder and short and thick in the walls of the blood vessels.

The cytoplasm of smooth muscle cells contains filaments known as **myofilaments.** They are about 100 Å thick and correspond closely in diameter and orientation to similar structures seen in the fibers of cardiac and skeletal muscle. In the smooth muscle cell, myofilaments occur in rather loose bundles, within which are clefts of varying sizes containing the mitochondria and other organelles. The nucleus itself may be considered as occupying a large cleft in the bundle of myofilaments. Electron-microscope study has given evidence of a truly syncytial arrangement for at least some smooth muscle (Mark, 1956; Thaemert, 1959); the cytoplasm (**sarcoplasm)** of several individual fibers can be seen to be continuous.

In the dermis (and a few other regions) of some vertebrates, smooth muscle cells may be scattered singly. However, the cells are usually placed close together in sheets, with the thick middle portion of one cell opposite the thin ends of adjacent cells. A reticular network of connective tissue penetrates into the spaces between the cells and binds them together. The pull of the contracting smooth muscle fibers is first transmitted to this sheath of reticular fibers that pass directly into the surrounding connective tissue. By such an arrangement the force of contraction of the entire layer is uniformly transmitted to the enclosed tissue, for example, as in narrowing the lumen of a blood vessel.

Physiologically, smooth muscle is slow in action and is found where rapid action is not required. These locations include the muscular coat of the alimentary tract, blood vessel walls, the tracheobronchial tree, the urinary bladder, and the iris and ciliary muscles of the eye.

Skeletal muscle. Striated muscle fibers are more complicated. They are elongated, with several spindle-shaped nuclei scattered through the cell. The muscle cell, or fiber, has a well-defined cell membrane, the **sarcolemma.** In its early development the fiber has only one nucleus, but it acquires others by repeated nuclear division without division of the cytoplasm.

In skeletal muscle the contractile material is concentrated in **fibrils** with a diameter of 1 to 2 μm (Fig. 6-2). Under the electron microscope these fibrils appear as long columns, or "cables," with the myofilaments gathered rather closely together as the individual strands of the cables (Fig. 6-3). The fibrils show a pattern of striations similar to that seen in the fiber at the lower magnifications of the light microscope. Generally the striations of the fibrils are in line with each other; thus under the light microscope, the sum of the striations of the fibrils appears as a striation of the fiber (Fig. 6-4).

In the pattern of striations seen in skeletal muscle, the repeating unit is known as the **sarcomere.** The length of a sarcomere is usually 2 to 3 μm, somewhat greater than the diameter of the fibril. It is bounded at each end by a narrow dark line—the Z membrane, or **Z line.** Midway between the Z lines is a dense, birefringent band called the **A band** (anisotropic band); the Z line itself is found in the center of the **I band** (isotropic band), a less dense area almost lacking in birefringence. Thus the A band is found in the middle of a sarcomere that contains one half of an I band above and one half below the A band. The question is only academic as to whether the Z line belongs to one or the other of the sarcomeres that it delimits. In the center of the A band there is a narrow zone of lesser density, referred to as the **H zone,** with a dark **M line** through its center.

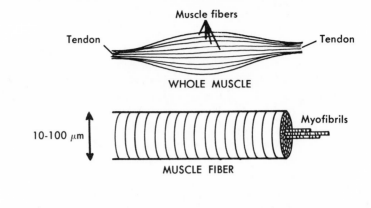

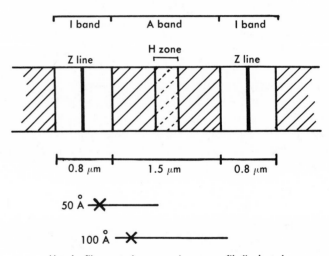

Muscle filaments (same scale as myofibril above)

Fig. 6-2. Structure of striated muscle as seen grossly and at different degrees of magnification with light and electron microscopes. The dimensions given are those for psoas muscle of the rabbit. (Redrawn from Huxley, 1959.)

Fine structure. Study of fine structure shows each fibril to be composed of many parallel myofilaments, each 50 to 100 Å in diameter. Whether the pattern of striations of the fibril is due to variations in these myofilaments, as such, or in some of the general sarcoplasm between them is an important question that has been at least partly resolved.

Actually, each fibril contains two kinds of myofilaments, differing in location, anatomic charac-teristics, and chemical composition. The **thick** filaments, about 100 Å in diameter and 1.5 μm in length, extend from one end of the A band to the other. The **thin** filaments, about 50 Å by 2.0 μm, extend in either direction from the Z line to the edge of the H zone. Thus, in the A band the thin myofilaments **interpenetrate** and overlap the array of thick filaments. It is not known whether there is some linkage between the ends of the thin filaments across the H zone. In cross section,

Fig. 6-3. Cardiac muscle of rat. Several fibrils with intervening sarcoplasm are shown. The A band, Z line, I band, and H line, as well as sarcoplasmic reticulum, *SC,* and mitochondria, *M,* are shown. (Perfusion with glutaraldehyde and postfixation by osmium tetroxide; ×47,500.) (Courtesy Dr. W. G. Forssmann, Geneva, Switzerland.)

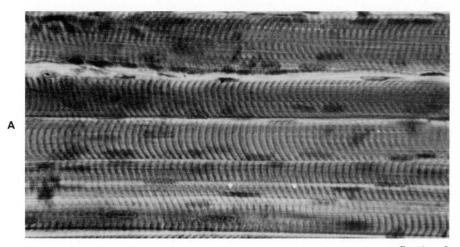

Continued.

Fig. 6-4. A, Striated muscle fibers of mammal (longitudinal section) showing cross striations and nuclei. **B,** Skeletal muscle of *Notophthalmus viridescens.* Individual fibrils of fibers have been partially separated out; each shows its pattern of striation. (**A,** ×800; **B,** iron hematoxylin and orange G, ×2000.)

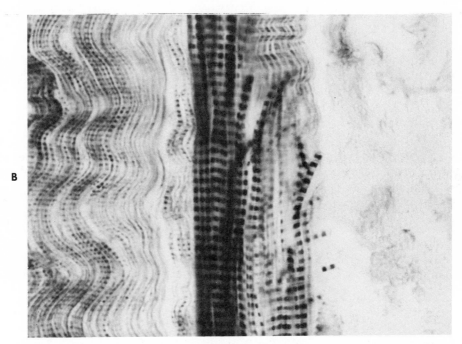

Fig. 6-4, cont'd. For legend see p. 95.

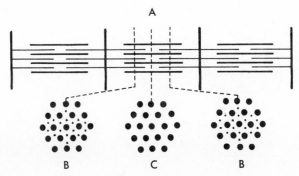

Fig. 6-5. Arrangement of thick and thin filaments of striated muscle. **A,** Longitudinal view, showing overlapping arrays. **B,** Cross sections through region of overlap, showing double hexagonal array. **C,** Cross section through H zone, showing simple hexagonal array of large filaments only. (Redrawn from Huxley, 1959.)

the regions where thick and thin filaments overlap show an hexagonal array in which the geometric relationship between thick and thin filaments is constant and in which there are twice as many thin filaments as thick ones (Fig. 6-5).

Chemically, the thick filaments are composed of the protein **myosin;** the thin filaments are the protein **actin.** Chemical linkages occur at regular intervals between thick and thin filaments, forming an **actinomyosin** complex.

Muscle contraction. The theory of muscle contraction is based partly on morphologic and partly on chemical evidence. When a striated muscle changes its length during contraction, the thin actin filaments are thought to slide farther into the array of thick myosin filaments. During this process, the length of each kind of myofilament apparently remains nearly constant. The resulting decrease in length of the fibril first brings the ends of adjoining groups of thin filaments together, obliterating the H zone. Then the thick filaments may also be brought into apposition, tending to eliminate the I bands.

The actinomyosin complex seems to be present only in small amounts in resting muscle. It is thought that adenosine triphosphate (ATP) maintains this condition in which few firm cross-links exist between myosin and actin components and in which the two kinds of filaments can slide easily past each other. In actively contracting muscle there appears to be a cyclic repetition of connection and disconnection; sliding of the actin filaments occurs at the times of disconnection (Huxley, 1959). Evidence for this concept of the mechanism of muscle contraction comes from various kinds of studies. These studies have been done with the light microscope, the electron microscope, x-ray diffraction, and chemical techniques. Approaches to the problem have included determinations of the location of muscle proteins and changes in the band pattern of muscle during contraction, relaxation, and stretch. Although the muscle studies give much support for the sliding-filament theory, it does not seem to make possible a clear comparison between the contractions of striated muscle and of unstriated smooth muscle.

It must be remembered that the three kinds of muscle fibers contain the organelles found generally in cells, as well as the specific contractile elements. Each fiber contains nuclei, often many in the larger skeletal muscle fibers; multitudes of mitochondria, especially in cardiac muscle; a Golgi apparatus, found at one or both poles of the nuclei; and the sarcoplasmic reticulum, first described in its ultrastructural aspects by Bennett and Porter (1953).

The **sarcoplasmic reticulum** is a delicate system within the muscle fiber, showing a complex architectural relation to the cross striations of the myofibrils. It appears to be a specially differentiated kind of endoplasmic reticulum. It consists of channels, tubules, and cisternae forming a set of fluid-filled, transversely oriented compartments, usually in a "triad" arrangement. Recent investigations have shown that contraction in a myofibril begins at the site of the triad of the sarcoplasmic reticulum. After stimulation of a muscle fiber, a depolarization of the sarcolemma is distributed to the interior of the fiber by way of the tubular system (the so-called T system) of the sarcoplasmic reticulum. The depolarization causes a release of calcium ions from elements of the sarcoplasmic reticulum to the myofibrils. These calcium ions activate the enzyme ATPase to break down ATP to ADP, with the release of energy. This is the energy that drives the contractile process. The ATPase is localized in the bridges between actin and myosin filaments. After contraction, the calcium ions return to the sarcoplasmic reticulum.

Modifications of striated muscle. Two distinct kinds of skeletal muscle fibers are found in most vertebrates. They are grossly distinguishable as red fibers and white fibers by their pigment contents. A band of muscle need not be composed exclusively of red or white fibers. Often it is made up of combinations of these fibers and a certain percentage of "intermediate" fibers as well.

Recently two workers have published similar details of histochemical and ultrastructural differences between these two kinds of fibers in two different groups of vertebrates. Gauthier (1969) worked on red and white bands in the semitendinosus muscle of the white rat (Fig. 6-6). Watanabe (1969) studied these muscle cells in the tails of tadpoles of two toads, *Bufo vulgaris* and *Bufo regularis,* and a Japanese frog, *Rana japonica* (Fig. 6-7).

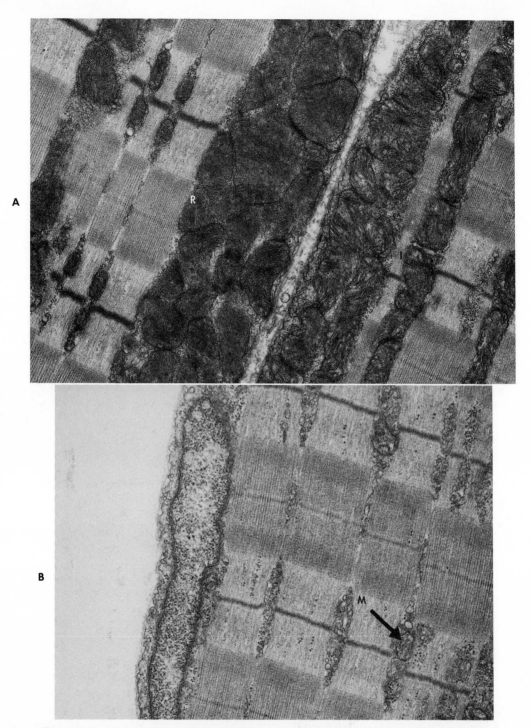

Fig. 6-6. A, Portions of a red fiber *(R)* and an intermediate fiber *(I)* from semitendinosus muscle of rat. Red fiber shows more numerous mitochondria with more closely packed cristae than intermediate fiber. Z lines are definitely thicker than in intermediate fiber. **B,** White fiber of semitendinosus muscle of rat. Relatively few mitochondria present. Z lines are narrower than in red or intermediate fibers. (**A,** ×19,800; **B,** ×18,800.) (**B** from Gauthier, G. F. 1969. Z. Zellforsch. Mikrosk. Anat. **95:**462-482.)

In general, the red fibers are smaller than the white ones and contain more lipid, as is readily shown by Sudan staining. Electron-microscope study shows a sparser distribution of sarcoplasmic reticulum and more numerous mitochondria in red fibers. The mitochondria lie in long chains between the myofibrils and just beneath the sarcolemma. They are longitudinally oriented and closely packed, as in cardiac muscle fibers (Fig. 6-8). The myofibrils themselves have thicker Z lines in the red fibers, and these Z lines run a more zigzag course than do the thinner ones of the white fibers (Fig. 6-9).

In higher forms, white muscle is characteristically found in structures requiring rapid contraction, and red and white fibers have come to be referred to as "slow" and "fast," respectively.

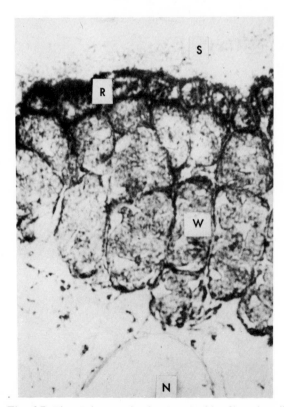

Fig. 6-7. Photomicrograph of red and white fibers in tail of *Bufo regularis* tadpole. Sudan black B stain. Red fibers are superficially located and contain more lipid than the white fibers. *N,* notochord; *R,* red fibers; *S,* skin; *W,* white fibers. (Courtesy Dr. K. Watanabe, Tokyo, Japan.)

Fig. 6-8. A, Longitudinal section of red fiber in tail of *Bufo regularis.* Chains of closely packed mitochondria, *M,* are seen between myofibrils. **B,** Longitudinal section of white fiber from same source. Mitochondria, *M,* are relatively scarce. (**A,** ×8000; **B,** ×8000.) (Courtesy Dr. K. Watanabe, Tokyo, Japan.)

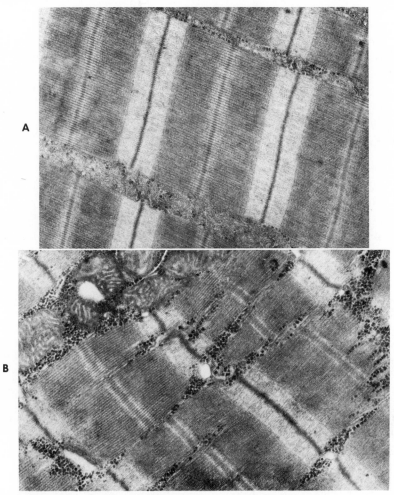

Fig. 6-9. A, Single myofibril of white muscle in tail of *Bufo regularis*. Note narrow and straight nature of the Z lines. **B,** Several myofibrils of red muscle from same source. Note thicker, less regular Z lines and tendency to branching of fibrils. (**A,** ×40,000; **B,** ×40,000) (Courtesy Dr. K. Watanabe, Tokyo, Japan.)

The exact relationship between the morphologic and physiologic differences is not entirely clear. The distinction between slow and fast fibers is related to the innervation and manner of response of the muscle. Vertebrate skeletal muscles are organized in units, each of which may consist of several hundred muscle fibers innervated by one motor neuron. The majority of muscles are made up of many fast units, with each fiber activated by a graded end-plate potential to give an all-or-none impulse. Some muscle fibers, however, receive slow nerve fibers in which variation in frequency of the nerve impulse serves to determine

the response. Gradation of movement of the fast muscles is effected chiefly by variations in the number of motor neurons activated in the ventral horn of the spinal cord.

Cardiac muscle. The fibers of cardiac muscle also show cross striations, but each fiber has only a single nucleus. The fibers may be branched, with adjacent attachments, and so arranged as to form a network of anastomosing fibers.

The arrangement of the fibrils in the fibers of cardiac muscle is of a less compact type than that in skeletal muscle. Around the central nucleus is a region lacking fibrils but containing mitochon-

Fig. 6-10. Cardiac muscle of rat. Fibrils are separated from each other by sarcoplasm containing long chains of mitochondria, *M*, with closely packed cristae. Note portions of two cell junctions, *CJ* (formerly called intercalated disks). The general correspondence in position of these structures with the Z lines, *Z*, of the fibrils is seen in this field. (Perfusion with glutaraldehyde, postfixation by osmium tetroxide; ×47,500.) (Courtesy Dr. W. G. Forssmann, Geneva, Switzerland.)

dria, fat and glycogen deposits, a Golgi net, and occasionally granules of pigment. The sarcoplasm between the fibrils contains rows of closely packed mitochondria (Fig. 6-10).

Each cardiac fiber has a sarcolemma, similar to that found in a skeletal muscle fiber, but more delicate. At the ends of each fiber, or cell, is a structure known as the **intercalated disk.** Elec-

tron-microscope studies indicate that these disks are actually cell junctions. Thus they are cell membranes, often arranged in a stair-step or zigzag manner, with considerable amounts of amorphous dense material associated with them (Fig. 6-10). This complex interdigitation of the cell membranes of cardiac fibers seems to permit the strong and complicated contractions of the heart

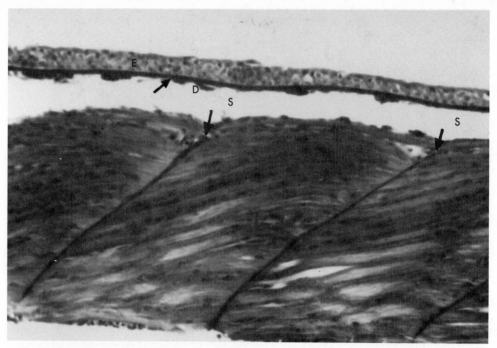

Fig. 6-11. Myotomal muscles of young rainbow trout, *Salmo gairdneri*. Individual myotomes are sharply defined by thin septa, *S,* of connective tissues. Above myotomes are epidermis, *E,* and dermis, *D.* (×100.)

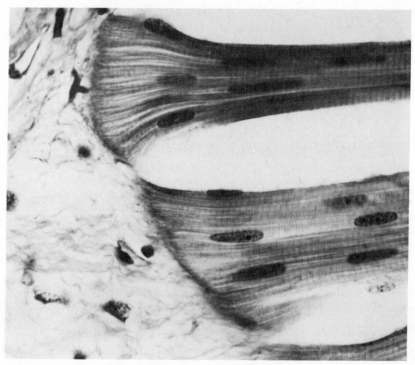

Fig. 6-12. Portion of myotomal muscle of lamprey, *Petromyzon marinus*. Attachment of fibers directly to connective tissue is shown. (×400.)

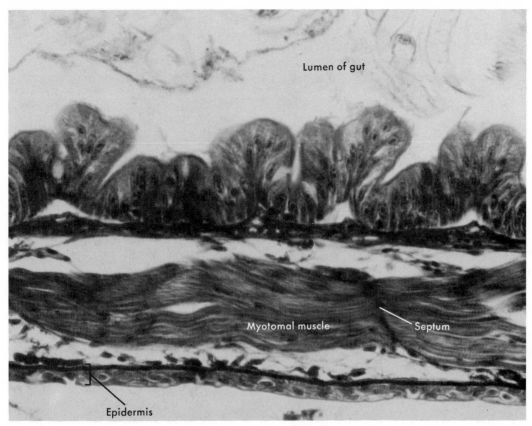

Fig. 6-13. Portion of longitudinal section through hatchling of *Salvelinus fontinalis.* Epithelium of gut is seen above, then parts of two myotomes with intervening septum, and fascia and integument below. (×100.) (Courtesy Mr. Alan Fletcher, Ithaca, N. Y.)

and yet maintain the integrity of the total muscle mass.

In cardiac muscle the sarcoplasmic reticulum is somewhat less complex than in skeletal muscle. Longitudinal connections are fewer, and the triads frequently are replaced by individual transverse channels. The pattern of striation is generally similar to that of skeletal muscle. However, **unstriated** or very weakly striated heart muscle can be found in certain animals, for example, the sloth.

Pigment is found rather commonly within striated muscle fibers, both cardiac and skeletal. Although we tend to associate the accumulation of pigment in muscle cells, as in nerve cells, with advancing age (Andrew, 1956; Strehler and co-workers, 1959), its presence is certainly not confined to older animals or persons. Long ago, Coghill (1929) described pigment masses in the muscle fibers of very young *Ambystoma.* In addition, Coghill believed pigment to be formed most abundantly in regions of higher metabolic rate. On this basis, the middle of the muscle cell in these young amphibians is such a region.

AQUATIC VERTEBRATES

Segmentation, or **metamerism,** of vertebrate musculature is seen clearly in the lateral muscles of the fishes. They are divided into **myotomes,** or muscle segments (Figs. 6-11 to 6-13), each of which is bent into a single V with the angle directed anteriorly. The function of these lateral muscles in almost all types of fishes is locomotion by swimming. Basically, alternate contractions passing posteriorly along the muscles on the two sides of the body result in the well-known undulating movements of trunk and tail.

Each myotome is divided into an upper

(epaxial) and a lower (hypaxial) half by a groove running along the side of the fish. Successive myotomes are separated by obliquely oriented connective tissue partitions **(myosepta).** The epaxial portion is separated from the hypaxial myotome by a fibrous septum running anteroposteriorly, parallel to the longitudinal axis. Removal of the skin reveals the zigzag arrangement of the myotomes with the myosepta.

The skeletal muscle fibers that make up the myotomes run parallel to the myosepta and thus are oriented obliquely to the long axis of the body. The fibers vary in thickness, ranging generally from 100 to 200 μm in diameter. Among them, however, are some small ones, only 10 to 20 μm thick. There is a longitudinally arranged capillary network around the fibers, derived from blood vessels that travel in the myosepta.

The body musculature is relatively simple in fishes. The fins are usually provided with individual small muscles, but the most complex organization is in the head region. Here individual, often very thin, muscles move the jaws, operculum, gill arches, eyes, and other structures. Many individual muscles show an arrangement of fibers into bundles separated from each other by connective tissue partitions.

The nerves generally penetrate a muscle on its side and branch out as they penetrate the connective tissue. Within a muscle bundle, **myelinated** fibers are seen, both singly and in small groups. The type of ending on the muscle fiber is variable. In many cases it is vinelike—a long stem with many small side branches, or buds, on the muscle fiber. Some of the endings, however, are more compact and form typical motor end plates, such as are characteristically found in the skeletal muscles of all the higher vertebrates.

In many amphibian larvae the lateral muscles continue to play a prominent role in locomotion, and myotomal arrangement is conspicuous. However, in adult amphibians there is great reduction of lateral muscles, and the process is carried farther in the reptiles, birds, and mammals. All higher vertebrates, however, including man, show a clear segmental arrangement of the lateral muscle masses at some stage of their embryonic development.

Some morphologic distinctions of fish muscle should be noted. Striated muscle of sharks shows no H zone or M line, but an uninterrupted, dark A band; also, intercalated disks have not been detected in shark cardiac muscle. In the fishes and larval amphibians the myofibrils of heart muscle are commonly located peripherally, whereas in higher vertebrates there is a more uniform distribution of these elements throughout the muscle fiber.

Electric organs. Several kinds of fishes, in rather widely different taxonomic groups, possess so-called electric organs (Fig. 6-14). These appear to have evolved from a combination of nervous tissue and skeletal muscle tissue. Such organs offer a rather difficult problem with regard to their evolution. What would be their usefulness in evolutionary stages before they had the capacity to deliver a shock? How would such stages add to the likelihood of survival of the species?

We do know, however, that the formation of currents in nervous, muscular, and even glandular tissues is common. The potentials involved usually are only a few millivolts. In some of the species of electric fishes, however, especially where the units are numerous and arranged in series—like a row of batteries—the potential may exceed 500 volts. A possible explanation of how the normally weak voltages in tissues might develop gradually to such proportions is offered by some weakly electric fishes that dwell in particularly murky waters. They may use their primitive electric organs to develop an electric field, a disturbance in which could be detected via their sensory **lateral line.**

The efficiency of the shock in many living species, however, is well known (Ballowitz, 1897, 1899, 1938). It serves both as a defense from predators and as a means of obtaining prey. Among the best known electric fishes are the electric torpedoes, of which there are several species, the electric catfish *(Malapterurus electricus),* and the electric eel *(Gymnotus electricus).*

The electric organ consists of a series of platelike structures, the **electroplaques.** A single electroplaque has three parts. The middle portion usually contains striated elements with alternating isotropic and anisotropic bands as in skeletal muscle (Fig. 6-15). The other two layers are

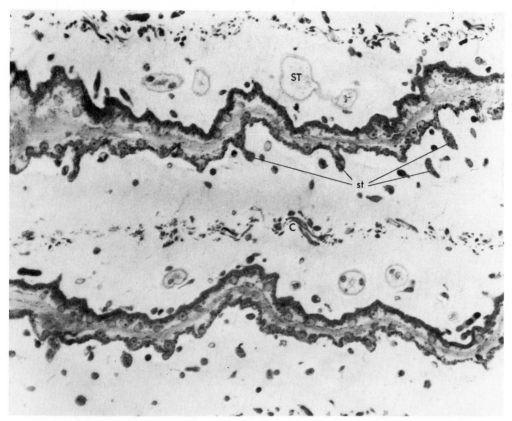

Fig. 6-14. Light micrograph of portions of two adjacent electroplaques of *Gnathostemus* sp. A number of small stalks, *st,* can be seen projecting from posterior (innervated) face. At *ST* are cross and oblique sections through stalks after they have penetrated back through body of electroplaque and before they have fused to form single stalk. Septum including some connective tissue elements, *C,* is seen in the extracellular space between the two electroplaques. These electroplaques are flattened into disks. Their dark edges reflect packing of mitochondria beneath each surface. Multiple nuclei are clearly visible. It is possible to distinguish band of organized filaments running through center of cell. (1% osmium tetroxide in phosphate buffer; ×845.) (Courtesy Dr. Ilsa R. Schwartz, New York, N. Y.)

nervous (electric) and nutritive, respectively. Each electroplaque is supplied by one or more myelinated nerves. The fiber or fibers branch profusely and spread out to form end plates that cover much of the surface of the electric layer. The relative positions of the electric and nutritive layers differ in various species. In *Mormyrus,* which has a well-developed striated layer, the electric layer of each electroplaque is located posteriorly. In *Gymnotus* it is anterior. In still a third teleost, *Astroscopus* ("the star-gazer"), the electroplaques are stacked in a horizontal plane, with the electric parts facing upward. Current flows from the innervated side of the electroplaque to the opposite side. Hence in fishes with the nervous layers placed anteriorly in the units, current will flow in the animal from front to back; in those with such layers posteriorly, it flows from back to front. In either event, the circuit will be completed in the surrounding water—and in the unwary prey or enemy!

Recent work on the ultrastructure of the electric organs (Wachtel, 1964) shows that a number of different modifications of the fundamental myofibrils have occurred in different species. All species studied seem to have retained the actin

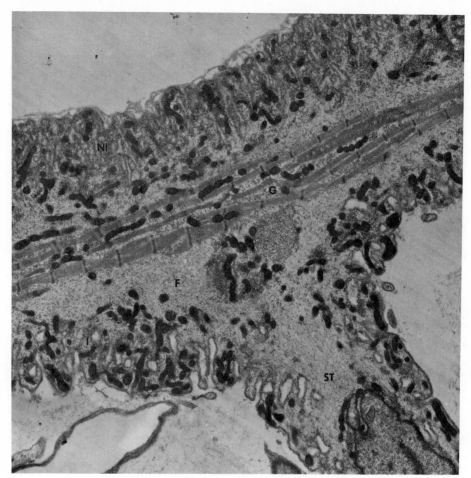

Fig. 6-15. Electron micrograph of portion of a *Gnathonemus petersi* electroplaque showing stalk, *ST*, projecting from the innervated face, *I*. Note slightly more extensive proliferation of the noninnervated face, *NI*, and interdigitation of mitochondria between invaginations from both surfaces. Organized filaments with prominent Z bands run in plane between the two surfaces. A few mitochondria and clusters of glycogen particles, *G*, are associated with the organized filaments. Disorganized filaments (as at *F*) fill the remainder of the cell. (1% osmium tetroxide in phosphate buffer; ×43,800.) (Courtesy Dr. Ilsa R. Schwartz, New York, N. Y.)

myofilaments (Figs. 6-16 and 6-17) and some also appear to have myosin myofilaments in their striated layer.

TERRESTRIAL VERTEBRATES

The tetrapod organization has produced changes in the muscular system. Fishes and tetrapods alike exhibit epaxial and hypaxial muscle masses separated by the transverse processes of vertebrae and innervated respectively by the dorsal and ventral rami of spinal nerves. But there have been certain modifications on this basic plan. For example, the tetrapod limb musculature would seem to be derived from the muscles that raise and lower the fins of fishes. However, such homology is not easily established, since in tetrapods these muscles do not originate from the early myotomes as in the fishes but develop from the mesenchyme in the limb itself. Generally speaking, homology between the muscles of fishes and those of terrestrial vertebrates is difficult to establish, although it is implied.

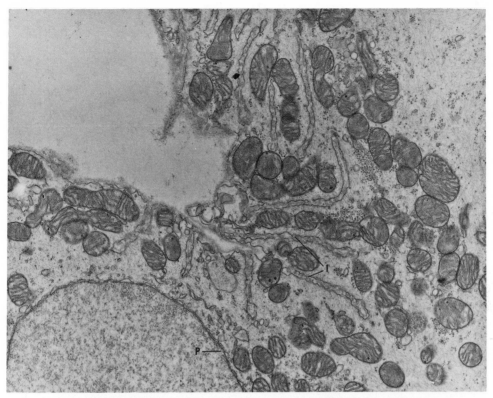

Fig. 6-16. Electron micrograph of innervated face of *Gnathonemus petersi* electroplaque. Note tubular invaginations, *I,* from this face and concentration of mitochondria immediately beneath the face and lined up between the invaginations. Clusters of free ribosomes and disorganized filaments are present, especially close to the cell surface. Several nuclear pores are clearly visible at *P.* (1% osmium tetroxide in phosphate buffer; ×40,000.) (Courtesy Dr. Ilsa R. Schwartz, New York, N. Y.)

Epaxial muscles lie along the vertebral column dorsal to the transverse processes and lateral to the neural arches, from the base of the skull to the tip of the tail. In most tetrapods the epaxial myotomes tend to fuse and form elongated bundles, with the deeper muscles retaining traces of the primitive segmental arrangement, at least in amphibians and reptiles. These epaxial bundles are often somewhat buried under the extrinsic limb muscles, which have become very well developed in the higher vertebrates.

Many changes in gross musculature occurred during the evolution of the various terrestrial forms. Such changes were particularly important in allowing successively more of the maneuverability essential to survival on the land. The limbs and the head and neck regions are significant examples of areas whose muscles underwent great evolutionary development. In birds, further modifications of the basic arrangement were necessary. To keep the body weight near the center of gravity, the powerful muscles of flight (pectoral and supracoracoid) are centrally located on the sternum. In the pigeon these muscles make up about a fourth of the total weight of the body. Also, most leg and wing muscles have their fleshy parts lying near or within the body mass. Such muscles operate through long tendons that run along the posterior aspect of the leg, inserting on the claw-bearing digits.

In the body wall, hypaxial muscles of tetrapods are disposed in three distinct layers: (1) an external oblique mass with fibers directed ventrally and caudad; (2) an internal oblique mass whose fibers are directed ventrally and cephalad; and (3) a transverse mass against the parietal peri-

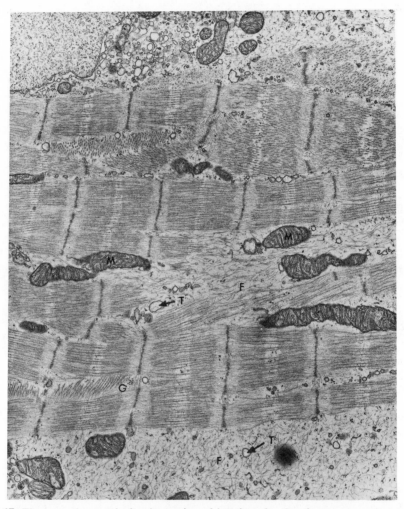

Fig. 6-17. Electron micrograph showing region of interior of a *Gnathonemus* sp. electroplaque with organized filaments posessing all band patterns of normal striated muscle, although the M band is not clearly defined. Not all fibrils run in same direction, and thus some are cut obliquely. Between the organized filaments are mitochondria, *M,* profiles of tubules, *T* (possibly remnants of the sarcoplasmic reticulum), glycogen particles, *G,* and disordered filaments, *F.* (1% osmium tetroxide in phosphate buffer; ×49,000.) (Courtesy Dr. Ilsa R. Schwartz, New York, N. Y.)

toneum, with ventrally directed fibers. These oblique and transverse muscle sheets are formed chiefly by the fusion of adjacent myomeres, and they tend to obliterate the segmented organization so prominent in fishes. The metameric pattern does persist in some muscles, for example, the intercostals.

References

Andrew, W. 1956. Structural alterations with aging in the nervous system. J. Chronic Dis. **3**:557-596.

Ballowitz, E. 1897. Zur Anatomie des Zitteraales (*Gymnotus electricus* L.); mit besonderer Berücksichtigung seiner elektrischen Organe. Arch. Mikrosk. Anat. **50**:686-750.

Ballowitz, E. 1899. Das elektrische Organ des afrikanischen Zitterwelses. G. Fischer, Jena. 96 pp.

Ballowitz, E. 1938. Elektrische Organe. Pages 657-682 *in* L. Bolk, E. Göppert, E. Kollius, and W. Lubarsch, eds. Handbuch der vergleichende Anatomie der Wirbeltiere, vol. 5. Urban & Schwarzenberg, Vienna.

Bennett, H. S., and K. R. Porter. 1953. An electron microscope study of sectioned breast muscle of the domestic fowl. Am. J. Anat. **93**:61-105.

Coghill, G. E. 1929. Anatomy and the problem of behaviour. Cambridge University Press, London. 113 pp.

Gauthier, G. F. 1969. On the relationship of ultrastructural and cytochemical features to color in mammalian skeletal muscle. Z. Zellforsch. Mikrosk. Anat. **95**:462-482.

Heidenhain, M. 1913. Über die Enstehung der quergestreiften Muskelsubstanz bei der *Forelle*. Arch. Mikrosk. Anat. **83**:427.

Huxley, H. E. 1959. Muscle cells. Pages 365-481 *in* J. Brachet and A. E. Mirsky, eds. The cell, biochemistry, physiology, and morphology. Academic Press, Inc., New York.

Mark, J. S. T. 1956. An electron microscope study of uterine smooth muscle. Anat. Rec. **125**:473-493.

Schwartz, I. R., and G. O. Pappas. 1968. The fine structure of electroplaques in some weakly electric fish. Anat. Rec. **160**(2):424. (Abstr.)

Strehler, B. L., D. D. Mark, A. S. Mildvan, and M. V. Gee. 1959. Rate and magnitude of age pigment accumulation in the human myocardium. J. Gerontol. **14**:430-439.

Thaemert, J. D. 1959. Intercellular bridges as protoplasmic anastomoses between smooth muscle cells. J. Biophys. Biochem. Cytol. **6**:67-69.

Wachtel, A. W. 1964. The ultrastructural relationships of electric organs and muscle. I. Filamentous systems. J. Morphol. **114**:325-360.

Watanabe, K. 1969. Observations on the types of muscle in the tadpole tail of the *B. vulgaris*. Acta Anat. Nippon. **44**(2):80. (Abstr.)

7 RESPIRATORY SYSTEMS

All vertebrate animals are aerobic. Oxygen is used in the energy metabolism of every living cell of the vertebrate body. Another gas, carbon dioxide, is a waste product of this cellular metabolism. Major structural adaptations in all animals have been concerned with getting oxygen from water or air into the body and then to the individual cells and with carrying carbon dioxide away from the cells to the outside medium.

The process of gas interchange between the outside and inside of the body is called respiration, or breathing. The respiratory system is closely allied with the circulatory system, to which it delivers oxygen and from which it removes carbon dioxide. The circulatory system, to be considered later, serves to carry the gases to and from the individual cells of the body.

Vertebrate breathing systems are specialized in diverse ways such as integument, gills, and lungs. Usually two parts have become specialized for different purposes. The first acts merely as oxygen-conduction passages and is represented in aquatic forms by gill chambers, gill clefts, gill arches, gill rakers, etc., and in terrestrial forms by the larynx, trachea, bronchi, and bronchioles. The other part of the system is a specialized region where the exchange of gases occurs by simple diffusion between the blood and water, as in gill filaments and gill lamellae, or between the air and blood as in respiratory bronchioles, alveolar ducts, atria, and alveolar sacs.

In addition to these devices, some fishes and most amphibians make use of the integument as an accessory respiratory organ. The rectum and cloaca may also be specialized for gaseous exchange. An excellent review of respiratory tissues is given by Bertalanffy (1964, 1965).

Any epithelium in contact with the external environment has the potentiality of functioning in gas exchange. One fundamental requirement of a breathing surface is that it have a film of liquid in which the gaseous elements can dissolve before they diffuse in or out. Water serves this function in aquatic forms; mucus in terrestrial vertebrates. Breathing surfaces must also be thin, readily permeable, and well vascularized (supplied with blood capillaries). The diffusion distance between blood and water or air is rarely more than 2 μm in respiratory epithelia and often much less (Hughes and Grimstone, 1965).

DEVELOPMENT OF RESPIRATORY SYSTEM

The respiratory system of vertebrates has arisen from the pharyngeal region of the anterior part of the alimentary canal. In the embryos of all vertebrates the endodermal wall of the pharynx develops a series of outgrowths that meet the ectoderm of the body wall, forming visceral pouches. In aquatic vertebrates this endoderm-ectoderm contact becomes perforated and forms a gill slit, which is a passageway from the pharynx to the outside. In early vertebrates these modifications of the pharynx probably served as food-gathering devices but later acquired respiratory function. Alternating with the pouches and grooves are the visceral arches on vertical columns of supporting cartilage, muscle, and blood vessels. Outgrowths of the epithelium on the margin of the cleft (mostly on the endodermal side) differentiates into gill filaments, gill lamellae, and gill rakers (Bevelander, 1935).

In most air-breathing vertebrates lung originate embryonically as a single median diverticulum from the floor of the pharyngeal region, posterior to the branchial region of the visceral pouches. This pocket deepens, becomes bilobed, and forms lung buds that give rise to the bronchi and lung tissue. The primordium retains communication with the pharynx by way of a duct, the trachea, whose opening to the pharyngeal region is the glottis. Lungs may have had a bilateral origin, since the last gill pouches in some primitive fishes and certain amphibians do not open to the exterior but retain air taken in through the mouth.

CYCLOSTOMATA

The general mechanism of respiration in the great majority of aquatic vertebrates is for the respiratory current of oxygen-containing water to pass through the mouth into the pharynx, over the gills, and back to the outside through the gill slits.

The respiratory system of *Petromyzon* consists essentially of the "water duct" and the gill sacs. The duct courses caudally between the pharynx and the branchial artery and ends at about the level of the sixth external gill opening. It is only about 2 mm wide at its beginning and narrows posteriorly to half that width. From the water duct the internal brachial openings lead into the gill sacs. There are seven of these rather flat, pocket-like cavities on each side of the organism. Within the sacs are the gill leaves or filaments, which in turn carry the many gill leaflets or lamellae (Fig. 7-1). Both filaments and leaflets are covered by respiratory epithelium. The gill region is supported by seven pairs of branchial cartilages, which by anastomotic bridges form a "branchial basket."

The water duct is lined by epithelium of 3 to 4 layers of flattened cells; the surface cells show a cuticular border. There is no basement membrane. Scattered cells can be seen among the connective tissue lamina propria. Outside is a smooth muscle layer that varies in its degree of development in different parts of the duct.

Histologically, each gill sac has a wall consist-

ing of connective tissue and skeletal muscle. The lateral portions of the wall are continuous with subcutaneous and intermuscular connective tissue. Medially, the gill sac wall is related to the septal mass that contains the large vessels, the pharynx, and water ducts. The skeletal muscle on the medial side shows thin fibers, which are similar in appearance to the sarcoplasm-rich border fibers *(Randfasern)* of the muscle of the body wall.

Each branchial filament bears an artery, a vein, and the capillaries that connect them. The capillaries are in close contact with the water, especially where they lie in the membrana propria of the leaflets. The vein runs along the free surface of each leaf, where it causes a swelling of that area. The veins unite to form larger ones between the sacs; the larger veins return the blood to the heart and aorta. Usually there are two

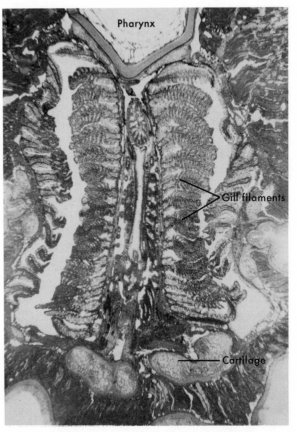

Fig. 7-1. Gills of a lamprey showing arrangement of gill filaments and lamellae. (×60.)

cell layers on the branchial leaf and only one of the leaflets. On a free ending of the leaf, the large vein is covered over by a two-layered epithelium. Between the epithelial cells in this area the **wandering cells** are especially numerous. A cuticular border covers all the epithelial surfaces.

CHONDRICHTHYES

Most sharks have five pairs of gill slits plus an anterior pair called the spiracles, back of the eyes without functional gills. Each gill slit opens separately on the sides of the neck (Fig. 7-2). Each opening leads into a scanty gill pouch. The connective tissue between successive gill chambers forms a thick intrabranchial septum of cartilage, muscles, nerves, and arteries. A series of cartilaginous gill arches lie at the edges of the septa and between the pharyngeal openings. Each arch has the form of a half-loop. There is a gill on the anterior and posterior walls of each gill pouch, except for the last where only the anterior wall has a gill. Each gill is called a hemibranch. The two hemibranchs facing each other across a cleft make a holobranch.

According to the recent study of Kempton (1969) gill filaments are attached in groups to the interbranchial septa separating the gill pouches. The interbranchial septum extends out to the skin and completely separates the two hemibranchs (groups of filaments) attached to it (Fig. 7-2). Each hemibranch consists of filaments extending from the base of the gill to its outer edge (Fig. 7-3). Basal arteries from the afferent branchial artery run in the interbranchial septum parallel to, and at the bases of, the filaments. Overlying each of these basal arteries is a parallel cavernous body with a connective tissue wall. Its lumen is provided with columns of connective tissue and endothelial cells (Fig. 7-4). Each cavernous body receives blood through several openings from its accompanying basal artery. Both the artery and its cavernous body extend out to the tip of each filament; the distal wall of the cavernous body is continuous with the connective tissue sheet in the center of the filament.

Secondary folds (lamellae), delicate extensions of the filaments, are the structures in which the gas exchange takes place between the water and the blood. Lamellae originate in the cavern-

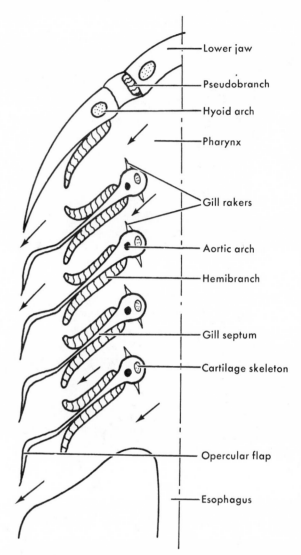

Fig. 7-2. Diagram of gill arrangement on one side of shark. Opercular flaps cover successive gill chambers when water is sucked into the pharynx. Water with oxygen passes from pharynx over gills and to outside through gill slits.

ous bodies; they consist of double sheets of epithelium (flat, polyhedral cells) held apart, or held together, by pillar cells (Fig. 7-5). The lamellae do not contain capillaries; blood moves freely in their spaces. Blood passes from the basal artery into the cavernous bodies and then into the lamellae, which empty into the distal artery, running around the free ends of the fila-

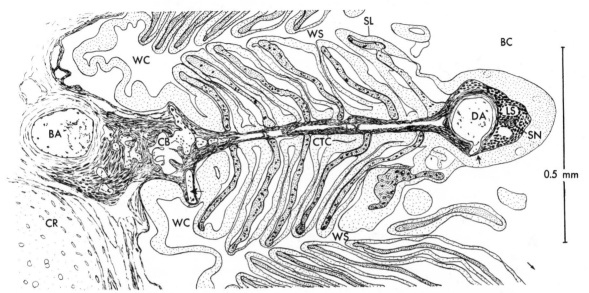

Fig. 7-3. Cross section of shark's gill filament from posterior hemibranch. Blood spaces are stippled more heavily than epithelium. Arrows indicate connections between secondary lamellae and both cavernous body and distal artery. *BA*, Basal artery running in gill septum; *BC*, branchial chamber; *CB*, cavernous body; *CTC*, connective tissue core of filament; *DA*, distal artery; *LS*, lymph space; *SL*, secondary lamella; *SN*, small nerve; *WC*, expanded water channel; *WS*, water space between filaments. (From Kempton, R. T. 1969. Biol. Bull. **136:** 226-240.)

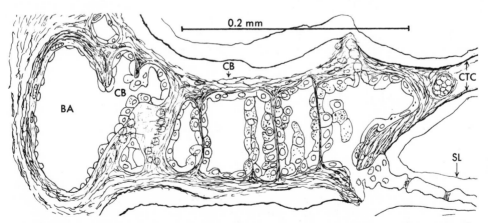

Fig. 7-4. Basal artery and cavernous body. Section is made perpendicular to filament (as in Fig. 7-3), showing opening between basal artery, *BA*, and cavernous body, *CB*, and the opening from cavernous body to one secondary lamella, *SL*. Cellular details of epithelium of secondary lamella and red blood cells omitted. *BA*, Basal artery; *CB*, cavernous body; *CTC*, connective tissue core of filament; *SL*, secondary lamella. (From Kempton, R. T. 1969. Biol. Bull. **136:**226-240.)

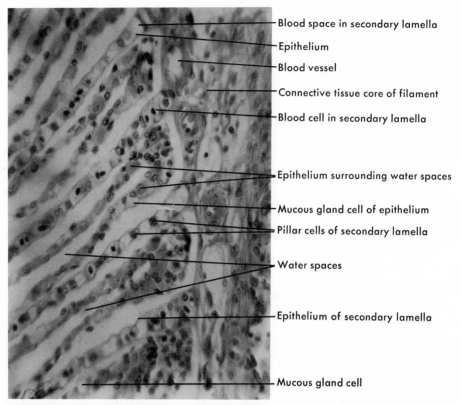

Blood space in secondary lamella
Epithelium
Blood vessel
Connective tissue core of filament
Blood cell in secondary lamella

Epithelium surrounding water spaces
Mucous gland cell of epithelium
Pillar cells of secondary lamella
Water spaces

Epithelium of secondary lamella

Mucous gland cell

Fig. 7-5. Cross section through portion of filament from hemibranch gill of *Squalus acanthias.* For relationship, compare with Fig. 7-3. (×640.)

ment. Both the basal and distal arteries have muscular walls for contractility.

The arrangement of the circulation is well suited for passing large amounts of blood through the gills. Water freely circulates in the spaces between adjacent cavernous bodies and between the lamellae. The possibility of a counter-current flow, by which the blood flows through the lamellae in an opposite direction to that of water (as in teleosts), is not clearly proven in the shark's gills.

OSTEICHTHYES

Gill breathing. In teleost fishes each gill is a hemibranch—the gill filaments on either the caudal or cranial edge of a gill cleft (Figs. 7-6 and 7-7). The lamellae of each filament have a rich supply of blood capillaries and a covering epithelium one cell thick (Fig. 7-8) in contact with the oxygen-laden water that is constantly

streaming through the pharynx and over the gills to the exterior. This thin epithelium of squamous or low cuboidal cells is thus a respiratory epithelium through which the diffusion of gases occurs, oxygen passing inward, carbon dioxide outward. The epithelial cells are characterized by ovoid or round nuclei and a cytoplasm less dense than is found in other cells. The capillaries of the gill lamellae, apparently because of their exposed and relatively unsupported position, show a peculiar structure. They are bridged across by many large cells, the supporting or pillar cells (Fig. 7-8). Evidently, these supporting cells are simply much modified endothelial cells, demonstrating again the versatility of individual cell types in relation to functional needs.

The blood is intimately associated with the functioning of the gill. The heart is located ventrally not far behind the gills. The ventral aorta passes forward from the heart and splits into right

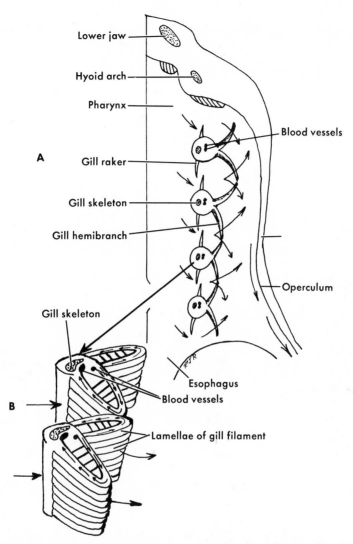

Fig. 7-6. A, Diagram of teleost gill arrangement and direction of water flow (only one side shown). **B,** Structure of teleost gills; four hemibranchs and their circulation. Large arrows indicate water flow. Small arrows in **B** indicate blood flow. (From several sources.)

and left branches ventral to the gills, running along each side of the body. Each branch divides into a series of aortic arches that pass dorsally through the tissue of the gill arches between the gill slits (Fig. 7-6, *A*). A rich capillary network from these aortic arches vascularizes the respiratory epithelia (Hughes, 1961). Each gill lamella receives a branch of an afferent branchial artery, and the blood is returned by an efferent branchial artery.

The flow of blood is in the opposite direction to that of the water over the lamella (Fig. 7-6, *B*). The blood pumped to the gills has been depleted of oxygen in the body tissues, and thus deoxygenated blood is exposed to oxygen-containing water and can pick up oxygen according to the laws of diffusion along a concentration gradient. The reverse is true for carbon dioxide, because the blood from the lamella has a higher carbon dioxide content than the water flowing over the gills (Scholander and associates, 1957). The aortic arches all rejoin on the dorsal side

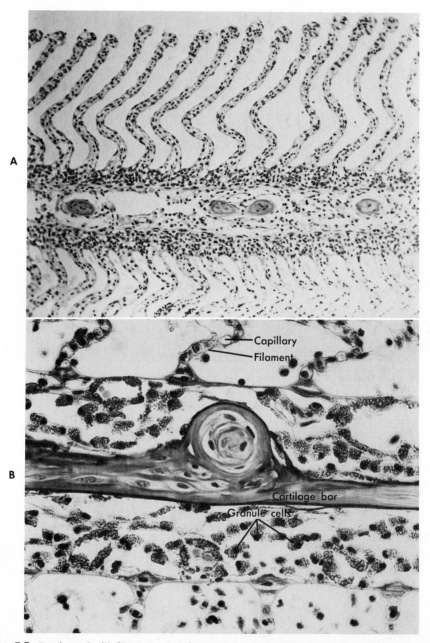

Fig. 7-7. Portion of gill filament of *Salmo gairdneri*. **A,** Thin bar of cartilage, portions of which are seen here, serves as skeleton for this structure; delicate lamellae, devoid of any such support, extend out on either side. **B,** High-power view of gill filament of *Salmo gairdneri*. Cells of cartilage bar are seen, and in connective tissue around it are many granule cells. Bases of a few lamellae are in the field. (**A,** ×125; **B,** ×400.)

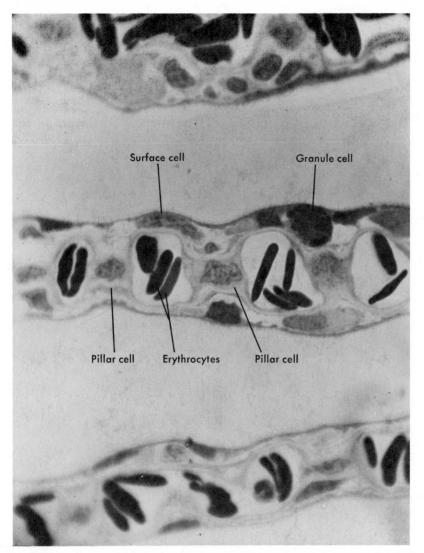

Fig. 7-8. One-micrometer thick section of several gill lamellae of *Salmo gairdneri*. Central section shows two large pillar cells and a migrating granule cell. (Toluidine blue stain; ×1000.)

of the body and form the dorsal aorta, which carries the oxygenated blood posteriorly to the body.

The diffusion of gases through the epithelium of the lamellae is apparently enhanced by the presence of microvilli on the surface cells (Fig. 7-9). These structures would serve to increase the surface area, much as they do in the intestine of vertebrates in general. Electron microscopy also reveals what could be a protective surface coating over these microvilli (Fig. 7-9). The coat-

ing appears to consist of innumerable fine filaments matted together.

Special cells for chloride excretion have been described by Philpott and Copeland (1963) in the gills of *Fundulus*. These cells are columnar and acidophilic and are located on the inside of the filament—the side supplied with afferent blood. In sea water or salt solutions, they form vesicles at their surfaces for expelling chloride anions.

On the branchial arches between the branchial

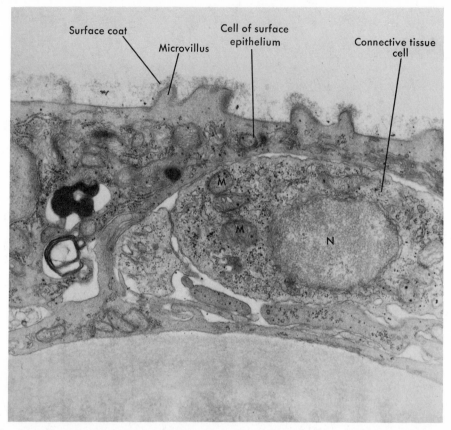

Fig. 7-9. Electron micrograph of portion of a gill lamella of *Salmo gairdneri*. Microvilli of the surface cells are covered with coat of material that appears similar to that found over microvilli of intestinal and some other epithelia, consisting of many very fine filaments. *N*, Nucleus; *M*, mitochondrion. (×28,800.)

filaments, the epithelium becomes stratified and thickened. Goblet cells may be found here, although such unicellular glands occur also at the tips of the gill filaments. Striated skeletal muscles occur at the base of each gill filament, making possible restricted gill movement. Basically, the respiratory organs in both aquatic and aerial breathers represent an adaptation of the pharyngeal epithelium to the different media of air and water. The fundamental respiratory pattern of epithelium, vascular supply, and general gaseous exchange are the same in each type of breathing. Most fishes cannot live out of water because their gill filaments and lamellae collapse in the air and their respiratory surfaces are thereby much reduced. Also, the lamellae become useless as the required film of water evaporates.

Some fishes overcome this handicap with a gill structure that traps sufficient moisture to prevent the drying out of respiratory surfaces. Such fish usually have small openings to the gill chambers, so that only small amounts of air can enter. Eels can respire through their skin as long as they are kept moist. Other methods for living out of water for considerable lengths of time are also known among fishes.

Lung breathing. Lungs are phylogenetically much older than tetrapods. They arose in the ancestors of the crossopterygian fishes that lived in shallow pools of water subject to drying up during drought. Some of these individuals were able to survive and move to more favorable pools because they developed vascularized outpocketings from the pharynx and limb skeletons, mak-

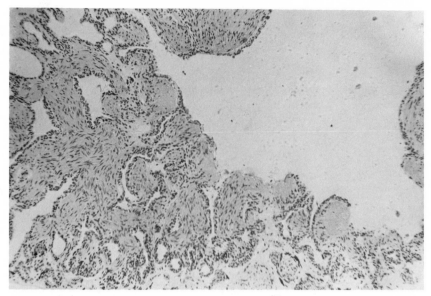

Fig. 7-10. Section through lung of *Protopterus,* a lungfish. Note large amount of smooth muscle in the walls. (×60.)

ing it possible for them to breath air directly.

The early vertebrates may have given rise to two main groups of descendents. Modern bony fishes come from forms that used the air pouch as a swim bladder; the second group used the pouch as a lung and gave rise to land vertebrates. In still others, this air-breathing device may have been lost altogether (elasmobranchs and some teleosts).

The complexity of the vertebrate lung varies enormously, from simple saclike lungs of primitive forms to the complex divided lungs of birds and mammals. The general evolutionary trend has been toward increased subdivision and thus greater surface area, so that the respiratory surface can come in contact with new volumes of oxygen-containing air. The surface of the lining of respiratory epithelium is usually many times the surface area of the animal's body.

As mentioned earlier, lungs may have had a bilateral origin. In one fossil placoderm suitable for analysis, the last pair of gill pouches did not open to the outside and appear to have been capable of retaining air drawn into the mouth. The other method of lung formation (in the tetrapods) involves a pocketlike evagination from the

floor of the pharynx. In the bichir *(Polypterus)* the lung is a bilobed sac that opens by a single duct into the floor of the gut. The lungfish *Protopterus* (Fig. 7-10) has a lung arrangement similar to that of *Polypterus,* but *Neoceratodus* possesses a single lung sac dorsal to the gut, with a duct passing down the right side and connecting to the ventral side of the gut. The nostrils of the lungfish are internal and appear to be for olfaction only, not for passage of air into the mouth; to get to the lungs, air must be gulped through the mouth.

AMPHIBIA

Amphibian gills are found only in the larvae and in those adults that never undergo metamorphosis but remain in a neotenous condition. Moving from the water to the land imposed on animals a new medium of oxygen supply for which the gill is not suitable. It is evident that delicate gill filaments would quickly dry out when exposed to air. Since gas exchange must take place through a moist membrane in order to be efficient enough for respiration, an internal organ that could retain moisture on its respiratory epithelial lining was a prime necessity. Of course,

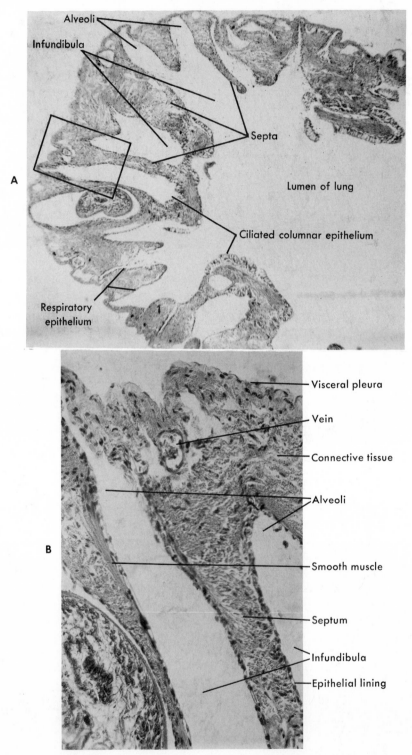

Fig. 7-11. A, Section through lung of frog *Rana pipiens*. Wall is folded into septa separating infundibula; these branch into blind pockets, the pulmonary alveoli. **B,** Boxed section of **A.** (**A,** ×65; **B,** ×260.)

lungs did not originate with the amphibians; they simply made use of lungs already established in some primitive fishes.

Land breathing has presented many problems to amphibians, and they have been solved in many ways. All amphibians make use of the integument and buccal cavity to help out; some use these means exclusively and have dispensed with lungs. It is in frogs that amphibian lungs have reached their greatest development.

Lungs are formed as evaginations from the ventral wall of the esophagus. In frogs they communicate with the alimentary canal through the laryngotracheal chamber by means of the slitlike glottis. The location of the lungs inside the body requires some pumping or sucking action for the exchange of gases. Since the frog has no ribs, it is unable to draw air in by enlarging the cavity as do higher vertebrates. Air is pumped in and out through the external nares by raising and lowering the floor of the mouth, with the mouth closed.

The larynx in urodeles is poorly developed and insignificant. It is lined with simple columnar epithelium with some connective tissue and smooth muscle lying beneath the epithelium. A pair of lateral cartilages from the last pair of gill arches are found near the glottis, the opening into the larynx.

In terrestrial amphibians (anurans) the larynx is better developed and contains the vocal cords, two lateral ridges of fibroelastic tissue projecting into the lumen. The larynx is reinforced by cartilages, and the internal surface is lined with simple squamous or cuboidal epithelium, ciliated in some regions. Striated muscle fibers attach the larynx to the hyoid apparatus and manipulate the glottis and larynx. The trachea is very short and branches directly to the anterior pole of the lungs without the intervention of bronchi.

The lungs of the frog are ovoid, thin-walled sacs. Their inner surface is divided by a network of septa into infundibula with small pulmonary alveoli, thus increasing their surface area (Fig. 7-11). Toads have more alveoli than frogs and depend less on their skin for breathing. The walls of the alveoli are richly supplied with blood vessels. This is where the gas exchange occurs. The alveoli are lined with a single layer of squamous epithelium (Figs. 7-12 and 7-13); on the edges of the septa, the ciliated epithelium is cuboidal to columnar. The blood vessels and lymph vessels are in a connective tissue layer beneath the epithelium. Smooth muscle is also found there. It is estimated that the inner surface of the frog lung is equal to about one half the total surface area of the skin.

Cutaneous respiration requires a highly vascular, thin, and moist skin. It usually serves as accessory to some other form of breathing (lung or gill), but it assumes great importance in many amphibians. For example, one family of salamanders has no other method, and some larval forms may breathe in this way until the gills are fully developed. In the hibernation stages, nearly all gas exchange occurs through the skin.

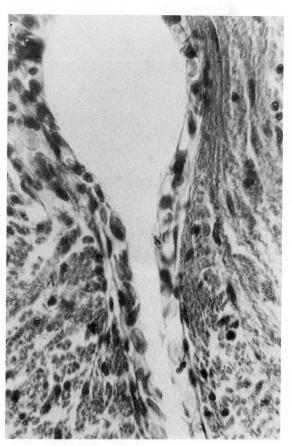

Fig. 7-12. Section through lung of *Rana pipiens* showing portion of one alveolus with its lining of squamous epithelium. (×640.)

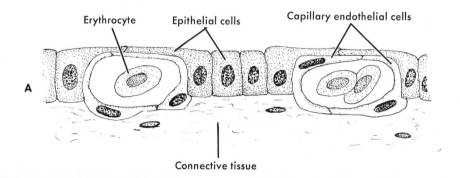

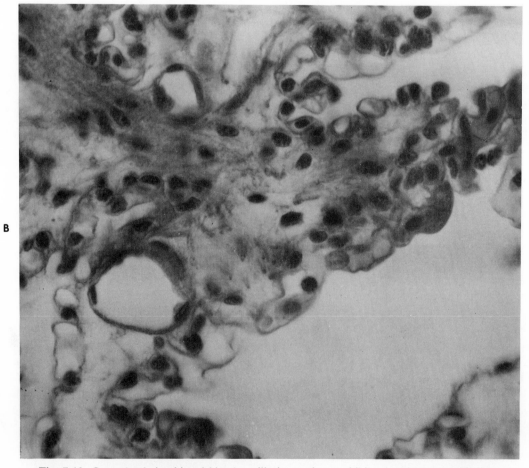

Fig. 7-13. General relationship of blood capillaries to the amphibian (respiratory) epithelium. **A,** Diagram. **B,** Frog's lung. (**B,** ×855.)

REPTILES

The lungs of reptiles are more adjusted to air breathing than amphibians; their respiratory surface is greater, and their air passageways are more complex. The lungs serve as the sole respiratory organs because the dry, scaly skin of reptiles is not adapted for this function.

Behind the tongue the slitlike glottis leads into a short chamber, the larynx, which is supported by cricoid and arytenoid cartilages and is lined with stratified ciliated columnar epithelium. The voice box is poorly developed, but there may be vocal cords in some lizards that can produce guttural sounds. Most, however, can produce only hissing sounds. The anterior wall of the larynx in male *Anolis* and some others is provided with a gular pouch, which can be inflated with air, thus causing the colored ventral surface of the neck to swell out for sex appeal during the breeding season.

The trachea is an elongated, cylindrical tube with its walls supported by rings of cartilage that may be complete or incomplete. These rings are found in the trachea of all forms above the amphibians. The trachea bifurcates into two bronchi. Each bronchus enters a lung on the median side, and gives off numerous bronchioles to the air chambers. The lung itself is a fusiform sac, the inner lining of which is folded into incomplete septa that separate it into several chambers. Higher lizards have the septa so constructed that the larger chambers are subdivided into many individual subchambers. The folds support a rich vascular supply covered by a lining of simple columnar or cuboidal epithelium (Fig. 7-14). The capillaries, with portions of the epithelial cells, may protrude out into the lumina of the alveoli.

Instead of raising and lowering the floor of the mouth as amphibians do in breathing, lizards breathe in by pulling forward the ribs with the oblique muscles that run from rib to rib. This action enlarges the body cavity and decreases the pressure around the lungs, which are then inflated by atmospheric pressure. For exhalation, the lungs are compressed by other muscles.

In lower lizards the lungs occupy the pleuroperitoneal cavity along with the other viscera, but higher forms have developed membranous folds of the dorsal and lateral parietal peritoneum that enclose each lung in an individual pleural cavity. The coelom is thus subdivided into a pericardial

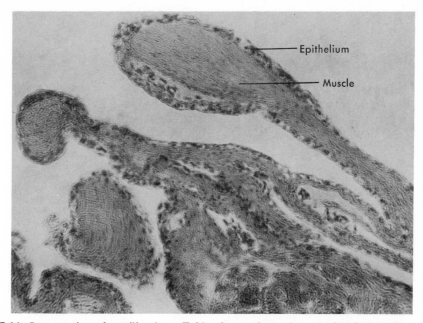

Epithelium

Muscle

Fig. 7-14. Cross section of reptilian lung. Folds of muscular and connective tissue walls extend into elongated saclike lung and increase surface area. Simple cuboid epithelium covers walls of these ridges. Note the prominent musculature. (×100.)

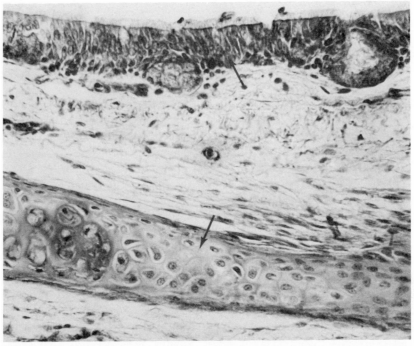

Fig. 7-15. Cross section through trachea of pigeon *(Columba).* Note ciliated pseudostratified epithelium (upper arrow) and hyaline cartilage (lower arrow). (×200.)

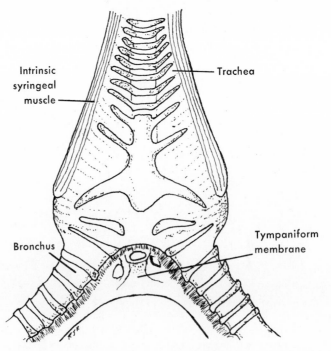

Fig. 7-16. Section through syrinx of pigeon. See text for discussion.

cavity enclosing the heart, the paired pleural cavities around the lungs, and the abdominal cavity containing other viscera. The lungs are surrounded by serosa under which lies a thin layer of muscular and connective tissue.

AVES

The bird respiratory system consists of a pair of lungs, a number of pairs of air sacs connected to air passageways, and accessory structures such as blood vessels and nerves.

The larynx of birds is rudimentary (Myers, 1917). No vocal cords are present. It is lined with stratified ciliated columnar epithelium and, in other ways as well, resembles the reptilian larynx (Marshall, 1960). The tracheal rings are complete and often bony in structure. The long trachea (Fig. 7-15) bifurcates to form the primary bronchi, one of which runs through each lung.

At the point of division that forms the bronchi, there is a membrane projecting into the trachea from the angle between the bronchi. This forms the **syrinx** (Fig. 7-16). The vibrations of the syrinx produce the sounds or song of the bird. It is actually composed of modified tracheal and bronchial rings and is expanded into a chamber of tympanum. Tympaniform membranes fashion the slitlike opening out of the bronchus, and air expelled from the lungs during expiration passes between these membranes. The tension of the membranes is controlled by intrinsic syringeal muscles, which alter the pitch of the sound (Myers, 1917). The variation in number of these muscles determines the versatility of the vocalist; the pigeon has a single pair, but the crow and catbird have seven pairs (Miskimen, 1951).

Secondary bronchi arise from the primary bronchi. The secondary bronchi, in turn, divide

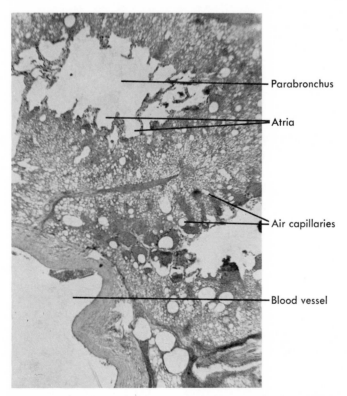

Fig. 7-17. Section of chicken lung, showing parabronchi and atria that give rise to air capillaries containing the respiratory membrane. Air capillaries arise as single tubules that branch often as they proceed from parabronchus and are entwined with blood capillaries. (×60.)

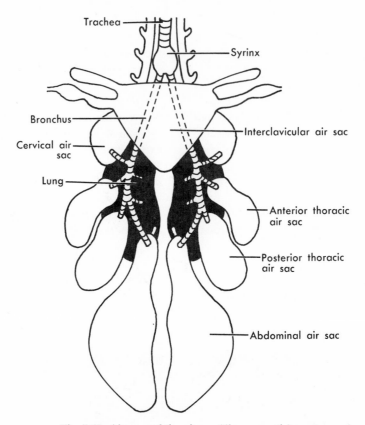

Fig. 7-18. Air sacs of the pigeon (diagrammatic).

into parabronchi (Fig. 7-17), which branch, coil, and anastomose freely. Thus the parabronchi are continuous, forming a network of air capillaries, the bronchial circuit, in which the blood capillaries are found. The lungs are small, spongy organs with little elasticity. They are attached to the ribs of the thorax and can dilate and contract only when the ribs and pulmonary diaphragm do so. The avian lungs (Fig. 7-17) do not have blind pouches (alveoli) as do those of mammals; the air goes through the complete circuit of passageways, with the numerous air capillaries serving as respiratory membranes.

The parabronchi are lined with simple squamous epithelium, with a layer of smooth muscle fibers in their adventitia. Air capillaries are lined by simple cuboidal epithelium near the parabronchi, but this epithelium becomes progressively thinner as the capillaries branch and finally fuse with the epithelium of the blood capillaries.

The larger air passageways (trachea and bronchi) all have an epithelial lining of pseudostratified ciliated epithelium, which contains alveolar mucous glands and goblet cells. A lamina propria with lymph glands is present, but a submucosa is lacking.

The lungs of birds are further peculiar in that they have a remarkable system of air sacs extending out between the visceral organs and into bones (Fig. 7-18). The walls of the air sacs are not vascular; therefore no exchange of gases takes place in them. They are lined by ciliated cuboidal epithelium except in the bones, where simple squamous epithelium is found (Ross, 1899). These sacs are reservoirs for air; they are filled at inspiration, and expiration sweeps the air out, forcing it back through the lungs. They also serve to counteract the excessive heat of the high metabolic rate of birds.

The course of air through the bird's lungs is

far from understood, and there are many puzzling aspects about their structure and functioning. The lungs are relatively small with a very modest respiratory surface. According to Gordon (1968) the lung surface of the crow is only 0.65 square centimeter (cm²) per gram of body weight, as compared with 7 cm² per gram for man. This would indicate an extremely efficient breathing system in birds, not equaled elsewhere among vertebrates.

MAMMALIA

The respiratory system in mammals consists of two portions: a conducting passageway of nasal tube, nasopharynx, larynx, trachea, bronchi, and bronchioles; and a place of gaseous exchange consisting of respiratory bronchioles, alveolar ducts, atria, and alveolar sacs.

The nasal tube is supported by skin, cartilage, and bone and lined with mucous membranes. It has two functions—smelling and conditioning the inspired air by warming and cleaning it. Moistening and warming of the air are performed by an extensive epithelium with rich vascular supply, supported on shelves and scrolls of bone formed by the maxillae and nasal and ethmoid bones. The epithelium is pseudostratified and ciliated, with numerous goblet cells. A current of mucus is carried toward the mouth. On its way, the air passes over a region of olfactory epithelium. In communication with the nasal cavity is a system of sinuses lying in certain bones of that region. These spaces also serve to lighten the bones and increase the resonance of vocal sounds.

The superior part of the pharyngeal chamber is the nasopharynx, lined with pseudostratified epithelium. On its posterior surface is a pair of pharyngeal tonsils. In general, the nasopharynx resembles the respiratory nasal cavities structurally. Leading into the pharynx is the slitlike entrance, the glottis, which is protected by the leaflike and highly elastic epiglottis. The larynx is a short cavity forming the first part of the respiratory tract.

In mammals the larynx reaches its best development and has a framework of cartilage (thyroid, cricoid, and arytenoid) and skeletal muscle. Its inner surface is continuous with the pharyngeal epithelium—stratified squamous and pseudostratified ciliated epithelium. Most of the cartilages are the hyaline type and are covered with perichondrium, which merges with the fibroelastic connective tissue of the larynx. The latter tissue contains much lymphoid tissue and some scattered lymph nodes.

The vocal cords are ridges extending from the lateral walls of the larynx into the lumen. Each vocal cord is provided with external skeletal muscles. The power of phonation, however, is phylogenetically a late development (Ham, 1969). An epiglottis is present in all mammals, preventing anything but air from entering the larynx. The epiglottis is made up of elastic cartilage, with a central zone of fat cells, and is covered by stratified squamous epithelium. In mammals it may also contain taste buds. It has both extrinsic and intrinsic muscles.

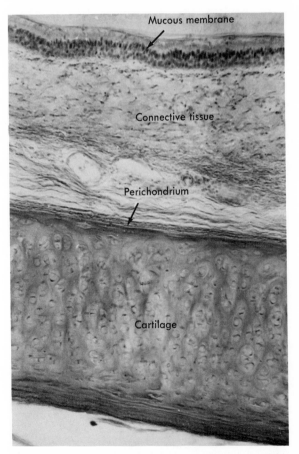

Fig. 7-19. Section through trachea of mammal. (×100.)

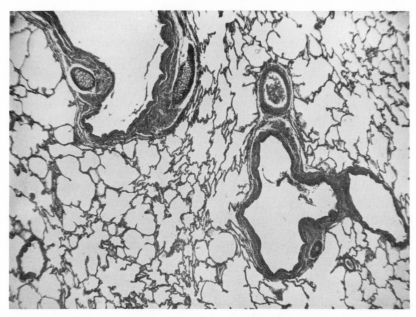

Fig. 7-20. Section through rabbit lung. Portion of intrapulmonary bronchus with some cartilage plates in wall (upper left). Blood vessel is found on top of bronchus near center; bronchiole is at extreme right. Alveolar ducts, alveolar sacs, and alveoli are represented as spaces in the mass. (×60.)

The trachea (Fig. 7-19) leads from the larynx to the place where the two bronchi are formed. It is kept open by incomplete (horseshoe-shaped) rings of cartilage. The space between the ends of the cartilage rings, adjacent to the esophagus, is filled by a band of smooth muscle and connective tissue. The cartilages are covered by perichondrium, which merges with the connective tissue between adjacent cartilages. The internal connective tissue forms a submucosa containing elastic fibers and lymph nodules. The ciliated pseudostratified epithelium often has many goblet cells; seromucous glands from the epithelium occur in the lamina propria and in the submucosa.

The trachea divides into bronchi, which are quite similar in structure in the three classes (reptiles, birds, and mammals) that have them. Each bronchus penetrates a lung and divides repeatedly into smaller bronchioles (Fig. 7-20). The bronchi and large bronchioles are lined with pseudostratified ciliated columnar epithelium with a distinct basement membrane. As the bronchi become smaller by successive divisions, they contain less cartilage, and the lining epithelium becomes a ciliated columnar type (Fig. 7-21, *A*). A bronchiole may be regarded as a small tube with little or no connective tissue and surrounded by respiratory tissue. As they become smaller, the bronchioles lose cartilage; the lining epithelium is reduced to simple, ciliated, cuboidal cells; smooth muscle and elastic fibers become more prominent; and glands, goblet cells, and lymph nodes tend to disappear (Fig. 7-21, *B*). Eventually, with still further branching, the bronchioles become terminal bronchioles, each of which divides into a number of respiratory bronchioles (Fig. 7-22).

The conducting system just described has certain functional correlations in respiration. One of the main functions of the tubes is to maintain patency, which is accomplished by the rigidity of the tubes, largely through the agency of the cartilages. The tubes also can undergo changes in diameter, caused by the smooth muscles under control of the autonomic nervous system. Elastic tissue, which is abundant throughout lung tissue, can expand and recoil in the processes of in-

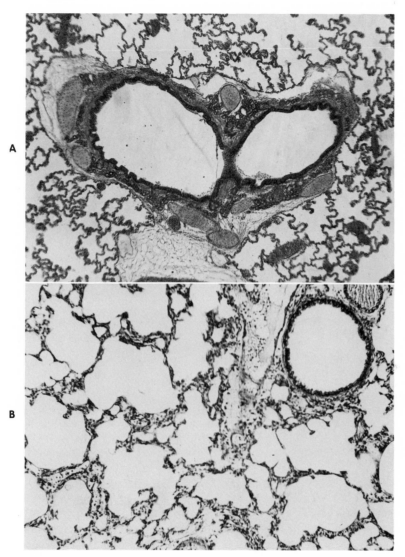

Fig. 7-21. A, Two bronchi at point of division in mammal lung. Note that cartilage rings are no longer found, but small scattered cartilage plaques are present. **B,** Section through mammalian lung showing bronchiole. Compare with cross section of bronchi in **A.** (×100.)

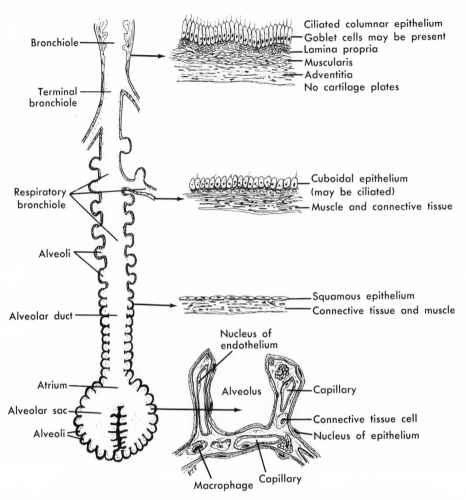

Fig. 7-22. Diagram of scheme of smaller air passageways from bronchiole to alveolus, inclusive. The characteristics of each of the pulmonary passages, as seen with the light microscope, is shown in figures at right. Structure of alveolus is based on electron microscopy. (From several sources.)

spiration and expiration. The cilia and the mucus produced by goblet cells and glands in the bronchial epithelium can trap foreign substances and eliminate them. The cilia in all respiratory passages propel toward the pharynx.

From each bronchiole many irregular alveolar ducts lead to the terminal alveoli, which are chambers wider than the bronchioles (Fig. 7-22). Each alveolar sac is lobed to form a number of alveoli. The epithelium of the alveolar duct is the simple squamous type, and the only muscle remaining is in scattered spirals in the wall between the alveoli. The mammalian lung is not a hollow sac, as is the lung of the lizard, but is subdivided into larger and larger numbers of increasingly smaller air spaces; so that it appears to be filled with spongy tissue in which blood vessels circulate, surrounded by the minute air spaces (Fig. 7-23).

Each alveolar sac consists of a variable number of thin-walled, minor compartments (alveoli), the walls of which are very thin allowing for a close relationship between air and blood (Fig. 7-24). These minor compartments are separated from each other and supported by a septum of dense reticular fibers and other connective tissue.

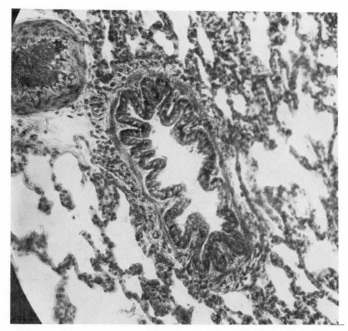

Fig. 7-23. Section through lung of rat showing alveolar sacs and alveoli. Note medium-sized bronchiole with longitudinal folds and prominent muscularis mucosae on left. (×220.)

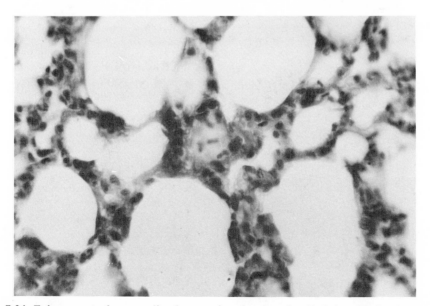

Fig. 7-24. Enlargement of mammalian lung section showing alveoli, their epithelial lining, and septal walls. Note that alveoli do not communicate with each other. (×860.)

This arrangement gives the interior of the alveolar sac a honeycomb appearance. In the fibrous septa, blood capillaries are anastomosed into a capillary net. The electron microscope shows that the free surface of each alveolus has a true epithelial lining (0.1 to 0.2 μm), which cannot be made out satisfactorily with a light microscope (Low, 1953). This lining is the respiratory surface and is closely applied to the endothelium of the capillaries.

Although the squamous epithelial cells show no evidence of phagocytic properties, macrophages are found in the interalveolar wall and as free cells in the alveolar lumen. These cells may contain carbon pigment that they have phagocytosed from smoke-containing air. Several types of cells can be recognized in interalveolar septa, for example, erythrocytes, lymphocytes, and monocytes. Epithelial cell usually occur in groups of three or four, the outer ones forming thin cytoplasmic flanges that arch over a blood capillary to form part of the air-blood barrier between air in the alveoli and blood in the pulmonary capillaries. This barrier consists of structures through which gaseous exchange is made and includes the thin cytoplasm of the pulmonary epithelial cells, the basement membrane of the capillary, and the thin cytoplasm of the capillary endothelium. The alveoli are lined with a fluid film containing lipoproteins. This fluid is secreted by the larger of two types of surface epithelial cells (the secretory cells).

The lungs of the mammal are covered over by the pleural sac consisting of an inner visceral layer and an outer parietal layer. These are serous membranes whose surfaces are of mesothelium. A thin film of pleural fluid is found between the two layers.

For lung inflation mammals rely on a negative-pressure pump. The dome-shaped diaphragm is flattened by its muscles and ribs are raised by the intercostal muscles, thus enlarging the thoracic space. At the same time, the pressure within the lungs (intrapleural pressure) decreases, and the air passageways and the alveoli are enlarged. Expiration takes place when the volume of the thorax lessens as the muscles of the diaphragm and ribs relax.

References

Bertalanffy, F. D. 1964. Respiratory tissue: structure, histophysiology, cytodynamics. Part I. Review and basic cytomorphology. Int. Rev. Cytol. **16:**234-328.

Bertalanffy, F. D. 1965. Respiratory tissue: structure, histophysiology, cytodynamics. Part II. New approaches and interpretations. Int. Rev. Cytol. **17:**214-297.

Bevelander, G. 1935. A comparative study of the branchial epithelium in fishes with special reference to extrarenal secretion. J. Morphol. **57:**335-350.

Gordon, M. S., ed. 1968. Animal function: principles and adaptations. The Macmillan Co., Publishers, New York.

Ham, A. W. 1969. Histology, 6th ed. J. B. Lippincott Co., Philadelphia.

Hughes, G. M. 1961. How a fish extracts oxygen from water. New Scientist **247:**346-348.

Hughes, G. M., and A. V. Grimstone. 1965. The fine structure of the secondary lamellae of the gills of *Gadus pollachius*. Q. J. Microsc. Sci. **106:**343-353.

Kempton, R. T. 1969. Morphological features of functional significance in the gills of the spiny dogfish, *Squalus acanthias*. Biol. Bull. **136:**226-240.

Low, F. N. 1953. The pulmonary alveolar epithelium of laboratory mammals and man. Anat. Rec. **117:**241-263.

Marshall, A. J., ed. 1960. The biology and comparative physiology of birds. Academic Press, Inc., New York. 2 vols.

Miskimen, M. 1951. Sound production in passerine birds. Auk **68:**493-504.

Myers, J. A. 1917. Studies on the syrinx of *Gallus domesticus*. J. Morphol. Physiol. **29:**165-216.

Philpott, C. W., and D. E. Copeland. 1963. Fine structure of chloride cells from three species of *Fundulus*. J. Cell Biol. **18:**389-404.

Ross, M. J. 1899. Special structural features in the air sacs of birds. Trans. Am. Microsc. Soc. **20:**29-40.

Scholander, P. F., L. Van Dam, J. W. Kanwisher, H. T. Hammel, and M. S. Gordan. 1957. Supercooling and osmoregulation in Arctic fish. J. Cell. Comp. Physiol. **49:**5-24.

8 CIRCULATORY SYSTEMS

Heart and blood vessels

In general, the circulatory system of the vertebrates is **closed,** in contrast to what is found in many invertebrates. Exceptions are to be found in the spleens of all vertebrates and perhaps in a very few other locations.

The heart, in both its ontogenetic and phylogenetic development, may be considered as an organ formed by modification of the wall of a blood vessel. It retains the three layers (Fig. 8-1) found in the blood vessel: (1) intima, the internal layer, (2) media, and (3) adventitia, the outer covering. The middle layer, however, has in-creased remarkably in the heart, and the muscle, once smooth, has differentiated into the cardiac variety (Chapter 6). The three layers of the heart have special names. The endocardium corresponds to the intima, the myocardium to the media, and the pericardium (both visceral and parietal) to the adventitia.

CYCLOSTOMATA

Heart. As in fishes generally, the heart of cyclostomes consists of a ventricle and an atrium, the latter partially surrounding the former. An opening leads from the atrium to the ventricle, and backflow is prevented by an atrioventricular

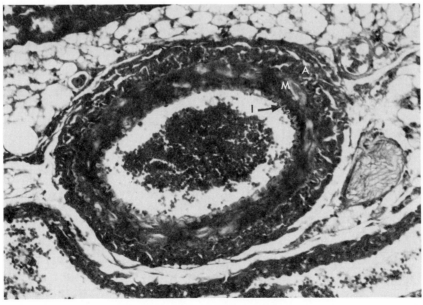

Fig. 8-1. Section through wall of an artery near kidney of the mouse. Three layers are shown: *A,* adventitia; *M,* media; *I,* intima. (×125.)

valve guarding the opening. The great veins empty into the atrium. From the ventricle an aortic bulb carries the blood into the ventral aorta, which actually may be considered as the "gill artery," since its branches are the afferent branchial arteries. Thus in a functional sense, the ventral aorta corresponds to the pulmonary artery of air-breathing forms, carrying **deoxygenated** blood to the respiratory organs. In the fishes, however, the blood does not return to the heart but passes by way of the efferent branchial arteries into the dorsal aorta for distribution throughout the body. Venous entrances to the atrium and the exit of the arterial bulb from the ventricle are guarded by valves.

The walls of the cyclostome atrium and ventricle consist of muscle fibers, with some organization into bundles that run in many directions. The myofibrils are located chiefly around the periphery of each fiber; the sarcoplasm and the nuclei occupy the interior. There is a well-developed connective tissue in both chambers of the heart.

A substantial layer of this connective tissue lies between the covering epithelium (epicardium) and the muscle layer (myocardium), penetrating between the muscle fibers. A similar, but not so strongly developed, connective tissue is also found beneath the endocardial epithelium. In this case, however, the ventricular cavity is not completely blocked off from the spaces among the masses of myocardial tissue, and the blood penetrates everywhere among the large muscle bundles, just as it does in the embryonic heart of higher vertebrates.

With the exception of a few minor features, the circulation of the cyclostomes either agrees with that of the fishes in general or represents a stage in its development. Not until we reach the lungfishes does the heart become more complex, beginning to separate into right and left halves.

One major aspect is peculiar to the circulation of one group of cyclostomes, the myxinoids, or hagfishes. The propulsion of their blood is due not only to the contraction of the heart but also to the action of a number of "accessory hearts" that are part of the venous system. For example, the "portal heart" is part of the portal vein leading to the liver. It has striated muscle in its wall and displays a spontaneous intrinsic beat.

The true heart of *Myxine* is also unusual. In adult life it is completely aneural: there are apparently no nerve fibers going to the organ (Greene, 1902; Augustinsson and associates, 1956). This condition is similar to the nerve-free heart of the embryos of teleost fishes. The heart of the lamprey, on the other hand, is definitely innervated.

It is likely that heart activity in *Myxine* is regulated by changing concentrations of catecholamines, adrenaline-like substances produced by **chromaffin cells.** Such cells are common in the adrenal medulla of other vertebrates and have been identified in hagfish heart tissue by Östlund and associates (1960). Catecholamines are probably involved to some extent in the regulation of heart activity in fishes, in general.

Blood vessels. The arterial walls of cyclostomes already show the intima, media, and adventitia. The intima consists of endothelium and connective tissue; the media contains considerable smooth muscle; the adventitia is connective tissue only. Smooth muscle is most abundant in the arterial bulb and in the ventral aorta, or gill artery. It becomes very scanty in the smallest arteries.

There are no elastic fibers in the arteries of *Petromyzon,* although some properties of elasticity seem to be present. These properties may be related to the chromaffin cells that are found in the adventitia of these arteries. The cells appear to constitute a **local** mechanism for control of vessel tonus, since they are in a position to discharge catecholamines directly into the tissues of the arterial wall.

The walls of the veins are made up of connective tissue that shows little or no distinction from surrounding tissue. Capillaries consist of endothelium with a minimum amount of connective tissue.

CHONDRICHTHYES

Heart. The elasmobranch heart has four chambers—sinus venosus, atrium, ventricle, and conus arteriosus—through which blood passes in sequence. The sinus venosus has very little fibrous tissue. It receives the large veins that return blood from all parts of the body. Among its chief veins are two common cardinal veins, two hepatic sinuses, and coronary veins from the heart.

From the sinus venosus the blood passes into

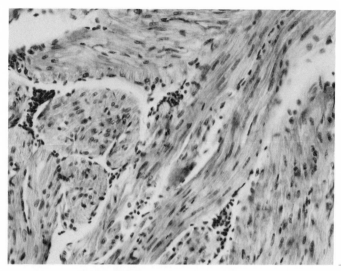

Fig. 8-2. Section through shark's heart, showing anastomosing groups of muscle fibers. Note nuclei located in centers of fibers. Striations not seen. (×300.)

the large, thin-walled atrium and then by way of an atrioventricular valve into the ventricle. The walls of the ventricle are thick and muscular (myocardium) (Fig. 8-2). From the prolonged forward end of the heart, the conus arteriosus arises. It is also well developed and muscular. In the shark the conus keeps a steady, forward arterial pressure to and through the gills.

In front of the conus, the base of the ventral aorta is enlarged into a noncontractile bulbus. The bulbus has smooth muscle in its walls, whereas the walls of the conus contain cardiac muscle. In higher bony fishes the conus tends to disappear and the bulbus is larger.

The heart lies in a pericardial cavity that is lined with a parietal pericardium. A visceral pericardium covers the heart. Lining the inside of the heart is the endocardium of endothelium and elastic tissue. The heart may be slowed by the vagus nerves running to its sinus venosus, but there is no evidence of sympathetic innervation.

Blood vessels. The arteries carry blood away from the heart. They have muscular and elastic walls, although in the shark, because of the intervention of the gills, they have no pulse. Arteries terminate in capillary beds; veins carry blood from the capillary beds back toward the heart. Although veins usually exhibit the same three coats characteristic of arteries, they have thin walls with little muscle or elastic tissue but a great deal of fibrous

tissue. Capillary walls are made up of endothelial cells only, and capillaries have very narrow lumina. The small arteries are connected to capillaries by arterioles, and venules connect capillaries with veins.

As in all gill-breathing vertebrates, blood is pumped from the heart to the gills where external respiration (gas exchange) takes place. From the gills the blood flows through arteries to capillaries throughout the body and is returned to the heart by way of the veins, which often take the form of large sinuses. A renal portal system from the capillaries of the tail carries blood through the kidney before returning it to the heart. The hepatic portal system carries blood from the digestive tract, pancreas, and spleen to the liver before the blood continues to the heart. Thus the shark really has a single circuit circulation, and only venous blood is pumped by the heart.

OSTEICHTHYES

Heart. In the bony fishes the heart consists of a sinus venosus, an atrium, and a ventricle. It is a two-chambered heart, the sinus venosus representing only the entranceway for the venous blood. Randall (1968) has reviewed the functional morphology of the heart in fishes. In general, he says, the heart is composed of typical vertebrate cardiac muscle (Fig. 8-3), but minor differences exist between species and between fishes and other verte-

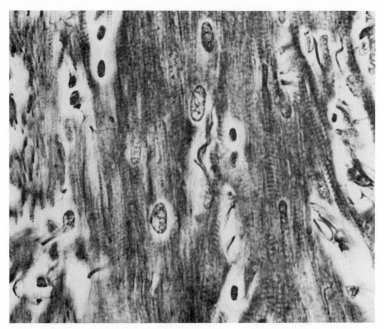

Fig. 8-3. Cardiac muscle from heart ventricle of rainbow trout, *Salmo gairdneri*. The centrally located nuclei are surrounded by an area of clear sarcoplasm. (Wilder's reticulum stain; ×400.)

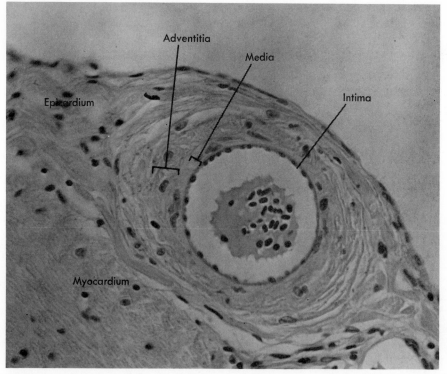

Fig. 8-4. Small artery in epicardium of *Salmo gairdneri*. Note endothelial cells bulging into lumen. Nucleated red cells are seen in the lumen. (×250.)

brates in the distribution of spontaneously active cells and in the rate and nature of the spread of excitatory waves.

The hearts of fishes lack sympathetic innervation. They are furnished with cholinergic stimulation by the vagus nerves, but the level of vagal tone shows a considerable degree of variation. Thus in some fishes the heart appears to be essentially uninfluenced by the vagi during exercise, and the stimulus to increased cardiac output appears to be the increased venous return as in the case of an isolated mammalian ventricle. As noted in the discussion of the aneural hagfish heart, changes in the levels of catecholamines, acting on special receptors in the myocardium, probably are important in regulating activity of the fish heart.

Blood vessels. The ventral aorta arises from the ventricle. The proximal end of the ventral aorta is dilated to form an aortic bulb, the **bulbus arteriosus.** This contains no cardiac muscle and is not rhythmically contractile. It differs from the conus arteriosus, which lies in a similar location but has cardiac muscle and rhythmic contraction. The conus is thus considered to be a part of the heart; the bulb is part of the arterial system.

In the elasmobranchs and in the most primitive teleosts, the bulb at the beginning of the arterial system is actually a conus arteriosus that contains striated cardiac muscle, undergoes pulsation, and is often provided with special valves. However, throughout their evolution the bony fishes show a shift from the contractile conus to the elastic, but noncontractile bulbus arteriosus.

The arteries of fishes in general show a poorer development of smooth muscle than those of other vertebrate classes (Fig. 8-4). The media and adventitia consequently are less sharply set off from one another.

AMPHIBIA

Heart. The amphibian heart is three-chambered, having two atria and a single ventricle

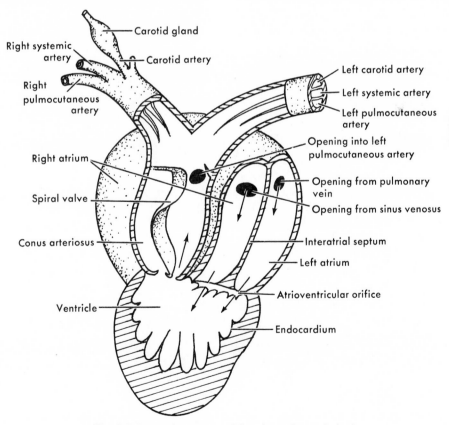

Fig. 8-5. Internal structure of frog heart (ventral view).

(Fig. 8-5). As in all vertebrates, the wall of the heart is made up of three layers: endocardium, myocardium, and pericardium. The endocardium consists of an endothelial lining continuous with that of the blood vessels, and a subendothelial collagenous layer including some smooth muscle.

The myocardium makes up the great bulk of the heart. It consists of spirally arranged cardiac muscle layers that are much thicker in the ventricle than in the atria. The myocardium is supplied with blood by a sinusoid circulation and by seepage through the spongy wall of the ventricle. The visceral pericardium is a serous membrane covered externally by a single layer of mesothelial cells. Underneath it are a thin layer of connective tissue and a subpericardial layer of areolar tissue that attaches to the myocardium.

The histologic aspect of a section through the frog's heart will naturally vary with the region through which the section is made. A section through the ventricle will reveal bundles of cardiac muscle with spaces of varying sizes, considerable amounts of connective tissue, blood vessels, and sinusoids. The tissues of the amphibian heart are not as compactly arranged as are those in higher vertebrates. The network arrangement of the cardiac muscles is shown by their branching (Fig. 8-6).

The myocardium is not a syncytium, but the intercalated disks along the branching fibers represent boundaries between cells as do those revealed in mammalian heart tissue by the electron microscope. The nucleus in a frog cardiac fiber is oval in shape and lies near the center of the individual fiber. Intercalated disks are very difficult to demonstrate; they stain very faintly, as do the cross striations. In the interstices between the fibers there can be seen, in some cases, the **endomysium** of the heart muscle: a network of reticular fibers with its lymphatics and blood sinusoids. Dyer (1967) has described the ultrastructural features of the cardiac muscle of the salamander *Amphiuma* (Fig. 8-7).

In the lungfishes and amphibians an increased use of lungs in breathing is correlated with a progressive increase in degree of separation between the pulmonary and systemic circulations (Johansen and Hanson, 1968). Member species of all genera of lungfishes and several caudate and acaudate amphibians show a method of distribution of blood that tends to decrease to a minimum any recirculation within the pulmonary and systemic circulations. The separation depends on a variety of factors. Among them are the partial ventricular septum in lungfishes and a massive trabeculation formed by cardiac muscle on the

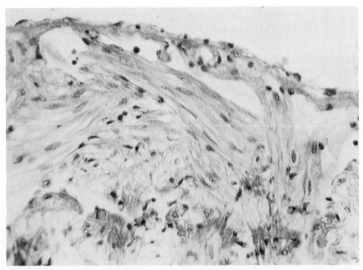

Fig. 8-6. Portion of frog heart, showing network of cardiac muscle (myocardium) with bundles of long fibers running in different directions. Striations of muscles not seen. (×350.)

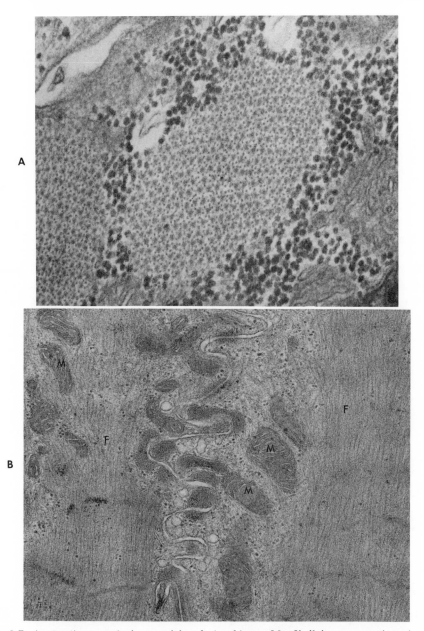

Fig. 8-7. A, Cardiac muscle in ventricle of *Amphiuma*. Myofibril in cross section. Arrangement of thick and thin myofilaments can be seen. In surrounding sarcoplasm are numerous granules of glycogen and several mitochondria. **B,** Cardiac muscle from atrium of *Amphiuma,* showing side-to-side interdigitation of adjacent muscle cells. Mitochondria, *M,* and longitudinal sections of myofibrils, *F,* are seen in each of the two cells. (Lead citrate and uranyl acetate stain; **A,** ×48,500; **B,** ×31,000.) (Courtesy Robert F. Dyer, New Orleans, La.)

inner walls of the ventricles in both lungfishes and amphibians.

In most amphibians the blood from right and left atria generally remains separate on its journey through the ventricle, as far as the base of the conus arteriosus (Simons, 1959). In the frog and toad this separation is continued in the conus arteriosus by the spiral valve, whereas in the urodeles the oxygenated and unoxygenated blood become mixed in the absence of a spiral valve. The blood entering the several arterial arches in the urodeles is therefore all of the same degree of oxygenation.

Blood vessels. In close relation to the chambers of the heart are the termination of venous system and the beginning of the arterial system. The sinus venosus, the largest venous channel, empties into the right atrium. At the beginning of the arterial system, a contractile **conus arteriosus** is generally present. Thus, whereas in the teleosts there has been evolution from muscular to elastic tissue in this region, the amphibians have retained the condition present in the less differentiated fish types from which they evolved. In addition to the well-developed conus, amphibians also possess a bulbus arteriosus. The bulbus, with a large amount of elastic connective tissue in its walls, exerts a steady pressure on the blood passing through it. This pressure serves to convert the pulsating motion imparted by the conus to a steady flow as the blood leaves it. Conus and bulbus together are called the truncus arteriosus and lead on to the ventral aorta.

The blood vessels of amphibians show a sharper definition of layers than do those of fishes (Fig. 8-8). The innermost coat, tunica intima, consists of a layer of flat endothelial cells, covered by an inner elastic membrane, found usually in a wrinkled condition. The middle layer, tunica media, is made up of circular smooth muscle and connective tissue. In the small arteries as well as large, the intima is set off from the media by a layer of elastic connective tissue, the **internal elastic lamina.** This layer remains as a conspicuous feature in the other classes of vertebrates. Smooth muscle is generally well developed in the media, alternating, in the large arteries, with lamellae of elastic connective tissue. The outer layer, tunica adventitia, is an elastic membrane over longitudinal smooth muscle and some white fibrous tissue.

The capillary bed has both muscular and nonmuscular vessels. The muscular type of capillary forms the main arteriovenous (A-V) bridges between the small arterioles and venules. The nonmuscular or true capillaries arise as side branches that are not in the direct path of blood flow from the arterial to the venous side (Zweifach, 1937). These capillaries have no smooth muscle cells but are capable of actively varying their caliber by the contractile power of the endothelium.

Some workers have explained the contraction of capillaries as due to the action of nonendothelial Rouget cells that partially encircle the endothelial layer. Others believe that the endothelium itself

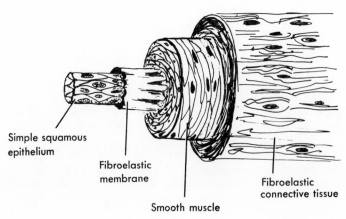

Simple squamous epithelium

Fibroelastic membrane

Smooth muscle

Fibroelastic connective tissue

Fig. 8-8. Structure of an artery. Arteries vary in proportions of various structural elements. See text for details.

is contractile, since contraction in mammalian capillaries seems to be independent of the presence of Rouget cells. In electron-microscope studies of capillaries in the mesentery and small intestine of two amphibian species, *Rana pipiens* and *Diemictylus viridescens,* Hama (1960) found fine filamentous structures, 60 to 80 Å in diameter, coursing through the cytoplasm of some endothelial cells. In places, these filaments were grouped into fibrils that ran roughly parallel to the cell surface. The filaments and fibrils resemble the structures of this type seen in smooth muscle cells and may account for the contractile property of amphibian capillaries. Such filaments have not been observed in mammalian tissue.

In general, veins have much thinner walls than arteries and are more easily collapsed. Whenever an artery and its corresponding vein are seen side by side in cross section, the vein nearly always shows the larger lumen. Although veins have the three characteristic coats, it is possible to distinguish the separate layers only in the medium and large veins; in other veins the components often obscure the exact relationships.

The inner coat (tunica intima) characteristically consists of endothelium and connective tissue fibers, both elastic and collagenous. The middle layer (tunica media) is circular smooth muscle and connective tissue, and the outer coat (tunica adventitia) is made of collagenous connective tissue fibers and some smooth muscle. Apparently to offset gravitational effects, the veins in the extremities and lower parts of the body tend to have more muscle. Common in the walls of the frog's veins are melanophoric pericytes. It is thought that these melanophores aid in the regulation of heat loss very much as they do in the skin.

REPTILIA

The gross differences between the reptilian heart and the amphibian one involve the separation (incomplete) of the single ventricle by a septum and a division of the conus (when present) to its base at the wall of the ventricle. With such an arrangement, the right and left systemic arches and the **pulmonary arch** now have independent openings from the partly divided ventricle. In lizards and turtles a "functional ventricular septum" may be said to occur during systole

(White, 1968), when the muscular ridge approaches the ventral ventricular wall. Low vascular resistance of the lungs favors pulmonary ejection of blood before systemic.

Thus the separation of oxygenated and unoxygenated blood has almost been achieved in the reptile. The condition is perfected in the birds and mammals. The loss of "skin breathing" in reptiles is indicated by the absence of cutaneous veins from the skin.

Microscopically, the medium-sized arteries of reptiles show a higher degree of development of smooth muscle and elastic membranes than do those of amphibians. However, the collagenous tissue is less well developed in the larger arteries.

AVES

In the birds the right and left sides of the heart are as completely separated as in mammals, but the separation develops differently from the manner seen in mammals and resembles more the development in the crocodile, reminding us of the archosaur (ancient lizard) ancestry of the birds.

In birds it is not only the large arteries that are of elastic type (Fig. 8-9). This condition holds for the so-called "arteries of flight" of the wings as well. The veins in birds show little separation into layers as compared with the arteries.

MAMMALIA

Heart. The muscular mammalian heart is four-chambered, with two atria and two ventricles. Thus there is complete separation of the systemic and pulmonary aspects of the circulatory system. Oxygen-poor blood returns to the right side of the heart; oxygen-rich blood, to the left.

The truncus arteriosus of the lower vertebrates is now split into two vessels. One of these (pulmonary) opens out of the right ventricle and goes to the lungs. The other vessel (aorta) leads from the left ventricle to the carotid arteries and the systemic arch. There is no sinus venosus in the heart of the adult mammal; the two superior venae cavae and the inferior vena cava open directly into the right atrium. The pulmonary veins open into the left atrium. The openings between the atria and ventricles are guarded by atrioventricular valves. The openings to the aorta and the pulmonary trunk have semilunar valves.

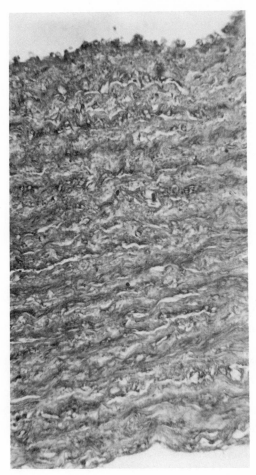

Fig. 8-9. Cross section of wall of the aorta of a pigeon. Note concentrically arranged elastic membranes (convoluted whitish strands) between which are fibroblasts and smooth muscle cells. The aorta is a highly specialized blood vessel, with the tunica media making up most of the thickness of the wall. (×640.)

The heart is surrounded by a double-walled sac, the pericardium, with fluid between the two layers. The heart in its pericardium lies in a septum, the mediastinum, which runs down the center of the thorax, separating the two lungs. The outer surface of the pericardium is attached to the mediastinum and to the diaphragm muscle. The surface of the heart and the inner surfaces of the pericardium are covered with simple squamous mesothelium.

The heart is a modified blood vessel with very muscular walls. On the inside next to the central cavity is the endocardium. As noted previously, this layer is homologous to the tunica intima of a blood vessel and is composed of a lining endothelium in contact with the blood flowing through the chambers, plus a deeper layer containing white fibers, fibroelastic tissue, and some smooth muscle. A connective tissue layer forms a sub-endocardial layer, which binds the endocardium to the myocardium.

The myocardium is the muscular middle layer of the heart, homologous to the tunica media. It is exclusively cardiac muscle, much thicker in the ventricles than in the atria. This cardiac muscle is arranged in sheets that wind around the atria and ventricles in a complex, spiraling manner. The sheets are columns of muscle cells forming end-to-end attachments, or intercalated disks, the numbers of which seem to increase with age (Zschiesche and Stilwell, 1934) (Fig. 8-10). Human cardiac muscle shows a clear distinction between perinuclear and peripheral cytoplasm (Fig. 8-11). Between the muscle fibers is connective tissue of reticular, collagenous, and elastic fibers.

The external coat of the mammalian heart is the epicardium, also called the visceral pericardium. This is the inner wall of the double-walled pericardial sac that surrounds the heart. The epicardium consists of a fibroelastic layer and a serous membrane of mesothelium bathed by pericardial fluid. Attaching the epicardium to the myocardium is a subepicardial layer of areolar connective tissue.

The heart has a supporting cardiac skeleton of dense fibrous tissue. This tissue forms a base on which cardiac muscle fibers are inserted and valves are attached. In some mammals this fibrous base becomes cartilaginous or ossified. The component parts of the base include (1) two sets of fibrous rings surrounding the origin of the pulmonary and aortic arteries and the atrioventricular orifices; (2) two fibrous triangles, or trigones, which are masses of fibrous tissue connecting the arterial openings and the atrioventricular orifices; and (3) the **septum membranaceum,** which forms part of the interventricular septum.

The tricuspid and mitral valves of the heart are folds of endocardium, reinforced with a core of dense fibroelastic tissue continuous with the fibrous rings. These valves connect with papillary muscles by fibrous cords (chordae tendineae) that

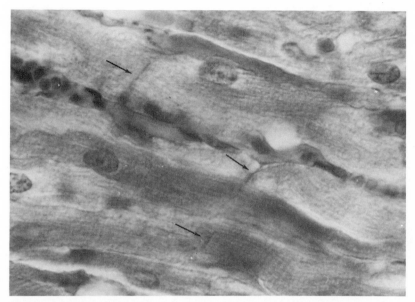

Fig. 8-10. Section through rabbit heart, showing portions of cardiac muscle fibers. Intercalated disks are indicated by arrows. (×475.)

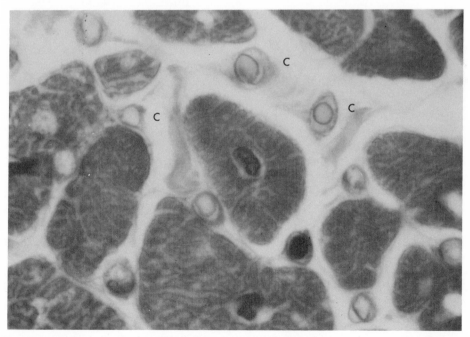

Fig. 8-11. Human cardiac muscle. Fibers shown are chiefly in cross section. The centrally placed nuclei have a small area of clear sarcoplasm around them, lacking myofibrils. Many capillaries *(C)* are present, each showing a single red corpuscle. (Hematoxylin-eosin; ×1000.)

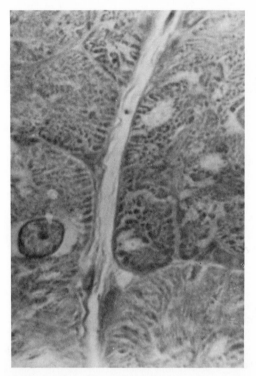

Fig. 8-12. Purkinje fibers of conducting system in mammalian heart. They are modified cardiac muscle fibers and conduct the impulse of contraction very rapidly. (×475.)

restrain the valves and prevent their eversion when the ventricles contract. The papillary muscles are really extensions of the trabeculae carneae (irregular projections of the myocardium).

To coordinate the heart beat, the rabbit, in common with other mammals, has a system of cardiac muscle fibers specialized for conduction. These fibers lie in the subendocardium, close to the myocardium. The pacemaker is the sinoatrial node at the junction of the superior vena cava with the right atrium. This is the site of the sinus venosus found in lower vertebrates and in mammalian embryos. The modified muscle fibers of this system, Purkinje fibers (Fig. 8-12), have a faster rate of conduction than ordinary cardiac muscle fibers. They are large fibers and are distinguished by having more sarcoplasm and glycogen than cardiac muscles. Impulses originating in the sinoatrial node pass diffusely over the atrial musculature to activate the atrioventricular node in the median wall of the right atrium. From this node a bundle of His communicates via branching trunks with each ventricle.

Blood vessels. As in all the vertebrates, blood vessels larger than capillaries follow a common pattern of organization (Fig. 8-8). This pattern may be altered in the blood circuit by emphasis of certain basic structures, reduction of others, or omission of certain features altogether. The innermost coat, the tunica intima (or interna), typically contains an inner endothelial lining of simple squamous epithelium, a subendothelial lining of fibroelastic connective tissue, and an external band of elastic fibers. This latter layer, the internal elastic lamina, first became conspicuous in the vessels of amphibia and may be reduced in many blood vessels. The middle coat, the tunica media, is chiefly circularly arranged smooth muscle, but it may also contain some elastic fibers. The tunica adventitia generally possesses an elastic membrane closest to the media and loose connective tissue, the fibers of which usually run parallel to the long axis of the blood vessels, linking the vessel to the surrounding tissue.

The largest artery, the aorta (Fig. 8-13), shows an intima with a definite connective tissue layer beneath its endothelium. The media of the aorta consists of alternating layers of elastic lamellae and circular smooth muscle. Collagenous fibers are intermingled with both the elastic and muscular elements (Fig. 8-14), and fine networks of argyrophilic reticular fibers surround the smooth muscle cells. The adventitia of these arteries is very thin and cannot be distinguished from the surrounding tissues. Other large arteries of the elastic type include the innominate, pulmonary, common carotid, subclavian, and common iliac.

In medium-sized arteries (Fig. 8-15) the muscular element of the tunica media predominates over the elastic. These are the so-called "muscular arteries." In the smallest arteries the media may consist of only a few individual smooth muscle cells.

The walls of large arteries and veins of vertebrates contain small vessels, the **vasa vasorum** (Fig. 8-16), which run to the adventitia where they terminate in a dense capillary network that may extend deep into the media. These vessels supply nutrients and oxygen to the tissues of the larger vessels and carry away wastes. They are

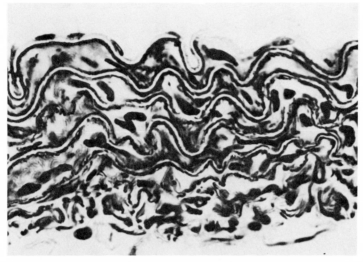

Fig. 8-13. Aorta of mouse. Elongated nuclei of endothelial cells are seen on inner (lower) surface. Elastic fibers appear as well-defined undulating strands. Between them are nuclei of smooth cells and fibroblasts. (Wilder's reticular stain; ×400.)

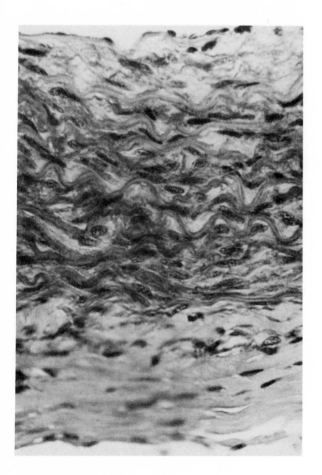

Fig. 8-14. Section through blood vessel of rabbit. Convoluted elastic membranes interspersed with collagenous fibers and smooth muscle are evident in the tunica media. (×475.)

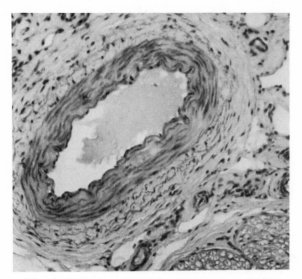

Fig. 8-15. Cross section of medium artery. (×200.)

comparable to vessels of the coronary system, which nourish the heart wall.

The simplest blood vessels are the capillaries (Fig. 8-17), which have an average diameter approximately equal to the diameter of a red corpuscle (7 to 9 μm). Capillaries exist in extensive networks or meshes of narrow canals. The meshes vary in size and shape in different capillary beds, according to the organ or tissue; rate of metabolism of the tissue is often an important factor. In general, capillary beds in mammals are of closer mesh than those found in the lower vertebrates.

The wall of a capillary is a simple layer of flat endothelial cells, which are really curving thin plates with ovoid nuclei. The cells are usually stretched along the long axis of the vessel and separated from a supporting bed of connective tissue. The cell margins are serrated or wavy. A delicate reticular network accompanied by fixed macrophages and fibroblasts surrounds the endo-

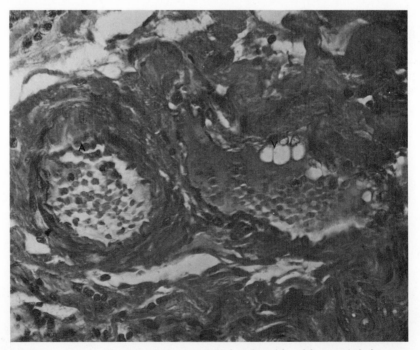

Fig. 8-16. Vasa vasorum ("vessels of the vessels") in adventitia of human inferior vena cava. Function of these vessels is to nourish tissues in the relatively thick walls of large vessels. *A,* Artery; *V,* vein. (×250.)

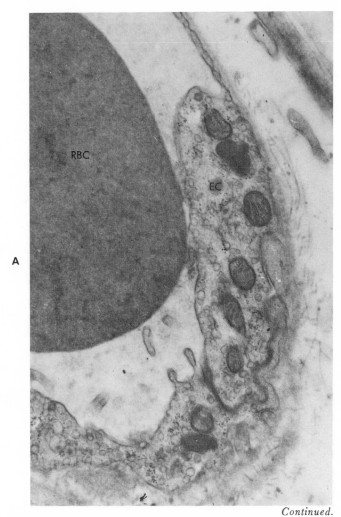

Continued.

Fig. 8-17. A, Portion of capillary in lamina propria of small intestine of mouse. Endothelial cells show a few processes projecting into lumen, mitochondria, and numerous vesicles in their cytoplasm. *EC,* Endothelial cell; *RBC,* red blood corpuscle. **B,** Capillary in lamina propria of duodenum of adult mouse. Vesicles, *V,* common in endothelial cells of capillaries, are seen. Infoldings, *I,* of cell membrane also are visible. A single red blood corpuscle is seen in the lumen. (×29,000.)

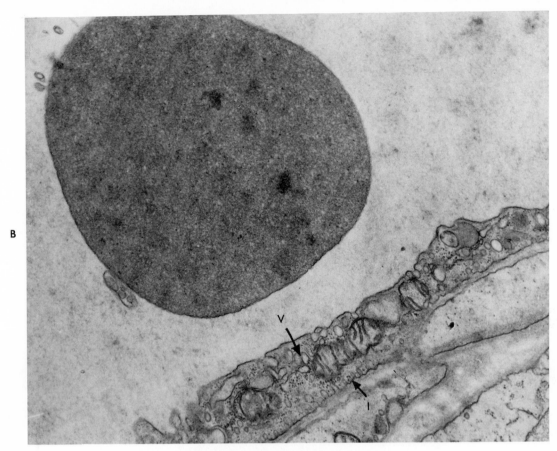

Fig. 8-17, cont'd. For legend see p. 147.

thelial tube. Rouget cells with branching processes may partially ensheath the capillary wall.

The fine structure of the capillaries shows some special features. The endothelial cells (Fig. 8-17) have many vesicles in the cytoplasm, probably indicating an active pinocytotic process.

In some organs such as the spleen and liver the connections between arteries and veins are not by way of ordinary capillaries but by less regular blood spaces, the sinusoids. These do not have a continuous lining of endothelium; the lining is an irregular one of reticuloendothelial cells, many of which are phagocytic elements.

The caliber of veins is usually larger than that of their corresponding arteries, but their walls are much thinner. They also communicate freely with each other, so that their channels are less regular than those of most arteries. The veins of

mammals show a more clear-cut distinction into layers than those of any of the other vertebrate classes. They usually have a reduced amount of muscular and elastic components in their coats. The poorly developed media usually contains only smooth muscle, whereas the adventitia, made up of fibroelastic tissue and often smooth muscle fibers, is usually relatively thick in comparison with the other layers.

Many of the small and medium-sized veins, especially those of the legs, have valves that block the backflow of blood away from the heart. The valves are semilunar pockets produced by local folding of the tunica intima. Valves usually face each other in pairs on opposite sides of the lumen. They are composed of connective tissue with rather large numbers of elastic fibers and are covered on both surfaces with endothelium.

Lymphatic circulation and lymphoid tissue

CYCLOSTOMATA

No distinct system of lymphatic channels has been described for the cyclostomes. White blood cells, which can be designated as lymphocytes, are found in the blood and in some places in the tissues. They are produced in masses of "lymphoid tissue" in the kidney and at other sites. Thus the cells and tissue of the lymphatic system appear to antedate the vessels in evolutionary development.

The cyclostomes studied by Murata (1959) show no lymphoid tissue in the core of the spiral fold of the intestine, but there is a mass of mesenchymal tissue that may be thought of as ancestral to lymphoid tissue. In the ammocoete larvae of *Lampetra* the spiral fold is the chief site of blood cell formation. By the adult stage, however, this tissue is almost completely replaced by adipose tissue, a very early and interesting incidence of the process of replacement of hemopoietic tissue by fatty tissue. Such a process is common and seen often in the human skeleton.

A longitudinal artery runs through this spiral fold of the cyclostome intestine. The densely aggregated lymphoid cells surround the artery much as such cells form a periarterial lymphoid sheath around the many "central arteries" in the white pulp of spleens of higher vertebrates. The peripheral part of the spiral fold shows many venous sinusoids filled with blood, giving an appearance similar to that of the red pulp of the definitive spleen. Because of its location, its tissue structure, and its circulatory apparatus, the spiral fold of the larval lamprey is believed to represent an initial stage in the evolution of the spleen (Kanesada, 1956).

A search for lymphocytes in a region where they are found constantly in higher vertebrates (i.e., in the epithelium of the intestine) led Fichtelius and associates (1969) to make an interesting distinction between the lamprey and the hagfish. In *Petromyzon,* the lamprey, they found cells that resemble lymphocytes in all respects. In *Myxine,* on the other hand, the cells occurring basally and between the regular intestinal cells have nuclei that, although rounded, resemble the epithelial nuclei in their chromatin pattern. This raises a

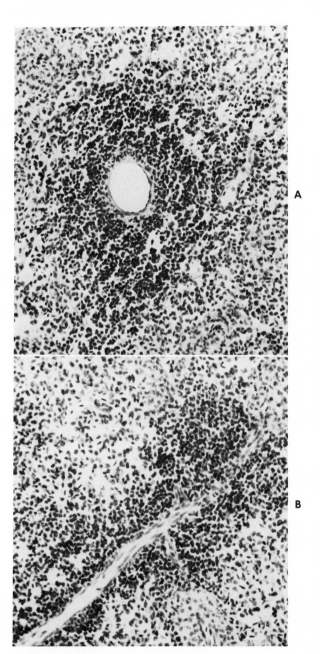

Fig. 8-18. Lymphoid tissue surrounding a central artery in spleen of elasmobranch *Scylliorhinus.* **A,** Cross section. **B,** Longitudinal section. As in higher vertebrates, mass of small lymphocytes forms a "sleeve" for the artery. (×400.)

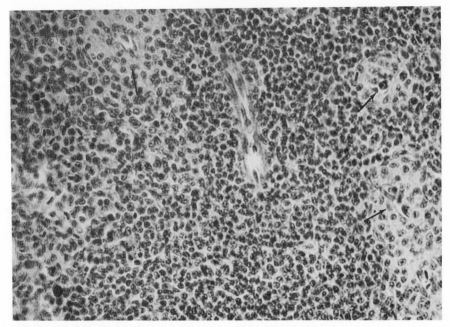

Fig. 8-19. Portion of shark's spleen. This lymphatic organ holds blood instead of lymph and communicates with arteries and veins. Islands of white pulp are indicated by arrows. (×500.)

question of considerable significance: does the hagfish possess an intestinal epithelium without lymphocytes? If so, this would be the only such case among the vertebrates!

CHONDRICHTHYES

Circulation. The lymphatic system is poorly developed in sharks and in many regions may not be distinct from the venous system. However, incomplete as this system may be, it definitely functions in collecting interstitial fluid lost from the bloodstream and returning it to the blood vascular system. Lymphatic valves have not been found in the elasmobranchs, and lymph nodes are absent.

Spleen. According to Osogoe (1953, 1954) and Kanesada (1956) true lymphoid tissue is seen for the first time in the spleen of the elasmobranchs (Fig. 8-18). Indeed, this spleen is well developed and is surprisingly similar to that of the mammal.

The shark's spleen is a darkish red, elongated body in the dorsal mesentery just to the left of the convex side of the U-shaped stomach. It is a large lymphoid organ but is not permeated by lymph capillaries and has blood instead of lymph in its sinuses. It is surrounded by a thin fibrous capsule, which may contain a few smooth muscle fibers. (Both muscle and elastic fibers are common in the mammalian spleen.) This capsule gives off septa or trabeculae, partitions that penetrate into the interior of the spleen, branch, interlace, and form a netlike framework. Within this framework is the parenchyma, made up of two kinds of lymphoid tissue known as white pulp and red pulp (Fig. 8-19). The white pulp is the chief source of lymphocytes. Each island of white pulp is an accumulation of small lymphocytes, a malphigian corpuscle. The red pulp contains great numbers of venous sinuses filled with blood. The white pulp frequently grades into the red pulp. A delicate network of reticular fibers is found throughout, enclosing the pulp cells and the sinusoids.

The splenic histology of the shark is not as clearcut as that of higher vertebrates, especially mammals. Arteries enter the interior by way of the trabeculae and give off branches to the lobules between the trabeculae. Some parts of the smaller arteries run through masses of white pulp where lymphocytes are proliferating. Some of the smaller arteries open into sinusoids, which become filled with red blood corpuscles while the plasma leaks

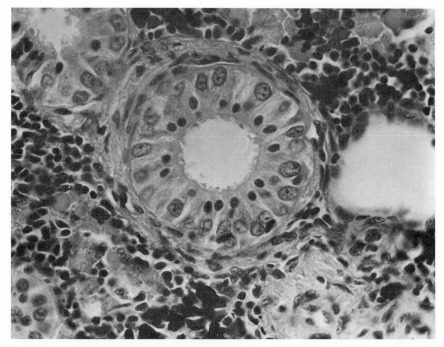

Fig. 8-20. Renal tubules of a teleost fish—*Salmo gairdneri,* the rainbow trout. Lymphocytes migrate into, and perhaps through, renal epithelium in a number of bony fishes. (×400.)

out into the red pulp. Some of the red corpuscles also go through the endothelial walls of the sinusoids into the red pulp. Besides the mature erythrocytes, the spleen also contains other types of cells found in the circulating blood.

The chief splenic artery and splenic vein enter the spleen at the hilus, a prominent indentation through which vessels and nerves pass. After blood has passed through the spleen, it is picked up by the splenic vein, which is a branch of the hepatic portal system.

The functions of the spleen vary with different vertebrates. Some produce erythrocytes, some may act as storage for reserve blood, and all may produce antibodies by the lymphoid cell types. The sinusoids are also well provided with phagocytic cells that destroy worn-out corpuscles and bacteria. All these functions are probably carried out by the shark spleen.

OSTEICHTHYES

Many of the bony fishes have a well-developed lymphatic system, with vessels beneath the skin and within the muscles, the walls of the viscera, and the glands. These vessels collect the fluid that

has passed out of the blood capillaries and diffused through the tissues, returning it through larger lymph vessels to the veins. "Lymph hearts," present in some bony fishes, are hollow pulsating structures, found generally at places where the larger lymph vessels open into the veins.

There are considerable amounts of lymphoid tissue in some organs, especially in relation to mucosal surfaces. The lymphoid masses vary in relation to species and stage of development, sometimes taking the more dense form of lymph nodules, such as in the spleen. However, these nodules do not show the pale, lymphocyte-producing germinal centers characteristic of such tissue in higher forms. Some teleosts have contractile lymph hearts, but these are far better developed in amphibians. Individual lymphocytes show migratory activity into epithelial layers. This activity is not confined to the sites involved in higher vertebrates but occurs in other locations as well, for example, in the kidney in many species (Fig. 8-20).

AMPHIBIA

Circulation. In higher vertebrates the lymphatic system, which is a part of the vascular system, is made up of lymph spaces, lymphatic capillaries, and larger vessels. It returns lymph from tissue spaces to the blood circulation by way of the large neck veins. In the frog the lymphatic system is remarkable for the number and large size of the lymph spaces and sinuses in the various parts of the body. The lymphatic vessels themselves are poorly defined in the frog, and the lymph flows in irregular spaces between and within the different organs. Some of the larger spaces are lined by flattened endothelial cells but are mostly devoid of other tissue.

The lymphatic system in frogs has four-lymph hearts that propel the lymph back into the blood. The paired anterior lymph hearts are just behind the transverse processes of the third vertebra and empty into the vertebral vein; the paired posterior lymph hearts lie on either side of the urostyle and empty into the transverse iliac vein. The lymphatic vessels lack valves, but some of the lymph hearts have them. The muscle tissue in the walls of these hearts is of the striated skeletal variety, arising from the myotomes.

Amphibians lack lymph nodes. Dense clusters of lymphoid tissue (lymph nodules) are common, but they appear to have no germinal centers.

Spleen. The spleen is the largest hemolymphatic organ in the frog's body. It is a rounded, reddish structure lying dorsal to the cloaca, supported by a mesentery. The amphibian spleen is structurally similar to that found in the shark and fishes generally. The organ is surrounded by a peritoneal layer beneath which there is a fibrous membrane. From this capsule, muscular trabeculae extend into the interior of the spleen and divide it into compartments, or lobules. The major part of the interior consists of the red and white spleen pulp. Red pulp contains lymphocytes, erythrocytes, and other elements of circulating blood; white pulp (splenic corpuscles) is made up of nodules of compact groups of lymphocytes held together by reticular tissue and usually with a central artery.

The red pulp is arranged as narrow zones of tissue between the sinusoids, which are lined with endothelial cells and associated with the netlike reticular cells. The sinusoids are connected to venous capillaries that carry blood to the trabeculae where veins convey blood out of the spleen. The spleen, among other functions, destroys worn-out erythrocytes by means of phagocytes in the sinusoids (Fig. 8-21). It also serves as blood-forming tissue in some frogs after their metamorphoses.

REPTILIA

Definite lymphatic vessels are well developed in reptilia. Thoracic ducts, as in higher vertebrates, open into the innominate veins in the neck region. A pair of posterior lymph hearts may pump lymph into the iliacs. The spleen is similar to that of birds and mammals.

The lymphatic system of reptiles resembles that of amphibians in a general way. Lymph sinuses, however, are not conspicuous. Notable features are the **lymph nodes,** which make their appearance in the mesentery of crocodiles. In some snakes the main lymphatic vessel forms a sheath around the dorsal aorta (Borghese, 1940).

AVES

In birds the lymphoid tissue tends to be dispersed rather than aggregated into nodes. Retention of the lymph heart mechanism is seen in the embryos of a number of birds, and it persists in the adults in some species. The important new feature in birds is the presence of **germinal centers** in the lymph nodules. Germinal centers are also abundant in mammalian lymphoid tissue. However, it it not known whether they may have been present in some of the reptiles that were the common ancestral stock of birds and mammals. If not, then it may be that this represents a case of parallel evolution in these two distinct vertebrate classes.

The spleen of the bird is specialized for filtering blood. Its sinuses are filled with blood instead of lymph. It has a collagenous framework within which there is a reticular network. Trabeculae pass from the capsule of the spleen into the interior of the organ. The chief cells (parenchyma) are of distinct types, white and red pulp (Fig. 8-22). The splenic nodules, each containing a small artery, and the lymphoid tissue surrounding them make up the white pulp. The white pulp is the chief site of lymphocyte production. The more abundant red pulp, containing many erythrocytes, is permeated with venous sinusoids.

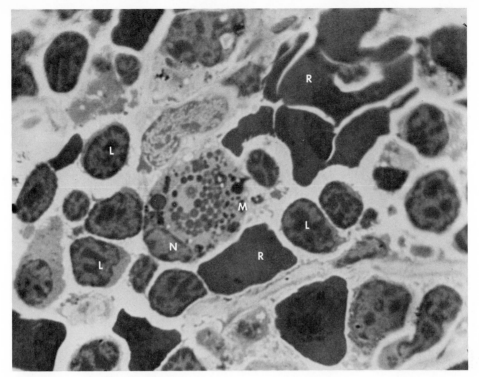

Fig. 8-21. Detail of splenic tissue of the axolotl, *Ambystoma mexicanum.* Large macrophage with eccentric nucleus, *N,* and many ingested particles, *M,* is at center of field. Portions of red blood cells, *R,* and a number of lymphoid cells, *L,* are in field. (One-micrometer epon section, toluidin blue stain; ×2000.)

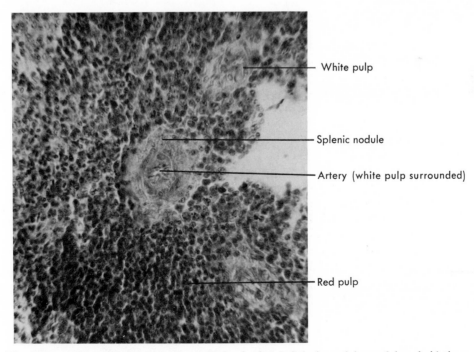

White pulp

Splenic nodule

Artery (white pulp surrounded)

Red pulp

Fig. 8-22. Section of spleen of pigeon *(Columba livia).* Splenic nodules and lymphoid tissue around central arteries make up white pulp. Red pulp, with numerous erythrocytes, is associated with sinuses and veins. (×640.)

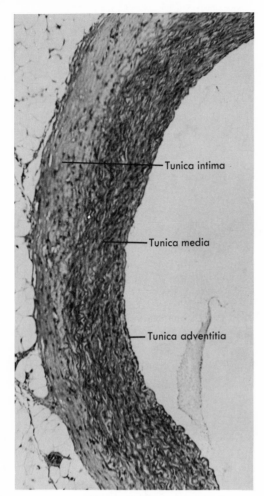

Tunica intima

Tunica media

Tunica adventitia

Fig. 8-23. Section through wall of a mammalian thoracic duct. This largest of the lymphatic vessels receives lymph from most of body and opens into bloodstream at junction of left internal jugular and subclavian vein. (×100.)

MAMMALIA

Circulation. The lymphatic system of mammals is quite elaborate. It consists of capillaries, forming networks in the tissues they drain, and larger and larger vessels that conduct lymph to its final destination. Lymph drainage is a one-way flow and is not really a circulation as is the blood system.

Lymphatic vessels have in their walls the same general tissue elements as blood vessels. In the larger ones the three layers, intima, media, and adventitia, can be distinguished (Fig. 8-23).

The intima consists of endothelium with a thin layer of connective tissue. Circular smooth muscle is the chief component of the media. The adventitia is well developed and shows, in addition to collagenous and elastic fibers, bundles of smooth muscle tissue.

The lymphatic vessels contain valves, which have a construction similar to those in the veins. These valves give the vessels a beaded appearance. The largest of the lymphatic vessels is the thoracic duct.

The smallest lymphatic vessels, the lymph capillaries, are lined by large, polygonal endothelial cells. They are usually broader than blood capillaries but are not uniform in caliber because of their dilations and constrictions. They are found in nearly all tissues, often forming extensive networks alongside blood capillaries.

Lymph nodes. Most of the lymph, before returning to the blood, passes through small round or oval lymph nodes that act as filters. Lymph nodes are not just clusters of lymphoid tissue as lymph nodules are. A node has a definite organ-type structure. It has a connective tissue capsule, projections from which (trabeculae) divide the cortex of the node into irregular subdivisions. Lymphatic vessels enter the node and empty their contents into a subcapsular sinus. The lymph then filters through the lymphoid tissue of the node, bathing the lymphocytes, reticular fibroblasts, and fixed and free macrophages that are the elements of this tissue. At the hilus, an indentation in one side, the lymph collects into the definitive efferent lymphatic vessels.

The architecture of the lymph node tissue shows generally two portions, cortex and medulla. In the cortex the aspect is that of a compact tissue with great hosts of small lymphocytes crowded together. Rounded nodules usually form a considerable part of the cortical tissue. Conway (1937) showed that these nodules are dynamic structures which come and go, probably in response to needs of the organism. The primary nodules often show secondary nodules that are the germinal or reaction centers (Fig. 8-24) where mitotic division and phagocytic activity may be observed. These germinal centers are found in all species of mammals. They are not present, however, in early postnatal life, and they

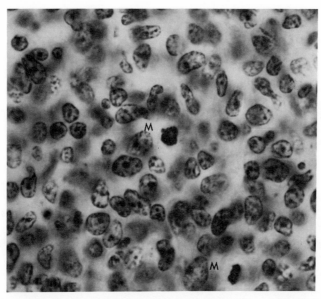

Fig. 8-24. Portion of germinal center in lymph node of rat. Nuclei are chiefly those of medium-sized lymphocytes. Mitotic figures, *M*, are seen. (×1000.)

often disappear from the nodules in old age (Andrew and Andrew, 1948).

The medulla is a loosely arranged tissue with wide sinusoids and anastomosing cords. The cords show reticular fibroblasts, macrophages, lymphocytes, plasma cells, eosinophils, and occasional mast cells. The lymph-filled sinusoids are lined by cells of the reticuloendothelial system. They often contain ingested particles of pigment or cellular debris. In the lumen of the sinusoids free macrophages occur in varying numbers.

Lymph nodes from different parts of different mammals, at different ages, and perhaps under various conditions that are not fully understood present differences in appearance (Figs. 8-25 and 8-26). For example, the relative masses of cortex and medulla may vary greatly, as may the degree of development of the trabecular system. In the most primitive mammals, the Monotremata, lymph nodes appear to be actually individual nodules with germinal centers. Like the ordinary, composite nodes, these nodules occur in specific groupings. Numbers of lymph nodes vary in different mammals. Man has approximately 465, dog 60, cow 300, horse 800 (Baum, 1925). The rabbit shows a very large mass of lymphoid tissue in the mesentery. This appears

to have arisen as an aggregation of mesenteric nodes. A similar "organ" is seen in some carnivores and rodents. In the **hemolymph nodes,** found in some species of mammals including the ox, the sinusoids contain blood. Otherwise the structure resembles that of lymph nodes in general. It seems reasonable to assume that the function of the hemolymph nodes may be much like that of the spleen and that they present a further opportunity for reticuloendothelial cells to remove debris from the blood.

Spleen. As in the lower forms, the spleen is the largest of the lymphoid organs. The mammalian spleen is fundamentally similar to those of the shark and frog, described earlier. Generally, in the birds and mammals there is a sharper demarcation between the white pulp and the red pulp. The red pulp consists of cords of primitive reticular connective tissue (with some wandering cells) and the venous sinuses, which contain all types of blood cells. The lymphoid tissue of the white pulp is present in well-organized nodules surrounding central arteries. In the spleens of many kinds of lower vertebrates (e.g., teleosts and amphibians) the lymphocytes are scattered diffusely, and nodules are difficult or impossible to distinguish.

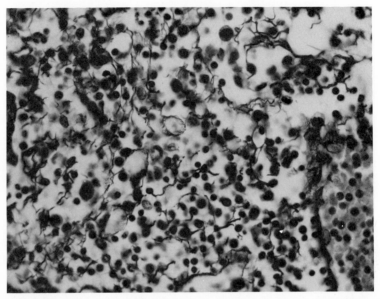

Fig. 8-25. Portion of human lymph node. Stain is combination of hematoxylin with silver impregnation, thus demonstrating both lymphocytes and reticulum fibers. (×660.)

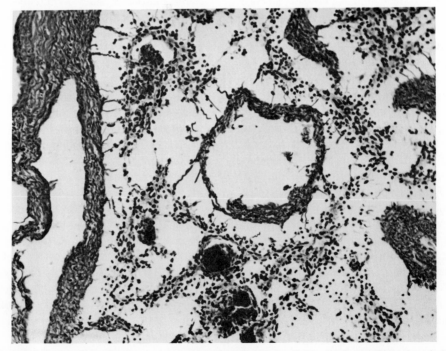

Fig. 8-26. Mediastinal lymph node of camel, *Camelus dromedarius*. Reticular fibers in relation to capsule and vessels are demonstrated. (Wilder's silver impregnation stain; ×125.)

The spleen is specialized for filtering the blood, just as the lymph nodes do for the lymphatic system, but it appears to perform several other important functions, as well. A primary function of the spleen in mammals is the elimination of old red blood corpuscles. It also serves as a temporary storage place for erythrocytes, releasing them into the circulation as needed. Under certain conditions the spleen may form red and white corpuscles; thus it has some of the characteristics of lymphoid tissue and others of bone marrow.

Hemopoiesis. Corpuscle-forming tissue is called hemopoietic. The formed elements of the blood are short-lived and are continuously being lost to the body. The process of forming new blood cells is called **hemopoiesis.** In lower vertebrates, various tissues—the general connective tissue, the kidneys, the subcapsular areas of the liver, and the spleen—carry out some hemopoietic functions. Lymphocytes and related corpuscles may develop in the lymphoid tissue wherever it is found (lymph nodes, appendix, etc.). However, in most reptiles, birds, and mammals, erythrocytes and granulocytes are normally produced exclusively within the bone marrow.

The marrow is a pulpy tissue filling the cavities of the bones. When it is relatively inactive and contains large numbers of adipose cells, it is called yellow marrow. When it is active in hemopoiesis and contains few or no adipose cells, it is called red marrow. The hemopoietic tissue of the marrow shows a framework of reticular fibroblasts and fibers, with fixed and occasionally free macrophages. In the interstices of this framework are vast numbers of developing erythrocytes and leukocytes. Here also are the giant cells with multilobed nuclei known as the megakaryocytes, the cytoplasm of which gives rise to the "noncellular" blood platelets.

A general term for the earlier stages in development of erythrocytes is erythroblasts. Erythroblasts are cells that are undergoing a continuous differentiation, during which they also undergo mitosis. The youngest erythroblasts have round, vesicular nuclei and a basophilic cytoplasm. They appear to be not far removed from the "stem cell" (hemocytoblast), which is very similar to, if not identical with, the large lymphocyte or lymphoblast.

Erythroblasts must produce and accumulate the respiratory pigment, hemoglobin. As this substance increases in amount in the cytoplasm, the latter becomes first polychromatophilic and then acidophilic. The nucleus loses its vesicular appearance and stains deeply. In this stage of acidophilic cytoplasm and deeply staining nucleus, the cell is a normoblast.

Until this stage, the history of a developing erythrocyte in the mammal is not unlike that of such a cell in fish, amphibian, reptile, or bird. Now, however, the differentiating cell undergoes a loss of its nucleus. Only in the mammals does this general phenomenon occur before the red blood cells are ready to enter the circulation. Thus the "red cells" of mammals are perhaps more correctly spoken of as red blood corpuscles. They still perform the active functions of taking up and transporting oxygen and releasing it to the tissues, and they carry on a metabolic activity probably necessary for maintaining their living condition and permitting them to carry out their functions.

The progenitors of the granular leukocytes are the myeloblasts. They have large vesicular nuclei and a small amount of nongranular cytoplasm. They undergo mitotic division and eventually differentiate into cells called myelocytes. These cells have nuclei that show a beginning of lobulation and some granules in the cytoplasm. The myelocytes also divide by mitosis, giving rise to metamyelocytes, which have a better developed granular content and more complicated, often ring-shaped, nuclei.

Cells of this type do not proliferate, but they do undergo further differentiation, acquiring a full complement of either neutrophilic granules to become **neutrophils** or of eosinophilic granules to become **eosinophils.** In those cells becoming neutrophils, the nucleus shows a complex lobulation, whereas in those becoming eosinophils, the nucleus may remain ring-shaped or may become bilobed. It is an interesting fact that myelocytic forms of the basophilic leukocytes are not found in bone marrow of the mouse. Some believe that the **basophils** develop by further differentiation of eosinophils, but this needs to be clarified.

In addition to the various stages of the red and white cells, all of which may occur together in the bone marrow, there are also large mononu-

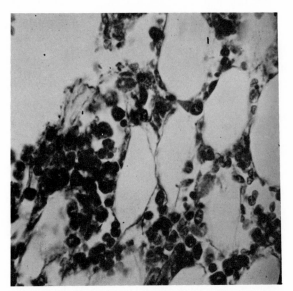

Fig. 8-27. Hemopoietic tissue in the wall of the digestive tract of the hagfish. (×800.) (From Good, R. A., J. Finstad, H. Gewurz, M. D. Cooper, and B. Pollara. 1967. Am. J. Dis. Child. **114**:477-497. Copyright 1967, American Medical Association.)

clear cells, lymphocytes, megakaryocytes, and fat cells. Blood vessels including arteries, veins, and sinusoids are abundant in the bone marrow. The cells that have become mature, both red and white, enter the circulation through the thin walls of the sinusoids.

Humoral and cellular immune mechanisms

The bodies of all vertebrates provide defenses against microorganisms and deleterious foreign substances. Such traits are present in varying degrees throughout the vertebrate series (Figs. 8-27 and 8-28); in the more advanced forms, including man, they have reached a high degree of complexity and efficiency.

Investigators in immunology have long been interested in the mystery of the two different kinds of immunity: **humoral,** through products of cellular origin set free into blood, lymph, and tissue fluids; and **cellular,** due to the direct action of

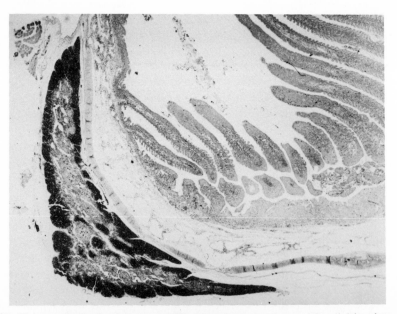

Fig. 8-28. Thymus gland of a baby guitarfish *Rhinobatos productus.* The division into darker cortex and lighter medulla is evident in this elasmobranch fish. The thymus is on the wall of the gut. (From Good, R. A., and A. E. Gabrielsen. 1968. The thymus and other lymphoid organs in the development of the immune system. Pages 526-564 *in* F. T. Rapaport and J. Dausset, eds. Human transplantation. Grune & Stratton, Inc., New York. Reprinted by permission.)

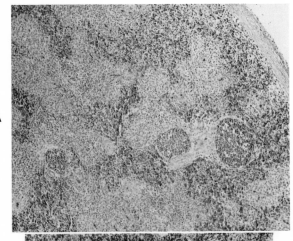

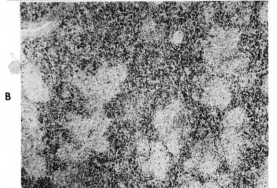

Fig. 8-29. A, Spleen of normal chicken, about 6 weeks of age. Two lymphoid tissue components are seen: the lighter germinal centers, which are bursa-dependent, and the clusters of small lymphoid cells, which are thymus-dependent. **B,** Spleen from a chicken that had been bursectomized and irradiated in the newly hatched period. Germinal centers are lacking but the small lymphocyte component remains. (Low power.) (From Good, R. A., and A. E. Gabrielsen. 1968. The thymus and other lymphoid organs in the development of the immune system. Pages 526-564 *in* F. T. Rapaport and J. Dausset, eds. Human transplantation. Grune & Stratton, Inc., New York. Reprinted by permission.)

cells. Studies have shown the lymphoid tissue of the chicken to be dependent on both the thymus gland and the bursa of Fabricius (an epithelial outgrowth of the cloaca of the young chick). The humoral and cellular aspects of immunologic function have thus been dissociated in the chicken (Warner and associates, 1962; Warner and Szenburg, 1964). The primordial tissue for the antibody-producing aspect was found to be in the

bursa, whereas the development of immunity to homografts was found to be due to the activity of the thymus. The bursa also controls delayed sensitivity reactions.

Bursectomy results in a clear suppression of the humoral system in chickens (Fig. 8-29). Repeated exposure of bursectomized chickens to agents that normally cause a marked humoral reaction leads to no detectable production of antibodies against protein or bacterial antigens and to no formation of immunoglobulins detectable by electrophoresis. There are no plasma cells and no germinal (reaction) centers in the lymphoid spleen. Thymectomy in chickens produces a defect in the cellular immune system, much like that of thymectomized mammals. Symptoms include poor rejection of homografts; impairment of delayed hypersensitivity reactions; reduction in number of small lymphocytes in the circulating blood; and a great depletion of such cells from the white pulp of the spleen.

PHYLOGENETIC ASPECTS OF THE IMMUNE MECHANISM

In the lowest vertebrates, the hagfishes, adaptive immunity appears to be lacking. Immunoglobulins are not produced, and no lymphoid system, lymphocytes, or plasma cells are present. There is a cell similar to the mammalian lymphocyte, which is believed by Good and associates (1966) to be a precursor of the thrombocyte, erythrocyte, and granulocyte and also to play a role in such inflammatory responses as will occur. An important addition to the knowledge of cellular defense in the hagfish has been made recently by Fänge and Gidholm (1968), who described active macrophages in these animals (Fig. 8-30).

According to Good and associates (1967), the lampreys show an important advance over the hagfishes; in *Petromyzon marinus* there is a thymus-dependent type of immune system. The "thymus" is not a paired organ such as occurs in higher fishes, but rather it consists of a number of small masses of lymphoid and epithelioid cells occurring in the gutters of the pharynx (Fig. 8-31). These tissue masses were described in an early paper (Salkind, 1915) as a lymphoepithelial thymus. *Petromyzon* also has splenic lymphoid tissue and what may be considered as three size-

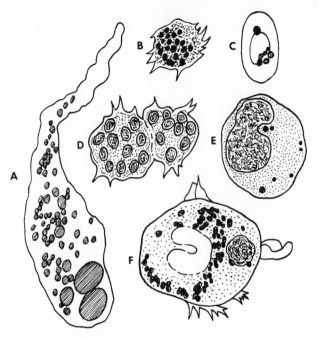

Fig. 8-30. Various kinds of phagocytic cells that occur in the hagfish, *Myxine glutinosa.* **A,** Pearshaped peritoneal cell after injection of Evans blue. **B,** Granulocyte of the blood, with ingested particles of carmine. **C,** Nongranulated cell, carmine. **D,** Agglutinated granulocytes of the blood, with ingested yeast cells that prior to ingestion had been stained with neutral red. **E,** Peritoneal cell with carmine particles, Giemsa-stained. **F,** Peritoneal cell with movable pseudopods and ingested carmine particles. (Redrawn from camera lucida drawings in Fänge and Gidholm, 1968.)

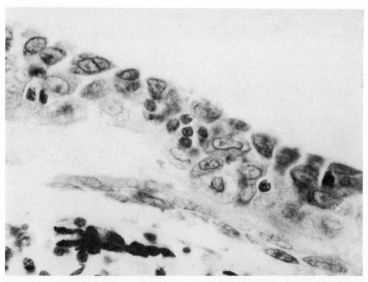

Fig. 8-31. Lymphoepithelial thymus of the larval lamprey. It consists of aggregates of lymphoid cells, five to twenty cells in a group, having an intimate association with the epithelium and lying in the pharyngeal gutters. (High power.) (From Good, R. A., and A. E. Gabrielsen. 1968. The thymus and other lymphoid organs in the development of the immune system. Pages 526-564 *in* F. T. Rapaport and J. Dauset, eds. Human transplantation. Grune & Stratton, Inc., New York. Reprinted by permission.)

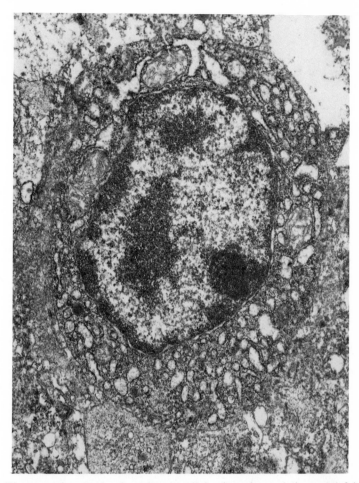

Fig. 8-32. Electron micrograph of a plasma cell in the spleen of the paddlefish, *Polyodon spathula.* It is typical of those found in pericardium and spleen of antigenically stimulated animals. The endoplasmic reticulum is well developed, as in this type of cells in the mammals. Embedment in Vestopal W; staining with lead citrate; ×26,000.) (From Good, R. A., and A. E. Gabrielsen. 1968. The thymus and other lymphoid organs in the development of the immune system. Pages 526-564 *in* F. T. Rapaport and J. Dausset, eds. Human transplantation. Grune & Stratton, Inc., New York. Reprinted by permission.)

classes of lymphocytes. No cells resembling plasma cells have been found. Although almost all the concomitants of the bursa-of-Fabricius type of immune system are lacking, two globulins of gamma type have been found, and some antibody activity has been identified. The presence of a rudimentary bursa-type immune mechanism in the lamprey may eventually make it possible to analyze the primitive condition of the two kinds of mechanisms and the manner of their separation during evolutionary development.

Among the cartilaginous fishes, the paddlefish, *Polyodon spathula* has been an object of study. It has been shown to have plasma cells (Fig. 8-32), which have been described in studies by both light microscopy (Good and associates, 1966) and electron microscopy (Clawson and associates, 1965). Vigorous antibody responses occur in the paddlefish. Lymphoid tissue is found in the wall of the alimentary tract, in the anterior portion of the kidney, and in the gonads. The thymus is well developed, and there is a considerable amount of white pulp in the spleen. Marked "cloning," or aggregation, of plasma cells occurs in the spleen during secondary antibody response.

Hildemann and Haas (1959) and Hildemann and Cooper (1963) have pointed out that in poikilothermic animals such as the fishes and amphibians, environmental temperature is critical to the immunity mechanism. They showed in homotransplantation and homograft experiments that a reduction of temperature can practically shut off demonstrable immunologic responses and adaptive proliferative responses by the lymphoid system. More studies are needed in some of the numerous genera of cartilaginous and bony fishes but it is clear that the appearance of plasma cells precedes the occurrence of germinal centers, recognizable tonsils, and bursae.

According to Good and associates (1966), it is in the Amphibia that plasma cells are first observed in considerable numbers in the lamina propria of the digestive tract. These cells have been observed in several species of amphibians, reptiles, birds, and mammals. It is believed that the plasma cells of the intestinal mucosa may represent the development of a local immuno-globulin-producing system, perhaps as a needed adaptation for terrestrial life. A major step was

probably the appearance of the "secretory system" of plasma cells, the products of which can bathe the surface of a mucosa with protective immuno-globulins.

In a number of species of amphibians lymphoid structures bearing some resemblance to lymph nodes are found in the juxtajugular and pharyngeal regions. These bodies may be of lympho-epithelial nature, but the first clear-cut appearance of follicular lymphoid tissue associated with epithelium of the alimentary tract is in the tonsils of the alligator (Good and associates, 1966). Birds have the much studied bursa of Fabricius (Fig. 8-33). Each of its follicles, or nodules, consists of a cortex of densely packed lymphoid cells and a medulla, or germinal center. A small amount of connective tissue separates the nodules from one another. In the marsupials we first find the "typical" mammalian lymph node structure. It has twenty or more nodules per node, each nodule having its own germinal center.

The paired thymus bodies of mammals are transient glands represented by a mass of lymphatic tissue in front of the pericardium at the base of the heart. Each is made up of many lob-

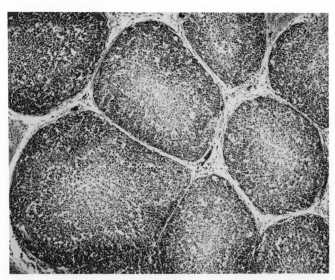

Fig. 8-33. Follicles (nodules) of bursa of Fabricius of a 60-day-old chicken. Note densely packed lymphoid cells of cortex and lightly staining germinal center as medulla. (×125.) (From Good, R. A., and A. E. Gabrielsen. 1968. The thymus and other lymphoid organs in the development of the immune system. Pages 526-564 *in* F. T. Rapaport and J. Dausset, eds. Human transplantation. Grune & Stratton, Inc., New York. Reprinted by permission.)

ules separated by fibrous connective tissue containing blood vessels and lymphatic vessels. The stroma consists of reticular cells and fibers; the parenchyma of lymphocytes (thymocytes). The gland lacks lymphatic vessels and nodules with germinal centers. Each lobule is sharply delineated into a cortex and medulla. The cortex is formed of dense accumulations of lymphocytes in a scanty stroma; the medulla contains more reticular cells and macrophages, with fewer lymphocytes. The medulla is much lighter in staining and often connects one lobule with another (Fig. 8-34). **Hassall's corpuscles** can be seen in the medulla. They are made up of concentrically arranged cells that may be remnants of the epithelial cells of the embryonic primordium of the thymus.

Recent studies indicate that the thymus, which reaches its maximum size in the young animal, produces lymphocytes and liberates them into the blood. These go to the spleen and lymph nodes and serve as the source of antibodies. The appendix of the rabbit may have a similar function. The presence of a hormone in the thymus is suspected but has never been demonstrated.

From these studies has come the concept of two main kinds of lymphoid tissue: central and peripheral (Good and Gabrielsen, 1968). The **central lymphoid tissue** arises from, or in close association with, the epithelium of the gastrointestinal tract. Long ago, Beard (1900) pointed to the thymus as this type of tissue. Later contributions in this field have come from a number of investigators. Recently it has been recognized that the central lymphoid tissues are not directly involved in the adaptive processes of immunologic reaction. Rather they appear to sustain and control the **peripheral lymphoid tissues** in such processes.

This idea is supported by the sequence of development of lymphoid tissue, which is always central first, then peripheral. For example, the thymus is highly lymphoid well before birth in the mouse, whereas the spleen shows no discernible lymphoid development at that time, and the full development of the lymph nodes comes even later.

In the rabbit the very large appendix has a structure generally similar to that of the avian bursa of Fabricius (Fig. 8-35). The rabbit ap-

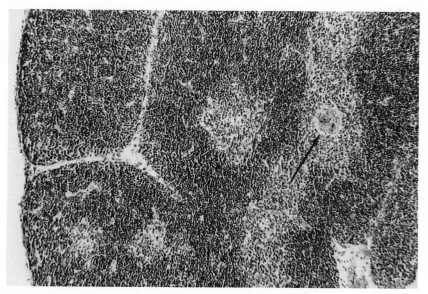

Fig. 8-34. Section of mammalian thymus (rabbit), showing parts of several lobules each with a dense cortex and a central, lighter-staining medulla. A Hassall's corpuscle, characteristic of thymus tissue, can be seen at right (arrow). Although the thymus is considered a transient gland, it is actually found in adult rabbits as well as in young animals. (×100.)

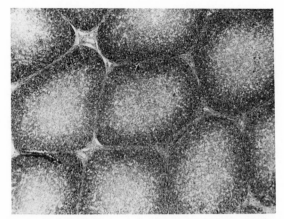

Fig. 8-35. Follicles (nodules) of appendix of a rabbit, 2 to 3 months of age. Note similarity to bursa nodules. (Low power.) (From Good, R. A. and A. E. Gabrielsen. 1968. The thymus and other lymphoid organs in the development of the immune system. Pages 526-564 *in* F. T. Rapaport and J. Dausset, eds. Human transplantation. Grune & Stratton, Inc., New York. Reprinted by permission.)

pendix, in postnatal life at least, seems to have a central lymphoid function in furthering the anatomic and functional maturation of peripheral lymphoid tissue; the spleen does not show such a function. According to the studies of Cooper and associates (1965, 1967), the central lymphoid complex of the rabbit includes, in addition to the thymus and appendix, the Peyer's patches of the small intestine and the sacculus rotundus, or intestinal tonsil.

References

Andrew, W., and N. V. Andrew. 1948. Age changes in the deep cervical lymph nodes of 100 Wistar Institute rats. Am. J. Anat. **82**:105-166.

Augustinsson, K. B., R. Fänge, A. Johnels, and E. Östlund. 1956. Histological, physiological, and biochemical studies on the heart of two cyclostomes, Hagfish *(Myxine)* and Lamprey *(Lampetra)*. J. Physiol. **131**:257-276.

Baum, H. 1925. Allgemeines über Lymphgefässsystem der Haustiere, insbesondere Unterschiede im makroskopischen Verhalten des Lymphgefässsystems verschiedener Tierarten. Z. Fleisch Milchyg. **36**:49-54.

Beard, J. 1900. The source of leukocytes and the true function of the thymus. Anat. Anz. **18**:550.

Borghese, E. 1940. Ricerche sul' apparecchio linfatico dei Rettili. II. I vasi linfatici paravertebrali degli Ofidi. Arch. Ital. Anat. Embriol. **43**(2):139-145.

Clawson, C. C., J. Finstad, and A. R. Good. 1965. Unpublished observations.

Conway, E. A. 1937. Cyclic changes in lymphatic nodules. Anat. Rec. **69**:487-513.

Cooper, M. D., B. R. Burmester, P. B. Dent, L. N. Payne, R. D. A. Peterson, and R. A. Good. 1965. Unpublished observations.

Cooper, M. D., D. Y. E. Perey, A. E. Gabrielsen, P. B. Dent, and R. A. Good. 1967. Further evidence that Peyer's patch tissues of mammals are similar to the bursa of Fabricius of birds. Fed. Proc. **26**:752. (Abstr.)

Cooper, M. D., R. D. A. Peterson, and R. A. Good. 1965. Delineation of the thymic and bursal lymphoid systems in the chicken. Nature (London) **205**:143-146.

Dyer, R. F. 1967. Ultrastructural features of *Amphiuma* heart. Anat. Rec. **157**:238. (Abstr.)

Fänge, R., and L. Gidholm. 1968. A macrophage system in *Myxine glutinosa* L. Naturwissenschaften **55**:1-2.

Fichtelius, K. E., J. Finstad, and R. A. Good. 1969. The phylogenetic occurrence of lymphocytes within the gut-epithelium. Int. Arch. Allergy Appl. Immunol. **35**:119-133.

Good, R. A., J. Finstad, H. Gewurz, M. D. Cooper, and B. Pollara. 1967. The development of immunological capacity in phylogenetic perspective. Am. J. Dis. Child. **114**:477-497.

Good, R. A., J. Finstad, B. Pollara, and A. E. Gabrielsen. 1966. Morphologic studies on the evolution of the lymphoid tissues among the lower vertebrates. Pages 149-168 *in* R. T. Smith, P. A. Miescher, and R. A. Good, eds. Phylogeny of immunity. University of Florida Press, Gainesville.

Good, R. A., and A. E. Gabrielsen. 1968. The thymus and other lymphoid organs in the development of the immune system. Pages 526-564 *in* F. T. Rapaport and J. Dausset, eds. Human transplantation. Grune & Stratton, Inc., New York.

Greene, C. W. 1902. Contributions to the physiology of the California hagfish, *Polistotrema stouti*. II. The absence of regulative nerves for the systemic heart. Am. J. Physiol. **6**:318-324.

Hama, K. 1960. On the existence of filamentous structures in endothelial cells of the amphibian capillary. Anat. Rec. **139**:437-441.

Hildemann, W. H., and E. L. Cooper. 1963. Immunogenesis of homograft reactions in fishes and amphibians. Fed. Proc. **22**:1145-1151.

Hildemann, W. H., and R. Haas. 1959. Homotransplantation immunity and tolerance in bullfrog. J. Immunol. **83**:478-485.

Johansen, K., and D. Hanson. 1968. Functional anatomy of the hearts of lungfishes and amphibians. Am. Zool. **8**:191-210.

Kanesada, A. 1956. A phylogenetical survey of hemocytopoietic tissues in submammalian vertebrates, with special reference to the differentiation of the lymphocyte and lymphoid tissue. Bull. Yamaguchi Med. Sch. **4**:1-35.

Murata, H. 1959. Comparative studies of the spleen in submammalian vertebrates. I. Topographical anatomy and weight of the spleen. Okajimas Folia Anat. Jap. **33**:1-9.

Osogoe, B. 1953. Phylogenetical study of bone marrow [in Japanese]. Symposium on Hematology **5**:1-19.

Osogoe, B. 1954. Phylogenetical study of spleen [in Japanese]. Symposium on Hematology **7**:1-35.

Östlund, E., G. Bloom, J. Adams-Ray, M. Ritzén, M. Siegman, H. Nordenstam, F. Lishajko, and U. S. von Euler. 1960. Storage and release of catecholamines and the occurrence of a specific submicroscopic granulation in hearts of Cyclostomes. Nature **188**:324-325.

Randall, D. J. 1968. Functional morphology of the heart in fishes. Am. Zool. **8**:179-189.

Salkind, J. 1915. Contributions histologiques à la biologie comparée du thyme. Arch. Zool. Exp. Gen. **55**:81-322.

Simons, J. R. 1959. The distribution of blood from the heart in some Amphibia. Proc. Zool. Soc. (London) **132**:51-64.

Warner, N. L., and A. Szenburg 1964. Immunologic studies on hormonally bursectomized and surgically thymectomized chickens; dissociation of immunologic responsiveness. Pages 395-411 *in* R. A. Good, and A. E. Gabrielsen, eds. The thymus in immunobiology. Harper & Row, Publishers, New York.

Warner, N. L., A. Szenburg, and F. M. Burnet. 1962. The immunological role of different lymphoid organs in the chicken. I. Dissociation of immunological responsiveness. Aust. J. Exp. Biol. Med. Sci. **40**:373-387.

White, F. N. 1968. Functional anatomy of the heart of reptiles. Am. Zool. **8**:211-219.

Zschiesche, E. S., and Stilwell, E. F. 1934. Intercalated discs of the heart muscle of the guinea-pig; counts of the number of discs present at eight ages of normal animals and in two experimental groups. Anat. Rec. **60**:477-486.

Zweifach, B. W. 1937. The structure and reactions of the small blood vessels in Amphibia. Am. J. Anat. **60**:473-514.

9 PHYLOGENETIC SURVEY OF THE BLOOD

In the vertebrates, the distinction between red cells, or erythrocytes, and white cells, or leukocytes, is a sharp one. The morphologic distinctions are based on important functional differences. The red cells are the more or less passive carriers of oxygen. As such, they show no active motion, only deformation due to contact with other cells or to being forced through vessels of narrow diameter by the pressure of the bloodstream. The white cells, on the other hand, have a role in the defense of the body against microorganisms, by means of phagocytic and other activities; and this role requires active ameboid motion.

Although all the details of function of the various kinds of white cells are not completely known, much has been learned about them. In all vertebrates (and indeed among the invertebrates also) there are two large categories of white cells: (1) the granulocytes, which contain specific granules in their cytoplasm, and (2) the agranulocytes, which in general lack granules. The nuclei of the granulocytes usually have two or more lobes, giving varied nuclear forms; hence the cells are also termed **polymorphonuclear.** The nuclei of the agranulocytes generally show no lobation (although they may be kidney shaped); these cells are termed **mononuclear.**

CYCLOSTOMATA

In the cyclostomes, lowest of the vertebrates, differences between red and white cells are already well established. There is also a clear distinction among the white cells between granulocytes and argranulocytes. However, the discoidal red cells (Fig. 9-1), which contain a relatively small amount of hemoglobin per erythrocyte, frequently show basophilic material in their cytoplasm.

In *Myxine,* reproduction of red cells commonly occurs in the bloodstream, as well as in the spleen and connective tissue. In fact, in these latter locations red cell production appears to be an intravascular process, apparently taking place in sinusoids where flow is very slow and is more of a streaming of fluid around and past the cells. The stem cells are medium-sized or large lymphocytes resembling the hemocytoblast that is generally accepted as the stem cell in higher vertebrates.

It has not been possible to identify thrombocytes as such in cyclostomes. The "spindle cells" resemble thrombocytes but lack the granules that contain thrombin and, in other forms, make the thrombocytes effective in the clotting process.

Although differentiation of red cells occurs, at least to some extent, in the circulating blood where hemocytoblasts also are found, such is not true of the granulocytic white cells. Their differentiation occurs in an extravascular location—chiefly in the spleen. The polymorphonuclear cells are of two kinds, named according to the staining properties of their specific granules: (1) the neutrophils, which are somewhat "neutral" in relation to affinity for acid or basic stain, and (2) the eosinophils, which have granules that stain well with eosin and other acid stains.

CHONDRICHTHYES

A blood smear from the dogfish reveals fairly large nucleated red blood corpuscles. They are

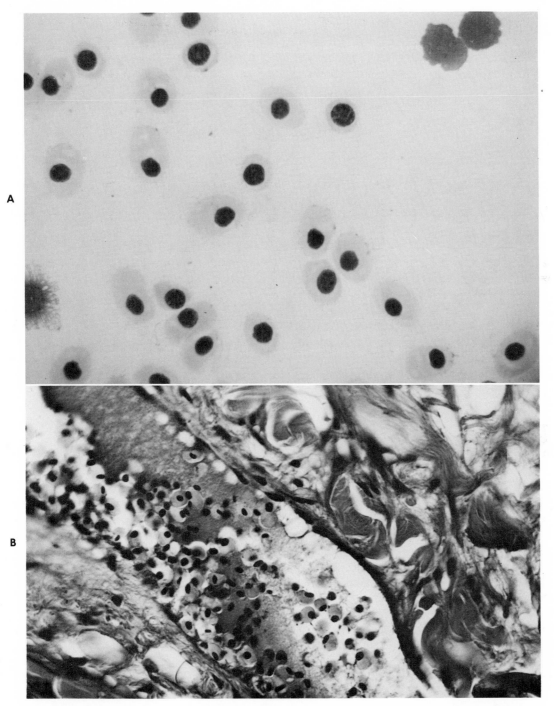

Fig. 9-1. A, Blood of brook lamprey, *Ichthyomyzon*. Erythrocytes generally have rounded nuclei. **B,** Blood in situ in a vessel of *Petromyzon*. Coagulated plasma fills large part of lumen. (**A,** Wright's stain, ×1000; **B,** ×250.)

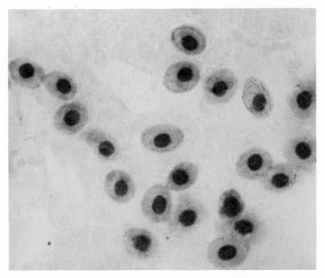

Fig. 9-2. Blood smear of shark. (×600.)

oval, variable in size, usually with diameters of more than 20 μm (Fig. 9-2), which is several times larger than mammalian erythrocytes. Their number varies from less than 100,000 to 600,000 per cubic millimeter, less than a tenth as many as found in mammalian blood. The percentage of hemoglobin in the whole blood of the shark is about 5%, compared with about 14% in human blood. Besides the erythrocytes, a variety of leukocytes are found: large and small lymphocytes, neutrophilic granulocytes, eosinophilic granulocytes, hemocytoblasts, thrombocytes, and certain transitional stages. Basophilic granulocytes are not present.

In the cartilaginous fishes, the spleen is of primary importance as a hemopoietic, or blood cell–forming, organ. Another structure of similar importance is the organ of Leydig (Fig. 9-3), a mass of tissue lying beneath the epithelium of the esophagus. Granulocytopoiesis (formation of granulocytes) occurs in the stroma of the subcapsular tissue of the testes and ovaries of elasmobranchs, in the intertubular tissue of the kidneys, in the portal connective tissue of the pancreas, and in portions of the intestinal wall. Thus the sites of formation of granular leukocytes (Fig. 9-4) are widely distributed. In many elasmobranchs there is active differentiation of red cells and thrombocytes from lymphoid cells in the circulating blood.

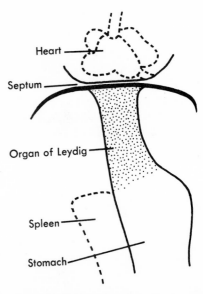

Fig. 9-3. Organ of Leydig in the ray *Raja radiata*. This structure lies in wall of the esophagus and consists of myeloid tissue in which granulocytes are formed. (Redrawn from Fänge and Gidholm, 1968.)

OSTEICHTHYES

Erythrocytes and leukocytes of teleosts are generally smaller than those of elasmobranchs. The kinds of leukocytes correspond to those in cartilaginous fishes. However, it is not clear whether basophils are actually lacking or whether they are represented by a cell that appears to be

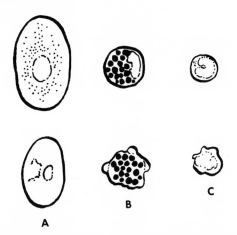

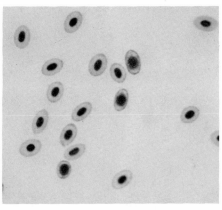

Fig. 9-4. Blood cells of cartilaginous fishes, stained supravitally with neutral red. Upper row, *Raja radiata;* lower row, *Chimaera monstrosa.* Erythrocytes, **A,** and nongranular leukocytes, **C,** have few structures that absorb the stain. Granulocytes, **B,** have numerous enzyme-rich cytoplasmic organelles, which are considered as lysosomes. (Oil immersion; ×1000.) (Redrawn from original, freehand drawing of freshly prepared samples in Fänge and Gidholm, 1968.)

Fig. 9-5. Blood of the bass, *Micropterus*—a teleost fish. Nucleated erythrocytes are small. (Wright's stain; ×810.)

an immature eosinophil. Again, the circulating blood contains many cells in stages of differentiation to erythrocytes and in the later stages of differentiation into mature leukocytes.

A curious situation is seen in certain bony fishes of Antarctic waters, the Chaenichthyidae, in which there actually are no red blood cells. The blood and tissues of these fishes have been studied in detail by Ruud (1958) and by Walvig (1958). Although some species of these fishes are very active at times, they apparently obtain enough oxygen by solution of the gas directly into the blood plasma.

The red blood cells of teleosts, although nucleated (Fig. 9-5), are in some cases as small or smaller than those of man, ranging in diameter from 6 to 13 μm. They are smaller than the red cells of elasmobranchs and considerably smaller than those of most amphibians. But these differences may be part of a general difference in cell size based on taxonomic considerations. The blood of bony fishes may contain nearly as many erythrocytes per cubic millimeter as human blood —for example, 4 million in the mackerel and about 5 million in man.

Leukocytes, especially eosinophils and neutro-phils, show considerable species differences in structure and size of their granules. In some fishes, including *Polyodon spathula* (the paddlefish) and *Ictiobus bubalis* (the buffalo fish), a peculiar kind of secretory granulocyte is found. In these cells the eosinophilic granules appear to undergo liquefaction and then to swell and fuse, forming large masses that leave the cell as a "secretion."

One group of bony fishes, the Dipnoi (lungfishes), are similar to the Amphibia in a number of respects, including various adaptations for terrestrial life. Their blood cells are of considerable interest in their resemblance to those of urodele amphibians. The African lungfish, *Protopterus ethiopicus,* has red cells in the shape of elliptical biconcave disks, measuring 40 by 30 μm (Fig. 9-6). Two kinds of thrombocytes occur, one with fine azurophilic granules, the other with coarse eosinophilic ones. They may represent different stages in function or in differentiation.

White cells are relatively numerous in this species (Fig. 9-6). Many of them may be classed as small lymphoid hemoblasts. It is interesting to note that degenerate forms of these cells are usually prominent; quite possibly, lack of success in differentiation leads to degeneration. Large lymphoid hemoblasts are present and may show mitosis. Monocytes are also seen.

Of the granulocytes in the Osteichthyes, the eosinophils show a graded series between those with small round granules and those with large

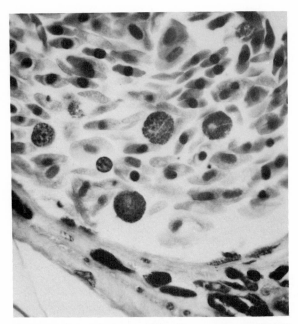

Fig. 9-6. Vein of the lungfish, *Protopterus*. Nucleated erythrocytes and leukocytes are seen in lumen. (×360.)

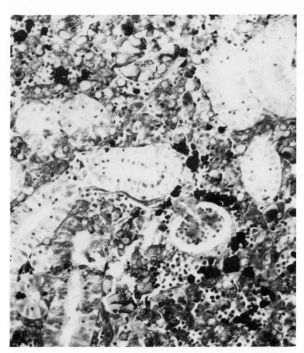

Fig. 9-7. Kidney of rainbow trout, *Salmo gairdneri*, showing large quantities of hemopoietic tissue among tubules, a condition common in many species of bony fishes. Many pigment cells are seen also. (×100.)

ones, either oval or rod-shaped. The nuclei of the small-granule eosinophils are generally rounded, those of the large-granule ones are bilobed or polymorphous. These differences seem to indicate a process of differentiation among the eosinophils. Granules of the different basophils also vary in size, but their nuclei are rather uniformly pale and ovoid. Duthie (1939) made the interesting statement that in the Triglidae the basophilic granules become eosinophilic when the wandering leukocytes reach epithelial layers.

In the bony fishes there is a tendency toward a shift from spleen to kidney (Fig. 9-7) as the primary hemopoietic organ. However, in some fishes (e.g., *Fundulus*) the spleen still has considerable activity, whereas in others (e.g., *Carassius*) the kidney has taken over most of the work. Yoffey (1929) described in detail the role of the spleen in fishes.

AMPHIBIA

The blood of amphibians consists of fluid plasma and cellular elements. Plasma contains water, proteins, salts, sugar, fats, oxygen, and the

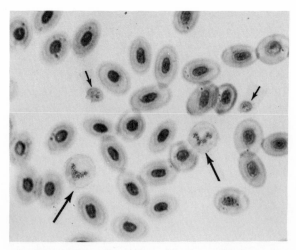

Fig. 9-8. Blood cells of frog. Mostly erythrocytes; two polymorphonuclear leukocytes (larger arrows) and two lymphocytes (small arrows) are also present. (×650.)

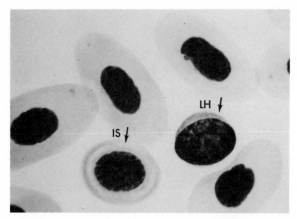

Fig. 9-9. Erythrocytes of *Amphiuma,* the largest of all red blood cells. Note characteristically uneven contours of nuclei. This photo shows evidence of differentiation of red cells within circulating blood of *Amphiuma.* Cell on right, with eccentric nucleus, is lymphoid hemoblast *(LH);* cell to its left is intermediate stage *(IS)* between hemoblast and mature erythrocyte. (Wright's stain; ×810.)

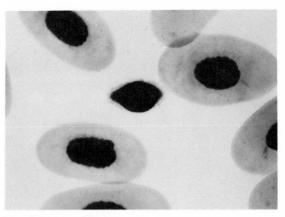

Fig. 9-10. *Amphiuma* erythrocytes with small thrombocyte in center of field. (Wright's stain; ×810.)

various products of metabolism. The cellular constituents include erythrocytes (red cells), spindle cells or thrombocytes, and leukocytes (white cells) (Fig. 9-8).

The large nucleated red cells of amphibians, like those of lungfishes, resemble the large primitive cells **(megaloblasts)** of the erythroid series found in the bone marrow of higher vertebrates. In most amphibians, differentiation of red blood cells occurs primarily within the sinusoids of the spleen. There are a few species in which the bone marrow is also active in this role.

The amphibian erythrocyte is an elliptical disk with a slight bulge in its center produced by the nucleus. In some species, a small percentage of red cells lose their nuclei. The frog *Rana pipiens* has these cells about 20 μm long, several times larger than mammalian erythrocytes. The largest of all red cells occur in *Amphiuma,* where they may be 70 μm in length (Fig. 9-9). The number of red blood cells varies with seasonal and nutritional conditions, but there are usually 250,000 to 400,000 per cubic millimeter, less than one tenth as many as found in mammalian blood. Erythrocytes live about a hundred days (Jordan and Speidel, 1925); when worn out,

they are removed by the spleen and by the Kupffer cells of the liver.

Thrombocytes (Fig. 9-10) are spindle-shaped ameboid cells. They possess a nucleus and are characterized by a clear ectoplasm and a granular endoplasm. Next in abundance to the erythrocytes, thrombocytes perform about the same functions as the platelets found in mammalian blood. They break down when in contact with foreign substances, providing the thrombin that acts on the plasma fibrinogen in the presence of blood calcium to form an insoluble clot. Being true cells, the thrombocytes do not correspond exactly to platelets, but they have been observed in frogs to give rise to similar bodies by the segmentation of their own pseudopods. These cells are usually formed in the spleen.

The leukocytes, or white corpuscles, consist of the agranular leukocytes (lymphocytes and monocytes) and the granular leukocytes (basophils, neutrophils, and eosinophils) (Figs. 9-11 and 9-12). The red cells outnumber the white cells by about 40 to 1. White cells vary in size and form as well as number, in response to nutritional and seasonal conditions. They probably live only two or three days. All unstained leukocytes are colorless and usually smaller than the red cells. The nuclei of the granulocytes often assume irregular forms or become divided into lobes. The monocytes are the most active in phagocytosis; they are able to wander throughout the tissues, devouring bacteria and debris.

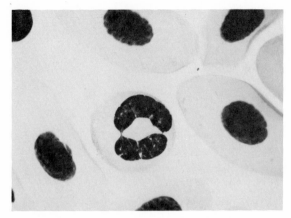

Fig. 9-11. Neutrophil in blood of *Amphiuma*. Sausage-shaped lobes of nucleus are connected by narrow nuclear strands. Specific granules are not well differentiated here. (Wright's stain; ×810.)

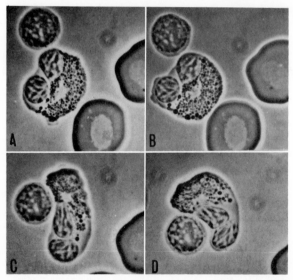

Fig. 9-12. Living cells of the blood of *Ambystoma mexicanum* as seen by phase microscopy. The large granular cell, an eosinophil, is seen in motion, as shown by the four time-lapse pictures taken at short intervals. The smaller round cell, with a narrow rim of cytoplasm, is a lymphocyte, the relative position of which becomes changed as the granular cell moves. It also is motile but was not moving at the time. (Courtesy Professors F. Fey and L. Kruger; from Andrew, W. 1965. Comparative hematology. Grune & Stratton, Inc., New York. Reprinted by permission.)

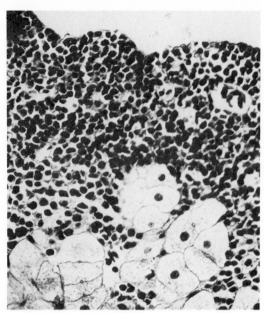

Fig. 9-13. Peripheral portion of liver of the axolotl, *Ambystoma mexicanum*. Beneath thin capsule is a thick layer of hemopoietic tissue in which neutrophils and other granulocytes undergo development. (Hematoxylin-eosin; ×125.)

Neutrophils and basophils of amphibians are not formed in the spleen with the red cells, but go through all their early differentiation in the liver, specifically in the subcapsular region (Fig. 9-13). Eosinophils seem distinctive when mature, but when "younger," they contain orthobasophilic granules that stain blue with Wright's stain. Thus eosinophils may be derived from basophils, or both may come from a common stock. A family of terrestrial salamanders, the Plethodontidae, differ from other amphibians in that their leukocytes originate and differentiate in the bone marrow rather than the liver. However, their red cells are still formed in the spleen, with some continued differentiation in the bloodstream.

REPTILIA

Blood cells in reptiles are formed in the spleen in some and in the red bone marrow in others. In *Phrynosoma,* the horned toad, the spleen is the primary site, although granulocytes are formed largely in the bone marrow (Fig. 9-14). In the turtles the process for both red and white cells occurs fairly equally in spleen and marrow,

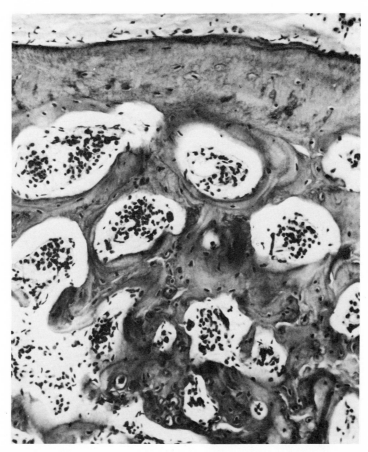

Fig. 9-14. Portion of the body of a vertebra of horned toad, *Phrynosoma*. The marrow of spongy bone is the primary site of formation of granulocytes. (Hematoxylin-eosin; ×100.)

whereas the lizards show the bone marrow as the major site. There are usually many primitive, relatively undifferentiated cells in the bloodstream of reptiles (Fig. 9-15, *A*). These include large lymphocytes (probably the hemocytoblasts from which all the various blood cells develop), medium-sized and small lymphocytes, and various developmental stages of red cells (erythroblasts). The thrombocytes of reptiles, are still **ameboid** cells but are relatively small, usually only about half the size of the red cells.

The erythrocytes of reptiles are elongated, flattened structures in which the ovoid nuclei are readily stained. The nuclei are centrally located, with long axes usually parallel to the axes of the cells, much as in the fishes and amphibians. Erythroblastic stages seen in the circulating blood include cells corresponding to those which in

mammals are found only in hemopoietic tissue—basophilic and polychromatophilic erythroblasts and normoblasts. Mitotic figures also occur.

In general, reptiles have erythrocytes intermediate in size between the large ones of amphibians and the smaller red cells of birds. (Mammals are not included in this comparison because the drastic change involved in loss of nuclei in mammals makes comparison of size less meaningful.) The erythrocytes of crocodiles and turtles are relatively large compared with those of lizards, and their nuclei are generally more rounded.

The blood of *Sphenodon punctatum* differs from that of other reptiles by the great size of its red cells. They sometimes are as long as 23 μm. *Sphenodon* (*Hatteria*) is the tuatara of New Zealand and is the oldest surviving lepidosaurian reptile, greatly resembling some extinct reptiles of the

Mesozoic Era. Very large red cells are an interesting microscopic feature to go with other primitive traits such as lack of a male copulatory organ, a well-developed third (pineal) eye, and amphicoelous vertebrae with intercentra.

Kamezawa (1955) made a comparative study of several features of cytoplasm and nuclei in the erythrocytes of various animals. He distinguished the following five main types based on variety and arrangement of mitochondria, as seen in living cells stained with Janus green B:

I. Diffusely arranged mitochondria
II. Perinuclear arrangement
III. Mitochondria forming coarse granules or locating in groups
IV. Transitional form between types I and III
V. Ringed and filamentous mitochondria

Among the granulocytes (Fig. 9-15, *B*) there

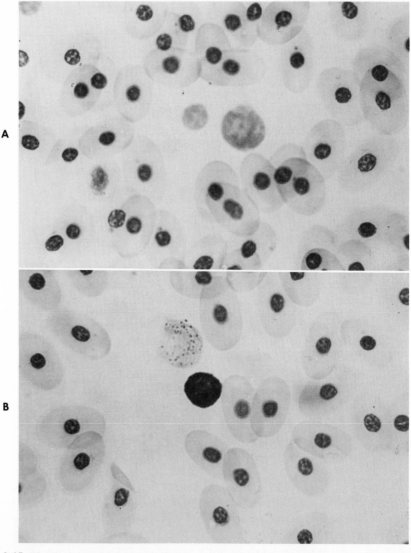

Fig. 9-15. Blood of painted turtle, *Chrysemys picta*. Nuclei in this smear are rather surprisingly spheroidal in elliptical cells. **A,** Near center of field is large "stem cell." **B,** An eosinophil with rather widely spaced granules and a basophil with densely packed granules are seen. (Wright's stain; ×1000.)

are fewer neutrophils and more eosinophils and basophils in reptilian blood than in mammalian blood. Usually two kinds of eosinophils are found: those with round granules and simple nuclei and others with rod-shaped, or bacillary, granules and complex polymorphous nuclei. In some species there are also two different types of basophils, based on the shape of the granules.

Snakes have generally the largest eosinophils, and lizards the smallest ones; those of turtles and crocodiles are of intermediate size. Their nuclei may be without lobes, bilobed, or multilobed. The basophils in a blood smear often resemble mul-

berries. The weakly staining nucleus is usually obscured by the large, deeply staining granules. Basophils are largest in crocodilians and turtles, smallest in the lizards, and of intermediate size in the snakes. Actually, the largest reptilian basophils are found in *Sphenodon,* which, as noted above, also has the largest erythrocytes. But *Sphenodon* has relatively small eosinophils.

The neutrophils are less prominent cells in reptiles (and in birds) than in mammals. The granules are small and azurophilic, the nuclei of irregular shape. The distinction between these and the so-called "azurophilic granulocytes" is not clear.

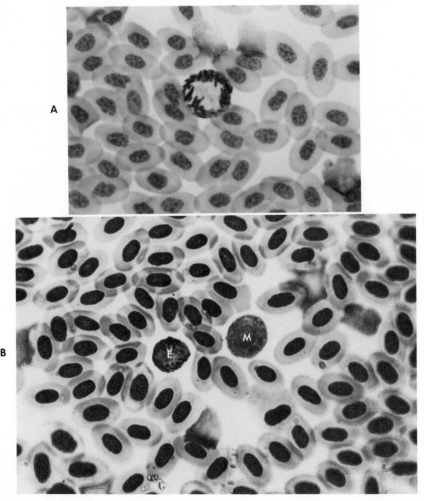

Fig. 9-16. A, Blood of chicken, *Gallus domesticus.* Red corpuscles have nuclei with relatively smooth outlines. Eosinophil near center has "granules" of bacillary form. **B,** Blood of sparrow, *Passer domesticus.* Large monocyte *(M)* and eosinophil *(E)* are shown among nucleated erythrocytes. (Wright's stain; **A,** ×1000; **B,** ×900.)

The latter, if they do constitute a distinct group, seem to vary considerably in size. The cytoplasm of such cells often contains lipid vacuoles, which appear to increase in number as the cell ages.

AVES

Which of the warm-blooded, or homeothermic, classes (birds or mammals) is "lower"? From the evolutionary standpoint, birds are considered to be one derivative of reptilian stock, and in many ways they seem highly "advanced." Mammals are another derivative and include in their numbers many advanced and specialized forms, including man himself. Whether birds, as a class, are more "reptilian" than mammals is a difficult question to answer. The presence of certain features found in reptiles, retained in birds, but lost in mammals led the English scientist and philosopher Thomas Henry Huxley to place birds and reptiles together in one large group that he called the Sauropsida. One of these features is the nucleated character of the erythrocytes (Fig. 9-16) that the birds have in common not only with reptiles but also with amphibians and fishes.

In the majority of wild birds the red cell count is somewhat over 3 million per cubic millimeter of blood—a bit lower than human blood. The erythrocytes tend to be larger than those of man and other mammals, but not much so. For example, the red cells of the chicken, *Gallus domesticus,* measure 10.6 by 6.6 μm (Lucas and Jamroz, 1961). In an extensive survey covering 50 species of North American birds, Bartsch and associates (1937) found the largest erythrocytes (16.5 μm) in the osprey, *Pandion haliaetus,* and the smallest in the Carolina chickadee, 1.0 μm. The nucleus of the erythrocytes in some birds, as in the starling, *Sturnus vulgaris,* has the chromatin in coarse clumps, giving a mulberry appearance.

In the blood of fishes, amphibians, and reptiles many cells can be found in various stages of differentiation, often including very early ones. This is **not** generally true of the blood of birds, although numbers of round erythroblasts with basophilic cytoplasm often are seen. The presence of the more advanced polychromatophilic erythrocytes is normal in the blood of birds of all ages.

Bird blood contains relatively large numbers of leukocytes, as compared to mammals—often above 25,000 per cubic millimeter or about four times the count in human blood. Polymorphonuclear forms have rather poorly defined identity. Kleineberger (1927), from his extensive quantitative studies, described cells with acidophilic granules of bacillary form as belonging to this class, constituting 40% to 65% of the leukocytes. Small and medium lymphocytes may account for 30% to 50%, monocytes 6% to 15%, and basophils and regular eosinophils (with spheroidal granules) usually about 2% of the leukocytes in the blood of birds. The morphologic and staining characteristics are similar to those of the same classes of cells in vertebrates in general.

The thrombocyte cells of birds are slightly smaller than erythrocytes, generally spindle-shaped but showing considerable variation in form. The cytoplasm is pale purple or blue containing pink to red granules. The cells undergo degenerative changes during clotting and finally break down completely. Thrombocytes are found in the blood of fishes, amphibians, and reptiles. But in birds we are seeing them for the last time; they will be replaced in mammals by the blood platelets. Thrombocytes have an enzyme similar to, if not the same as, the thromboplastin found in human platelets. Under conditions of hemorrhage and tissue damage, it is released as the cells break down. In the free state this enzyme acts as a catalyst for transformation of prothrombin to thrombin. It is thrombin that brings about the change of fibrinogen, in solution in the plasma, into the needlelike crystals of **fibrin** that form the main support of the "blood clot."

MAMMALIA

Erythrocytes. The outstanding distinction of the mammalian red blood "cell" is its lack of a nucleus (Figs. 9-17 and 9-18). The elimination of the nucleus from the cell is a complex process (Skutelsky and Danon, 1967) and occurs during the late reticulocyte stages of the erythroblast. There is little evidence of differentiation of red corpuscles in the circulatory blood of mammals. However, cells in the reticulocyte stages are common in the blood but vary among the species of mammals. In human blood, less than 1.0% of the erythrocytes are of this form, but higher reticulocyte percentages have been reported for others: rabbit,

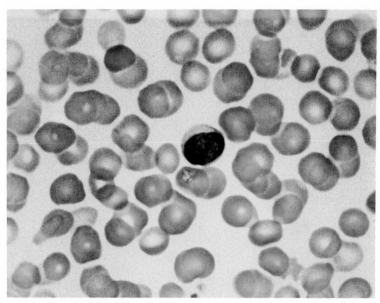

Fig. 9-17. Blood of human subject. Red corpuscles lack nuclei but show light center because of their biconcave shape. Lymphocyte in center of field has eccentric nucleus and small amount of cytoplasm. (Wright's stain; ×1000.)

5.0%; rat, 4.5%: cat, 3.5%; guinea pig, 2.25%; and dog, 1.92%. None has been found in pig, goat, horse, or ox. The nuclear material apparently becomes dispersed during this stage, since reticulocytes show polychromatophilia with Wright's stain.

The red blood corpuscles have the same shape in man as in the majority of other mammals—a biconcave disk. They are flexible structures, however, and in passing through capillaries of smaller diameter than themselves, they become cup-shaped. The average diameter of a red blood corpuscle of the mouse is somewhat under 6.0 μm, smaller than that of man and the rabbit (7.0 to 7.5 μm). The red cell count varies among the mammals: in human blood, 4.5 to 5.5 million per cubic millimeter of blood; in rabbit blood, about 7 million; in mouse blood, 7.5 to 12 million. In the goat the count ranges from 11.2 to 20 million (Wells and Sutton, 1915). The corpuscles in that animal are small, only 4.25 μm in diameter.

As a class, mammals have the smallest and most numerous erythrocytes. However, the numbers of red corpuscles can hardly be thought to follow an upward trend in the evolutionary ascent

within the class of mammals. In several species of marsupials, generally considered to be primitive mammals, the average is about 8 million red cells per cubic millimeter of blood (Ponder and associates, 1928).

There are some interesting features of the red corpuscles in various kinds of mammals. The average count for the cat, *Felis domestica,* is 7.5 million, ranging from 5.5 to 10 million (Schalm, 1961). A sharp rise in the red cell count under certain conditions suggests a correlation with the excitable nature of the cat. The increased number is caused by release from the spleen, which serves as a reservoir of red corpuscles and platelets.

Some of the changes in erythrocytes of the cat from birth to maturity seem to follow the general evolutionary pattern in the phylogeny of vertebrates. At birth the individual red corpuscles are about twice the volume of those of the adult and are fewer by several million per cubic millimeter. The adult number is reached by the third or fourth month, although the adult level of hemoglobin is not attained until the fifth or sixth month (Windle and associates, 1940). A common feature in red corpuscles of the cat is a small spheroid structure, staining deeply blue and lying

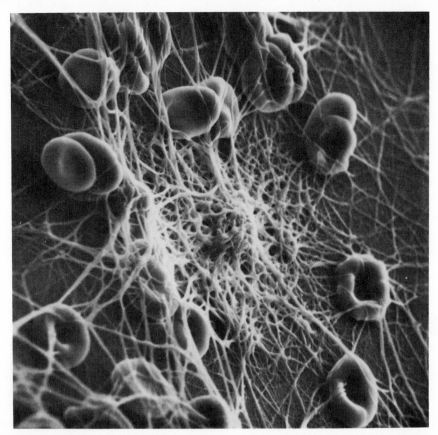

Fig. 9-18. Scanning electron micrograph of human red blood cells entrapped in fibrin clot. Clotting initiated by disintegration of platelets in the blood (after tissue damage). This disintegration leads to complex series of reactions that convert the plasma protein fibrinogen into long, tough, insoluble polymers of fibrin. Fibrin and entangled erythrocytes form blood clot that arrests bleeding. (×5180.) (Courtesy N. F. Rodman, Iowa City, Iowa; from Hickman, C. P., C. P. Hickman, Jr., and F. H. Hickman. 1974. Integrated principles of zoology, 7th ed. The C. V. Mosby Co., St Louis.)

close to the edge. These bodies are called the Howell-Jolly bodies. Their exact nature and function are obscure. They are found in the red corpuscles of man in some forms of blood disease.

Some mammals have red corpuscles of markedly different shape from the biconcave disk of man and of most of the familiar species. In the small mouse deer, or chevrotain, of Asia, *Tragulus javanicus,* the cells are almost spherical and have an average diameter of only 1.5 μm. The blood of the camel has ellipsoidal red corpuscles, shaped somewhat like those of the lower classes of vertebrates. Small groups of ellipsoidal nucleated red corpuscles have been found

in the blood of the dromedary camel (Andrew, 1965). Conceivably, these could represent a thrombocytoid type of cell, since they show a tendency to aggregate whereas the nonnucleated types do not.

The camels differ from other artiodactyls in not being cud-chewers. They evidently have constituted a distinct stock since the Eocene and have even been placed by some in a separate suborder, Tylopoda. It is not too surprising, then, to find unusual features in the blood of these animals.

Leukocytes. Leukocytes in mammals show the same general types as in other vertebrates: the agranulocytes, or mononuclears, consisting of lymphocytes and monocytes; and the granulocytes, which are the neutrophils, eosinophils, and basophils. Any cubic millimeter of human blood usually contains 5000 to 9000 white blood cells. In the mouse, however, the leukocyte count may vary from 3000 to 18,500 cells per cubic millimeter, depending on the part of the body sampled.

The monocytes are the largest of the leukocytes. Their nuclei often are kidney-shaped and eccentrically placed. The cytoplasm is abundant and somewhat less basophilic than that of the lymphocyte. Monocytes in the mouse constitute 9% to 14% of the total number of leukocytes as compared with 3% to 8% in human blood. Lymphocytes are by far the most numerous white cells in the mouse, making up 60% to 80% of the total as compared with 20% to 35% in humans.

As in the lower vertebrates, the granular leukocytes are named according to the staining affinities of the specific granules in their cytoplasm. The neutrophils make up 65% to 75% of the total number of leukocytes in human blood, but they are far less numerous in the mouse (8% to 25%). The eosinophilic leukocytes range from less than 1% to 4%, and the basophils constitute about 0.5% of the leukocytes in the blood of both humans and mice.

The total number of leukocytes and the differential count of the various types vary considerably among the mammals. These species differences would seem to indicate some greater role of particular kinds of leukocytes in certain species than in others. As more is learned about the functions of the leukocytes, the reasons for the differences may be clarified.

Biochemical, ultrastructural, and physiopathologic studies are helping to make progress in the study of **function** of the leukocytes. All white cells, although carried about passively in the bloodstream, are actively motile in the tissues. In man at least, it has long been known that the neutrophils are the most important defenders against bacteria in acute infections; that the eosinophils increase in allergic reactions and in cases of parasitism; that the monocytes, like the neutrophils, are highly phagocytic but act more in the role of scavengers, consuming portions of dead cells; and that the lymphocytes play a role during chronic disease and in immune reactions of the body.

An example of the importance of comparative study of normal animals is seen in the laboratory rat. This animal generally has very large numbers of eosinophils in the subcutaneous tissues, some 25 times as many as in the circulating blood. It has been suggested that the high eosinophil count is due to the presence of parasites (Spry, 1969), which indeed often abound in these animals. Yet the high level of the numbers and the constancy of the finding would tend to indicate that the great numbers of tissue eosinophils are not out of the ordinary for the rat.

The electron microscope has given us an increased knowledge of the structure of the specific granules in the various kinds of granulocytes in the mammals and also in the other classes of vertebrates (Fig. 9-19). It has revealed the granules of the eosinophils as membranous sacs, each of which contains a crystalloid. The shape of the granule is apparently dependent on the growth of the crystalloid. Under high magnifications, the fine granules of the neutrophils reveal a considerable variety of aspect, which may well represent stages either in their development or function. It does seem clear that the granules have much more significance as a convenient means of classifying these cells. The secrets of their specific roles in the body probably lie to a large extent in their type of granules.

It would seem that the lower chordates would provide a better view of blood in its more primitive condition. In fact, the blood of *Amphioxus* is almost colorless, and the cells appear to be of a relatively primitive kind, approaching that generally considered as the "stem cell" of vertebrates.

• • •

In the tunicates, however, the blood cell picture is not at all a simple one. The blood in both the free-swimming larva and the sedentary adult seems quite different from that found in any of the vertebrates. The various kinds of cells include lymphocytes, macrophages, green cells, blue cells, orange cells, and bladder cells (Fulton, 1920; George, 1926, 1930; Andrew, 1961, 1962). These types do not include any with specific granules such as seen in the neutrophils, eosinophils, and basophils.

The erythrocytes, contrary to some recent at-

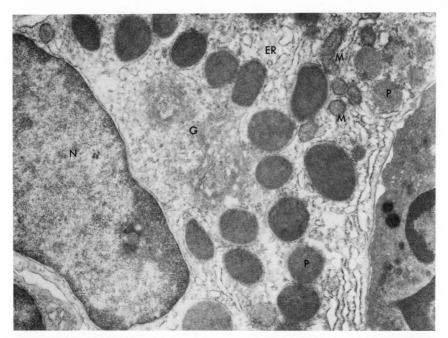

Fig. 9-19. Portion of basophilic myelocyte in bone marrow of guinea pig. *ER,* Endoplasmic reticulum; *G,* Golgi apparatus; *M,* mitochondria; *N,* nucleus; *P,* granule. (×30,000.) (Courtesy Dr. G. Winqvist; from Andrew, W. 1965. Comparative hematology. Grune & Stratton, Inc., New York. Reprinted by permission.)

tempts at generalizations in comparative vertebrate histology (Patt and Patt, 1969), are not typically vertebrate structures. Nonameboid, hemoglobin-containing cells are found in various species of holothurians and gephyrean worms; some brittle stars actually have nonnucleated red corpuscles. This is not to say that such cells are necessarily homologous to the red blood cells of vertebrates.

The lack of nuclei in the erythrocytes of mammals seems to be a specialization by which the corpuscle becomes hardly more than a hemoglobin-containing sac, thereby assuring a maximum content of the respiratory pigment per unit of volume. This phenomenon is perhaps foreshadowed in some lower vertebrates in which corpuscles without nuclei occur in varying, although usually small, percentages. In one amphibian, *Batrachoseps,* the great majority of the red cells lack nuclei.

The difficulty of being certain which of the various kinds of granulocytes are homologous from class to class of vertebrates is a great one.

Functional studies on the leukocytes of lower vertebrates are needed; it is dangerous to assume functional roles simply from comparison of morphologic aspects of these cells with similar cells in man and the better known mammals. Studies such as those of Kato (1957) offer interesting comparisons of enzyme content of blood cells and should be of assistance in functional and evolutionary considerations.

References

Andrew, W. 1961. Phase microscope studies of living blood-cells of the tunicates under normal and experimental conditions with a description of a new type of motile cell appendage. Q. J. Microsc. Sci. **102**(1): 89-105.

Andrew, W. 1962. Cells of the blood and coelomic fluids of tunicates and echinoderms. Am. Zool. **2**(2): 285-297.

Andrew, W. 1965. Comparative hematology. Grune & Stratton, Inc., New York. 188 pp.

Bartsch, P. W., W. H. Ball, W. Rosenzweig, and S. Salman. 1937. Size of red blood corpuscles and their nuclei in fifty North American birds. Auk **54**:516-519.

Duthie, E. S. 1939. The origin, development and function of the blood cells in certain marine teleosts. I. Morphology. J. Anat. **73**:396-412.

Fulton, J. F. 1920. The blood of *Ascidia atra* Lesueur; with special reference to pigmentation and phagocytosis. Acta Zool. Arg. **1**:3, 381.

George, W. C. 1926. The histology of the blood of *Perophora viridis* (ascidian). J. Morphol. Physiol. **41**: 311-331.

George, W. C. 1930. The histology of the blood of some Bermuda ascidians. J. Morphol. Physiol. **49**: 385-413.

Jordan, H. E., and C. C. Speidel. 1925. Studies on lymphocytes. IV. Further observations upon the hemopoietic effects of splenectomy in frogs. J. Morphol. **40**:461-477.

Kamezawa, S. 1955. A contribution to the morphological study of nucleated erythrocytes in man and various animals. Histochemical study of peroxidase; 24th annual report. Okajimas Folia Anat. Jap. **27**(6):383-400.

Kato, K. 1957. Comparison between histochemical peroxidase and alkaline phosphatase reactions of animal blood cells. Histochemical study of peroxidase; 26th annual report. Okajimas Folia Anat. Jap. **29**(6): 387-402.

Klieneberger, C. 1927. Die Blutmorphologie der Laboratoriumstiere. Barth, Leipzig. 136 pp.

Lucas, A. M., and C. Jamroz. 1961. Atlas of avian hematology. U. S. Dept. Agr. Monogr. 25. 234 pp.

Patt, D. I., and G. R. Patt. 1969. Comparative vertebrate histology. Harper & Row, Publishers, New York. 438 pp.

Ponder, E., J. F. Yeager, and H. A. Charipper. 1928. Studies in comparative hematology. III. Marsupialia. Q. J. Exp. Physiol. **19**:273-283.

Ruud, J. T. 1959. Vertebrates without blood pigment: A study of the fish family Chaenichthyidae. Section VI, paper 32, pages 1-3 *in* H. R. Hewer and N. D. Riley, eds. Proceedings of Fifteenth International Congress of Zoology, London, 1958. The Congress, Burlington House, London.

Schalm, O. W. 1961. Veterinary hematology. Lea & Febiger, Publishers, Philadelphia. 386 pp.

Skutelsky, E., and D. Danon. 1967. An electron microscope study of nuclear elimination from the late erythroblast. J. Cell. Biol. **33**:625-635.

Spry, C. J. F. 1969. Personal communication.

Walvig, F. 1958. Blood and parenchymal cells in the spleen of the icefish *Chaenocephalus aceratus* (Lönneberg). Nytt Magasin Zool. **6**:111-120.

Wells, J. J., and J. E. Sutton. 1915. Blood counts in the frog, the turtle and twelve different species of mammals. Am. J. Physiol. **39**:31-36.

Windle, W. F., M. Sweet, and W. H. Whitehead. 1940. Some aspects of prenatal and postnatal development of the blood of the cat. Anat. Rec. **78**:321-332.

Yoffey, J. M. 1929. A contribution to the study of the comparative histology and physiology of the spleen, with reference chiefly to its cellular constituents. I. In fishes. J. Anat. **63**:314-344.

10 URINARY SYSTEMS

Waste products are formed as a natural consequence of physiologic processes in the tissue cells of the body. The energy-yielding metabolism of food products in cells produces metabolic products that must be gotten rid of by the body. Excess water, nitrogenous wastes, and salts are the principal by-products of the breakdown processes of organic compounds, but there are a variety of others as well. Some are toxic to the body and must be removed; others perform useful roles in the body processes. Specific substances may be excreted by some vertebrates but retained by others, depending on individual circumstances and requirements. Therefore, it is best not to define excretion in terms of materials excreted but to call it a process of elimination of metabolic by-products no longer needed by the body. Of course, various other substances get into the body and must be eliminated along with the waste products of metabolism. These include particulate matter, drugs, and inorganic salts.

Many excretory devices are used by vertebrates. Carbon dioxide and some excess water are eliminated by way of the respiratory organs. The sweat glands of mammals discharge sweat containing water, salts, and some nitrogenous waste. Many marine fishes get rid of salt through their gills. Some marine birds and others eliminate salt by way of their orbital (nasal) glands. The rectal gland of elasmobranchs may be considered an excretory organ, since it is involved in the active transport and excretion of salt. The intestinal tract may excrete some inorganic ions and excess water, but the fecal discharge mostly represents what has not been digested (metabolized).

The chief organs of excretion are the kidneys, which also have several functions other than ridding the body of waste materials. They regulate the fluid balance of the body and the electrolytes in urine. They assist in maintaining a proper salt balance between blood and tissue fluid. Kidneys are involved in regulating the sugar content of the blood, and they exert an influence on blood pressure. To do this work, the kidneys have the power to conserve some substances and eliminate others.

The urinary system typically consists of two kidneys, two conducting ducts (called wolffian ducts, mesonephric ducts, or ureters), and, in some vertebrates, the urinary bladder and urethra. In the kidneys, waste products from the blood become separated and concentrated as a fluid (urine), which is then conveyed from each kidney by a tube to the urinary bladder, or directly to the cloaca, or outside. Variations of this basic plan of excretion are found among the different vertebrates.

Whatever type of kidneys particular vertebrates may have, all kidneys consist of morphologic and functional units called nephrons. The history of all kidneys is simply a history of a basic pattern made up of these specialized tubular structures. Structural modifications of nephrons have enabled vertebrates to solve the problems of their environmental conditions, whether in fresh or salt water or on land (Smith, 1953).

DEVELOPMENT OF THE VERTEBRATE KIDNEY

The kidneys develop from paired embryonic ridges representing the portion of the mesoderm called the mesomere, or intermediate mesoderm. This ribbon of nephrogenic tissue extends from about the level of the heart to the cloaca and lies between the dorsal and lateral mesoderm of the developing embryo. Beginning at the anterior end, most of this intermediate mesoderm produces the nephrons, and as differentiation proceeds, new units are added caudally. The intermediate mesoderm becomes segmented, at least in the anterior portion, each somite being called a **nephrotome;** one tubule or nephron develops from each segment.

In its early development each nephrotome contains a coelomic cavity, the nephrocoel. This nephrocoel is converted into a kidney tubule, or **nephron,** by the formation from its dorsolateral wall of a tubular outgrowth that communicates with the main coelom by way of a peritoneal funnel, the **nephrostome** (Fig. 10-1, *A*). Each tubule

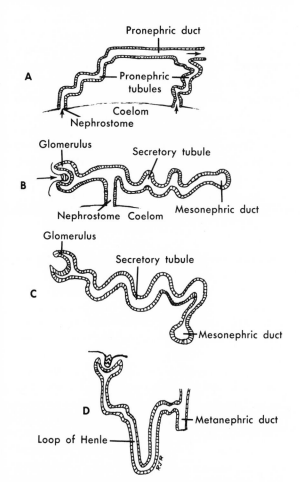

Fig. 10-1. Diagram of evolution of renal units (nephrons). See text for general description.

sends a dorsal extension that fuses with the tubule behind it to form a continuous duct (the **archinephric duct).** This duct continues to grow caudally, eventually reaching the cloaca.

Behind some posterior tubules, the medial wall of the nephrotome bulges inward, and the cavity so formed is invaded by a knot of capillaries (Fig. 10-1, *B*). The capillaries form a **glomerulus;** the cup-shaped investing wall of the nephrotome is the **renal,** or **Bowman's, capsule.** The complete structure is called the **renal, or malpighian, corpuscle,**

This excretory unit, the **archinephros,** may be considered as the basic unit of renal functioning in all vertebrates. By repetition in each segment the archinephros is supposed to have extended the entire length of the intermediate mesoderm. With its glomerulus and nephrostome, this primitive kidney could remove waste both from the blood (via the renal corpuscle) and from the coelomic fluid (via the nephrostome). Such a kidney is not entirely hypothetical; a similar one is found in embryonic myxinoid cyclostomes (hagfish). It has been further suggested that the original vertebrate nephros was made up of similar repeating nephrons throughout the length of the organ, each nephron opening independently to the exterior. But all known vertebrates have their nephrons emptying into the common duct (the mesonephric duct, wolffian duct, or ureter), which is called the archinephric duct because from such a hypothetical archinephros may have arisen the kidneys of later vertebrates.

How is this more or less hypothetical archinephros related to the kidneys found in present vertebrates? Embryologic evidence indicates that vertebrates pass through successive stages of development of three different regions of the archinephros, from the anterior to the posterior area. The same phenomenon seems to have occurred in the evolutionary development of vertebrates; the ancestral archinephros may be divided into three regions that developed into kidneys in different animals and came to have different functions. Such development always occurs in a head-to-tail direction, with general tendencies toward loss of segmentation and degeneration of the nonfunctional anterior portions. The various functional kidneys are known as pronephros, mesonephros (and opisthonephros), and metanephros.

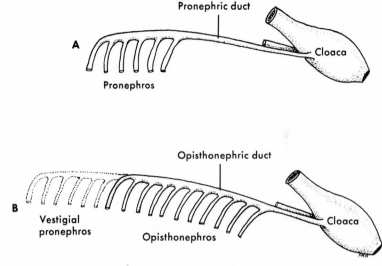

Fig. 10-2. Scheme of structure of pronephros and opisthonephros. **A,** Pronephros. **B,** Opisthonephros. Often used as a synonym for mesonephros but includes both mesonephros and metanephros portions of the embryonic nephrotome. Its duct is really the opisthonephric duct. Only tubules and ducts represented in scheme; glomeruli not shown.

The **pronephros** (Fig. 10-2, *A*) is that region of first embryonic kidney tubules arising from the anterior portion of the intermediate mesoderm. The pronephros is functional in all species that produce free-living larval forms, including cyclostomes, some bony fishes, and amphibians (Jaffee, 1954). It develops, but probably never functions, in amniotes. Only in the adult cyclostomes and a few teleosts does it persist throughout life, and then chiefly as lymphoid tissue. Its **pronephric duct** later becomes the mesonephric duct when the pronephric region disappears.

Posterior to the region of the pronephros, additional tubules (the **mesonephros**) are formed in the intermediate mesoderm. These new tubules establish connections with the pronephric duct. They are quite similar to the pronephric tubules; however, the primitive segmentation is mostly lacking, and there is a tendency for the development of tubules without nephrostomes (Fig. 10-1, *C*). The **collecting duct system** thus appears for the first time. The mesonephros is the functional kidney of fishes and amphibians and of the embryonic forms of reptiles, birds, and mammals. In addition, it may function for a short time after birth in some reptiles and monotremes.

There are many variations of the mesonephros, especially among the fishes. In dogfish sharks and caecilians it is a long, ribbonlike body extending throughout the coelom. Here it is really an **opisthonephros;** that is, it also includes the posterior metanephros (Fig. 10-2, *B*). In male sharks the most anterior tubules are typically part of the reproductive system and are used as a pathway for sperm. The nephrons in fish are often provided with nephrostomes, and the caudal end of the mesonephric duct may enlarge to form seminal vesicles or urinary bladders.

The **metanephros** is the kidney of the higher vertebrates (reptiles, birds, and mammals). It is formed from the caudal end of the nephrogenic intermediate mesoderm, but it is displaced anteriorly and laterally during development. This is the same mesoderm that forms the caudal part of the mesonephric or opisthonephric kidney of fishes and amphibians. It actually has a dual origin. The conducting mechanism of ureter, renal pelvis, calyces, and collecting tubules arises from the **metanephric bud,** whereas the nephrons are formed from undifferentiated intermediate mesoderm (the **metanephric blastema**).

The number of nephrons may be very large in some amniotes. In man it is estimated that there are more than a million in each kidney. The kidneys of amniotes are more or less compact, with the ureter coming from a notch or hilus, on their median surface. The inner end of the ureter is expanded into a sinus from which branches extend into the lobe or lobes of the kidney and into which collecting ducts open. The vertebrate kidney varies from a bean-shaped to an elongated, lobulated structure.

GENERAL FUNCTIONING OF THE VERTEBRATE NEPHRON

As the blood passes through the capillaries of the glomerulus, fluid and various materials filter out through the capillary walls into the capsule. This filtered material, the **glomerular filtrate,** passes into the tubule of the nephron. The filtrate differs from blood in that there are no blood cells or macromolecules; they are too large to pass through the glomerular capillary walls. The amount of filtrate is directly proportional to the blood pressure and extent of dilation of the capillary walls in the glomerulus.

Many blood vessels and capillaries are closely associated with the walls of the nephron tubule. As the filtrate moves along the tubule, which varies in length among the vertebrates, there is differential reabsorption of many constituents, via the cells of the tubule, back into the blood. The rest passes into the ureter and is excreted from the body as **urine.** Various substances are reabsorbed in different concentrations; some are not reabsorbed at all. Thus the urine of all animals has a different composition from the glomerulus filtrate and from the blood.

The pronephros and mesonephros are mostly adapted for filtration, not for reabsorption. Since vertebrates evolved in fresh water, their gills and alimentary canal continually take in water by osmosis, and this water must be continually eliminated by production of copious **hypotonic** urine. Thus bony marine fishes have inherited from their freshwater ancestors a mesonephros or opisthonephros designed to eliminate water. But they live in a **hypertonic** environment (sea water) that tends to draw water out of their bodies. They meet this problem by reabsorbing as much water as their opisthonephros permits, by swallowing large amounts of sea water for blood fluid needs

and excreting excess salt through specialized salt-excreting glands in the gills, and by producing a scanty hypotonic urine.

The chondrichthyes solve the problem by reabsorbing much of their nitrogenous waste product, urea, into their blood and tissues. This makes their interior body slightly hypertonic to the surrounding sea water, allowing them to draw in their required water osmotically. But they still excrete a hypotonic urine and thus lose more water then they should. The evolution of the metanephros in reptiles, birds, and mammals made possible the production of urine that is hypertonic to the blood. This capability is responsible for the conservation of water so necessary to their terrestrial existence.

CYCLOSTOMATA

The mesonephric kidneys of cyclostomes are two elongated organs lying in a retroperitoneal position. They begin about the center of the body and extend caudally to a position not far from the cloacal opening. They increase in size gradually, attaining a thickness of about 3 mm in the adult, then gradually narrow as they pass caudally. The lateral surface of the kidney is in contact with the body wall, and the ureters lie on either side. The medial surfaces are closely related to the sex glands.

The renal (malpighian) corpuscles of cyclostomes are relatively enormous (Fig. 10-3). They range from 0.5 to 1.5 mm in diameter and are thus visible to the naked eye. They are correspondingly few in number, about 70 in the entire mesonephros, arranged segmentally. In *Petromyzon* several tubules thus arise from each corpuscle. The first part of each tubule is a ciliated segment some 150 to 200 mm in length (Fig. 10-4). The second part is a convoluted segment. A definite constriction marks the transition from the ciliated segment to the second segment.

Each ciliated cell of the first segment bears a number of long cilia that join together to form a whiplike process (Fig. 10-5). Since these processes are over twice as long as the diameter of the tubule, they extend along the channel of the tubule in a direction away from the malpighian corpuscle. The ciliated cells are directly continu-

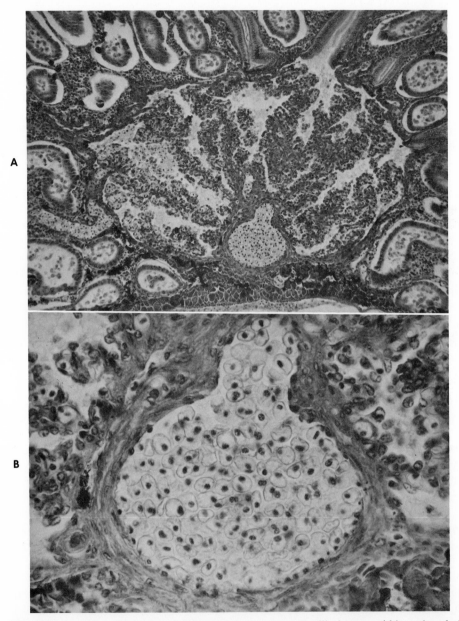

Fig. 10-3. A, Giant glomerulus of kidney of *Petromyzon*. Capillaries are within rather darkly stained tissue masses; clear spaces represent lumen of Bowman's capsule. At upper right is ciliated neck of one of the exits from this great malpighian body. **B,** Higher magnification of renal artery in **A.** Lumen is filled with nucleated erythrocytes. Endothelial nuclei are prominent, and smooth muscle of media is seen. There is poor development of elastic membranes. (**A,** ×100; **B,** ×400.)

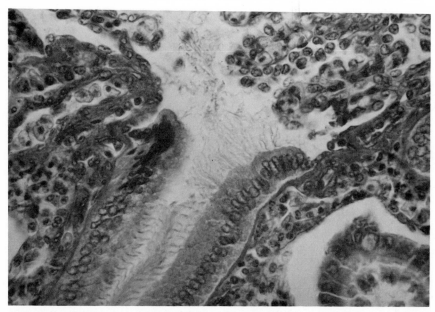

Fig. 10-4. Detail of neck of nephron of *Petromyzon*. Epithelial cells just at exit from lumen of Bowman's capsule have cilia several times the length of cell body; those farther along tubule have shorter cilia. (×400.)

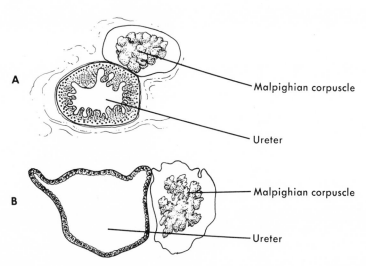

Fig. 10-5. Transverse sections through middle portion of *Myxine* kidney (mesonephros). **A** shows kidney with its large malpighian body and an empty ureter. **B** shows ureter distended with urine. Sections are from two different specimens of *Myxine*. (Redrawn from Fänge and associates, 1963.)

ous with the flattened cells of the parietal layer of Bowman's capsule.

In the second, convoluted segment the cells are columnar and contain deeply colored granules in their basal portions. They have well-developed brush, or striated, borders, the thickness of which seem to vary considerably, possibly in relation to functional state. The brush border is due to the presence of microvilli, generally characteristic of absorptive cells of vertebrates. Transition from the convoluted to the third, or end, segment takes place near the lateral side of the kidney. The epithelium loses its brush border, and the tubule decreases in diameter and courses ventrally toward the ureter, the mesonephric duct.

The cleftlike lumen of the ureter increases in size as it passes caudally, until ﬁ finally comes to occupy the greater part of the mesonephros. The ureter wall is lined by a pseudostratified, high columar epithelium. The shapes of the cells appear to change with the degree of distention of the ureter (Fig. 10-5). Outside the epithelium is a thick connective tissue layer, the lamina propria, containing many longitudinally coursing connective tissue fibers, numerous blood vessels, sympathetic nerves, and groups of ganglion cells. Outside the lamina propria is a layer of smooth muscle, primarily circular in arrangement and increasing in thickness caudally.

The mesonephros of *Petromyzon's* cousin, *Myxine* may be considered as almost "atubular." The tubules connecting the malpighian corpuscles with the ureters are very short. The epithelial cells of these short tubules are columnar and contain brownish-black pigment granules. These cells apparently change from high columnar, when the tubule contains little fluid, to cuboidal when it is distended, much as does the lining of the ureter. Surprisingly, the epithelium of the *Myxine* ureter has a brush border and also contains many pigment granules (Fig. 10-6). The wall of the ureter has a rich capillary net derived from efferent glomerular arteries and other branches of the renal arteries, and the cells of the epithelium contain many vacuoles. A common view is that the pigment granules probably represent material reabsorbed from the urine. Hence the ureters in the cyclostome would be serving a function carried out by the various tubules of the nephron in

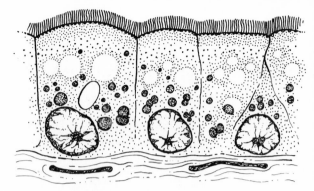

Fig. 10-6. Epithelium of the ureter of *Myxine*. Epithelium is not transitional as in higher vertebrates, but rather of pseudostratified type. Brush-border and pigment granules are conspicuous. (Redrawn from Fänge and associates, 1963.)

Petromyzon and higher forms. The functional aspects of the *Myxine* ureter have not been studied yet in great detail. However, the fact that urine seems to be retained in the tube for fairly long periods of time probably means more than simply a matter of storage. It may afford a real opportunity for resorptive activity, including probably the resorption of glucose and other substances. Some of this resorptive activity may be by a special type of phagocytosis (athrophagocytosis) by which cells are able to take solid particles out of suspension (Gerard, 1933, 1943).

The large malpighian corpuscles of marine cyclostomes may have significance. They are similar in size to those found in freshwater amphibians. Smith (1960) believes that such glomeruli are particularly well adapted to the elimination of excess water that has been taken up osmotically in fresh water. Although cyclostomes are generally marine in habitat, many of them do spend some time in fresh water; it may be that their ancestors were freshwater animals. In some species of lampreys (e.g., *Lampetra fluviatilis*) the glomeruli are fused into one large **glomus** (Fig. 10-7), and several renal tubules, or nephrons, emerge from the common urinary space. A glomus is usually provided with a single afferent arteriole but has several efferent arterioles (Hickman and Trump, 1969).

The pronephros of adult cyclostomes is largely transformed into hemopoietic tissue. Here lym-

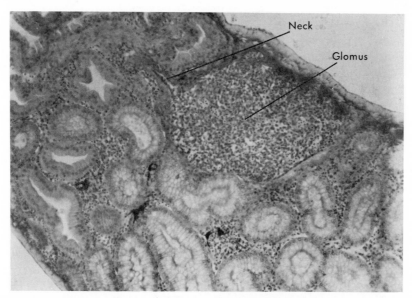

Fig. 10-7. Section through primitive kidney of adult lamprey. Renal corpuscles are relatively few in number. Note the dense nature of epithelium of the renal capsule, thickness of parietal epithelium, and brush borders of the proximal convoluted tubules. (×210.)

phocytes, erythrocytes, and other blood cells are formed. The organ seems to be active only at certain periods and, in comparison with hemopoietic tissue of the wall of the intestine, is probably of minor importance (Jordan and Speidel, 1930; Holmgren, 1950).

The pronephros also shows a large aggregation of lymphoid cells in its central portion. Willmer (1960) believes that it may serve as a filtration apparatus similar to the lymph nodes of higher vertebrates. Fänge and associates (1963) comment that the degenerate remnants of tubular epithelium show a resemblance to Hassall's corpuscles of the thymus gland of higher vertebrates. The "lymphatic" nature of the pronephros is of particular interest in relation to the supposed primitive condition of immune mechanisms in those forms; the lymphatic tissue may represent at least a beginning in development of an immune system.

CHONDRICHTHYES

There are some differences between the male and female excretory systems in sharks. The chief excretory organs, the paired opisthonephroi, occupy most of the length of the body cavity. (Recall that the opisthonephric kidney includes portions of both the mesonephric and metanephric

types.) The anterior ends are quite narrow and are sexual in male sharks, not excretory. But caudally, each opisthonephros has an excretory function. In males the anterior tubules are used as a pathway for sperm from the testes to the wolffian, or opisthonephric, duct. These delicate tubules are called **vasa efferentia.**

The wolffian duct, much larger in males, is highly convoluted and runs posteriorly along the ventral face of the kidney (Fig. 10-8). Posteriorly, the duct widens to form a seminal vesicle. In the region of the vasa efferentia, the wolffian duct lies on Leydig's gland, which may secrete a fluid beneficial to the sperm. At its posterior end in the cloaca, the seminal vesicle terminates in a forward-projecting sperm sac. The two sperm sacs, one from each wolffian duct, join and open to the outside at the tip of a urogenital papilla in the median part of the cloaca. The seminal vesicle receives no kidney tubules from the portion of the kidney on which it lies; this part of the opisthonephros is drained by accessory urinary ducts.

The female kidneys are similar to those in the male. Their smaller opisthonephric ducts, however, are entirely urinary. In some sharks (e.g., *Squalus*) there are small accessory urinary ducts

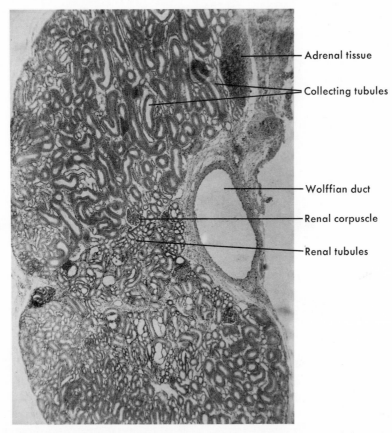

Fig. 10-8. Section through kidney of male shark showing large wolffian (opisthonephric) duct and general features of nephrons. (×70.)

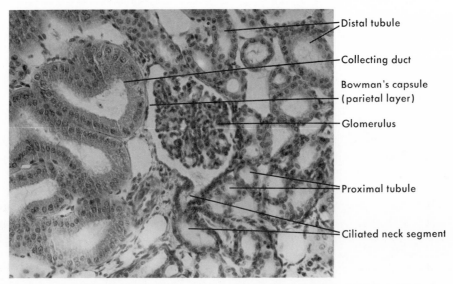

Fig. 10-9. Section through kidney of shark showing the structure of renal (malpighian) corpuscle and renal tubules. (×380.)

for draining the posterior parts of the opisthonephroi. In the female the anterior portion of the kidney is vestigial. The wolffian ducts pass to the cloaca, where they join and open externally through a urinary papilla.

The structural and functional units of each opisthonephros are the nephrons. Each nephron consists of a renal, or malpighian, corpuscle and a more or less convoluted renal tubule (Fig. 10-8). The malpighian corpuscle is located at the beginning of each nephron and is composed of capillaries of the glomerulus resting in the cup-shaped, blind end of the excretory tubule, Bowman's capsule (Fig. 10-9). The double-layered Bowman's capsule contains an inner visceral layer, in contact with the glomerulus, and an outer parietal layer.

Each renal tubule consists of the neck segment, the first and second proximal segments, the distal segment, and the collecting duct. The collecting ducts join to form several large ducts that open into the opisthonephric duct. The histologic structure of the different segments of the renal tubule varies, as indicated by the characteristic cell types shown in kidney sections. The long and constricted neck segment is thin-walled and lined by cuboidal epithelium. Cells near the origin and at the end of the neck are ciliated, but most of the other cells of this segment are free of cilia.

Sections through the neck region show many closely packed cross sections of the neck, indicating many convolutions of this tubule. Some of the neck loops have taller cells with round or elongated nuclei. They may also have faint striations, representing an area of the neck tubule that some authorities have believed responsible for the reabsorption of urea, so important to the well-being of the elasmobranch. However, Kempton (1943, 1962) believes he has shown that urea reabsorption does not take place through a special segment but is performed by regular tubular cells that do not have this capacity in the kidneys of other forms.

The first proximal segment of the renal tubule originating at the end of the neck has a lining of low columnar or tall cuboidal cells with characteristic brush borders and many mitochondria. The second proximal segment is convoluted and makes up most of the renal tubule. It is lined with tall columnar cells that show striations, many mitochondria, and brush borders. The distal segment is lined by a cuboidal epithelium that has no brush border. This tubule empties into the collecting duct, which has a larger diameter and a lining of tall columnar cells. The collecting ducts have distinct cell boundaries and are often found on the dorsal side of the kidney.

In addition to the various tubular segments, a great deal of reticular connective tissue and lymphocytes are present in the shark kidney. Hemopoietic tissue is also found among the tubules.

OSTEICHTHYES

All adult bony fishes have mesonephric or opisthonephric kidneys (Figs. 10-10 and 10-11), except for a few species that have a functional pronephros. The kidneys of the teleosts vary considerably (Fig. 10-12). Some are elongated; others are compact. Some have long, convoluted nephrons; others have rather abbreviated ones. Many marine teleosts have small glomeruli or none at all (aglomerular). In this dramatic adaptation to the hypertonic conditions of sea living, water is conserved by not losing it in the first place rather than by reabsorption. Aglomerular forms must rely on excretion of waste products into their kidney tubules from the surrounding capillary beds (Edwards, 1928).

Urinary bladders are found in some bony fishes. They form by enlargement of the caudal portions of the wolffian ducts. The kidneys of teleosts often show accumulations of lymphoid or hemopoietic tissue; migration of lymphocytes into the epithelium of the tubules is also a common phenomenon.

AMPHIBIA

The paired frog kidneys, or opisthonephroi, are flattened, elongated, dark red bodies lying dorsal to the peritoneum of the posterior part of the coelomic cavity. Each is a compact structure of about 2000 opisthonephric tubules, or nephrons. The distal ends of the nephrons unite with the wolffian ducts, which run along the outer lateral edges of the kidneys and continue posteriorly, emptying into the cloaca.

Each nephron has several regions (Fig. 10-12).

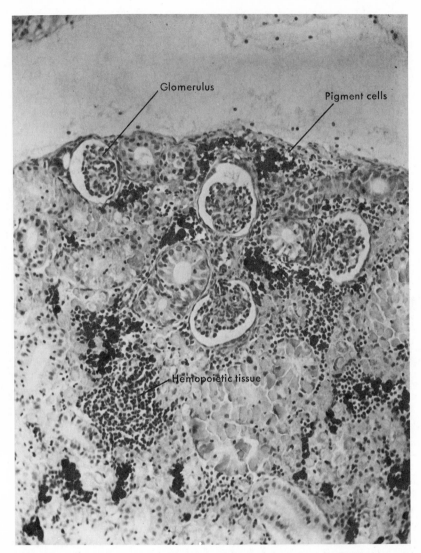

Fig. 10-10. Kidney of *Salmo gairdneri*. There are four glomeruli in the field. Large pigment cells occur normally in the trout kidney, and hemopoietic tissue is abundant. (×125.)

At the proximal end is the renal corpuscle, made up of a glomerulus of arterial capillaries surrounded or encapsulated by the double layer of thin, flat, epithelial cells. This is Bowman's capsule (Fig. 10-13). The region of the tubule just distal to Bowman's capsule, the neck segment, is lined with ciliated cuboidal cells. The cilia apparently function to sweep the glomerular filtrate into the proximal tubule.

The tubule continues as a proximal convoluted portion formed by low columnar cells with brush borders. Next, there is a short constricted portion (the intermediate segment) with a ciliated cuboidal epithelium. Then the tubule widens and becomes convoluted again, forming the distal convoluted segment lined with cuboidal cells. Just before joining the larger collecting duct, the tubule forms a straight connecting piece. A number of these collecting ducts fuse into the opisthonephric duct, which runs along the outer edge of the kidney.

The opisthonephric duct (Fig. 10-14, *A*) runs

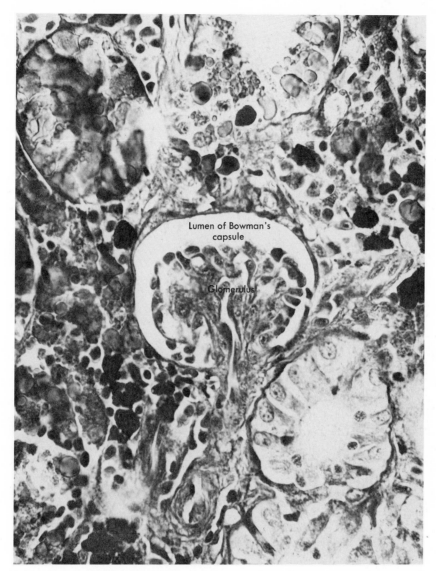

Fig. 10-11. Single malpighian corpuscle of *Salmo gairdneri*. Bowman's capsule surrounds the glomerulus. (×400.)

to the cloaca, not directly into the bladder, which also opens into the cloaca. The duct is lined with ciliated transitional epithelium (Fig. 10-14, *B*). A lamina propria of connective tissue is thrown into longitudinal folds, thus giving the lumen a stellate appearance. A muscular coat and a fibroelastic adventitia form the outer wall.

Attached to the ventral side of the cloaca is a thin-walled bilobed sac, the urinary bladder. It is so placed that it is below the opening of the urinary duct. The urine is discharged from the duct into the cloaca and then backs up into the bladder (Fig. 10-15). The bladder originates as an outpushing of the cloaca and is surrounded by peritoneum that attaches to the rectum and body wall. It is lined with a transitional epithelium that may be three to five cells thick, depending on the degree of distention of the bladder (Fig. 10-15). Goblet cells may be present in the epithelium. The electron microscope shows the mucosa lining with microvilli and with mitochondria distributed through the cells (Bentley, 1966). The

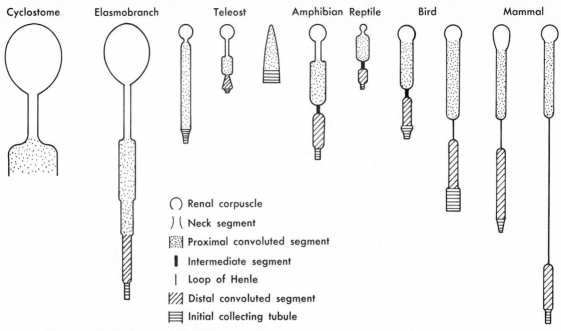

Cyclostome	Elasmobranch	Teleost	Amphibian	Reptile	Bird	Mammal

◯ Renal corpuscle
)(Neck segment
▦ Proximal convoluted segment
▮ Intermediate segment
│ Loop of Henle
▨ Distal convoluted segment
▤ Initial collecting tubule

Fig. 10-12. Schemes of renal units (nephrons) in various vertebrates. (Based on several sources.)

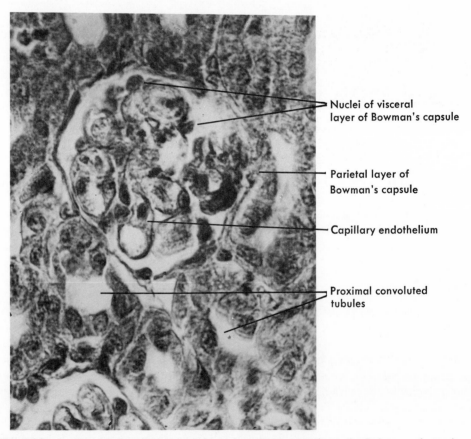

Nuclei of visceral layer of Bowman's capsule

Parietal layer of Bowman's capsule

Capillary endothelium

Proximal convoluted tubules

Fig. 10-13. Section through renal corpuscle of frog. Various regions of tubules are shown in cross section. (×700.)

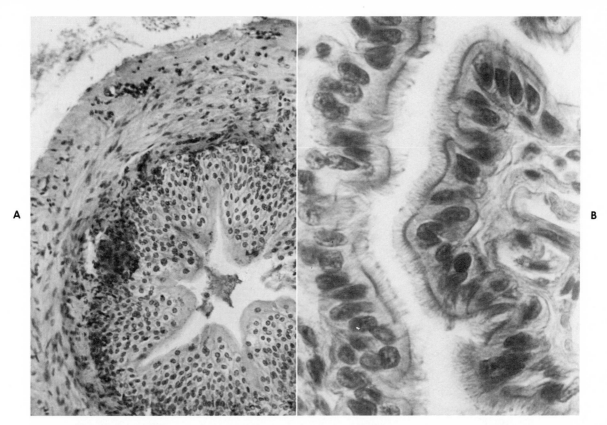

Fig. 10-14. Wolffian duct (mesonephric or opisthonephric duct), often called erroneously the ureter, of frog. **A,** Transverse section. Note mucous membrane of transitional epithelium, thrown into folds. **B,** Transverse section cut toward its entrance into the cloaca. Note that the squamous transitional epithelium of **A** is transformed into the columnar transitional epithelium of **B.** This change often happens in long ducts.

middle layer of the bladder is a network of smooth muscle, outside which is a thin sheet of connective tissue covered externally by the peritoneum.

Nephrostomes, or peritoneal funnels, which open to the body cavity, are found in the frog's kidney. Rugh (1938) identified 200 to 250 of these in *Rana pipiens* on the ventral facet of the kidneys, which is covered by peritoneum. These funnels are highly ciliated and connect not to the capsule or urinary tubule but to venous blood sinuses that eventually empty into the posterior vena cava. Rugh found that these funnels had a diameter of 50 to 80 μm on the average. They apparently function in picking up waste debris from the body cavity and passing it to the blood. It is then eventually discharged through the renal corpuscle along with the other waste.

Numerous physiologic experiments have been performed to determine how the nephron functions, but many aspects are still obscure. Blood flows through the glomeruli in spurts or intermittently. Some of the glomeruli are permeable to all the constituents of blood plasma except the blood proteins and corpuscles. The plasma is forced out through the capillary walls as a result of the high blood pressure caused by constriction of the efferent vessel carrying blood away from the glomerulus. The diameter of this vessel decreases in comparison to the afferent vessel carrying blood to the glomerulus and thus causes filtration.

As the filtrate passes down the tubule it undergoes many changes of reabsorption. The proximal convoluted tubule concentrates the glomerular filtrate by reabsorbing water (Walker and Hudson, 1937). Since the blood in the vessels envelop-

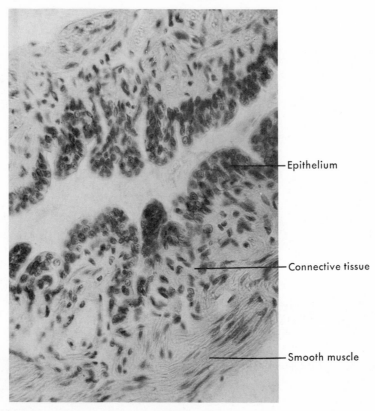

Fig. 10-15. Section of frog bladder. Epithelial lining varies from columnar to stratified or pseudostratified layer; often called transitional type. (×500.)

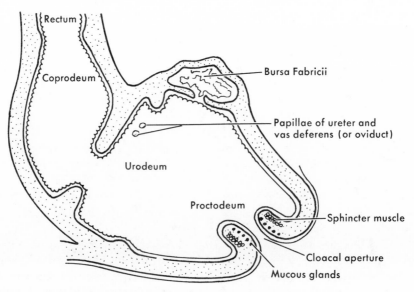

Fig. 10-16. Diagram of cloaca in reptiles and birds. Note the three incomplete compartments. (Modified from many sources.)

ing the kidney tubules has already passed through the glomerulus and lost a great deal of water, an osmotic gradient is produced that then causes water to flow out of the tubule and back into the blood. The other substances are reabsorbed at varying threshold concentrations. Some are regained by passive diffusion, and others by active (energy-requiring) reabsorption.

Blood arrives at the kidneys from two sources. Branches of the renal arteries from the dorsal aorta supply blood mainly to the glomeruli, whereas branches of the renal portal vein carry it to the tubules. However, blood from the glomeruli, after filtration, also passes to the vascular system of the renal tubules, where reabsorption of some of the tubular contents occurs. All blood returns to the heart by way of the postcava. In mammals, there is only one source of blood supply to the kidneys: the aorta via the renal arteries.

REPTILIA

The mesonephros functions during the embryonic lives of all amniotes (reptiles, birds, and mammals). However, in the adult forms, the most posterior regions of the archinephros become active—the metanephric kidney.

In the lizard the kidneys are paired, dark red bodies, each double-lobed, situated in close contact with the dorsal wall of the posterior part of the abdominal cavity. They are covered with peritoneum on their ventral face, and therefore they lie outside the peritoneal cavity. A ureter extends from each, coursing posteriorly into the urodeum part of the cloaca (Fig. 10-16). The ureter, or metanephric duct, forms as an outgrowth of the mesonephric (wolffian) duct. The latter persists in male amniotes as sperm-carrying structures.

The thin ureter is lined with simple columnar cells resting on a **tunica propria.** The submucosa is surrounded by a circular smooth muscle tunic with an external adventitia. A urinary bladder of endodermal origin opens into the cloaca on its ventral side. As in the amphibians, ureters do not enter the bladder directly, so urine tends to accumulate in the **urodeum** area of the cloaca. It is in the urodeum that most of the water is reabsorbed into the blood. To prevent the return

of toxic urea into the blood, lizards excrete most of their nitrogen as uric acid. This insoluble substance crystallizes as water is reabsorbed and is discharged with the feces as a white, semisolid mass. It is thus possible for the lizard to conserve most of its water without risking the reabsorption of the toxic nitrogenous waste product of the lower forms, urea.

Many reptiles, for example, snakes, crocodiles, and some lizards, do not have bladders. Some turtles, on the other hand, have two of them.

As in other amniotes, the lizard kidney has malpighian corpuscles with glomeruli and renal tubules (Fig. 10-17). The tubules, however, have no loop of Henle, an important segment of the mammalian nephron; the glomeruli are reduced in size, with only two or three short vascular loops within Bowman's capsule. The neckpiece of the nephron is ciliated, and the cells of the proximal convoluted tubule have a brush border. An intermediate segment passes to the distal convoluted tubule, which is lined with striated cells. The collecting tubule in lizards often has many mucus-secreting cells.

As in the lower forms, the capillaries and other vessels around the renal tubules still receive blood from the renal portal system, as well as from efferent arterioles from the glomeruli. This double arrangement ensures an adequate blood supply to the tubules of cold-blooded animals that otherwise might be subjected to the fluctuating conditions of sluggish flow in low temperatures.

AVES

The paired kidneys are located dorsally in the abdominal cavity, snugly embedded in a concavity of the synsacrum. Each kidney is three-lobed with each lobe subdivided into smaller lobules. As in all amniotes the kidneys of the adult are metanephric. The major lobes are of little structural significance, but each of the smaller lobules has a branch of the ureter into which the collecting tubules of their nephrons open. Each lobule has two parts, a medulla containing mainly nephron loops, and a cortex consisting of the other parts of the nephrons and some entire nephrons.

There are two types of nephrons. One type is restricted to the cortical part of the lobule; the

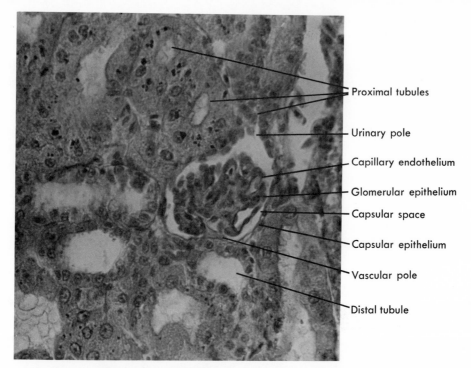

Proximal tubules

Urinary pole

Capillary endothelium

Glomerular epithelium

Capsular space

Capsular epithelium

Vascular pole

Distal tubule

Fig. 10-17. Section through kidney of *Anolis.* (×200.)

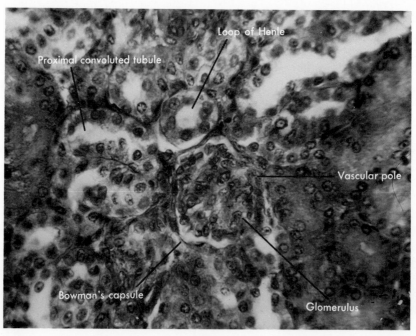

Loop of Henle

Proximal convoluted tubule

Vascular pole

Bowman's capsule

Glomerulus

Fig. 10-18. Section of chicken kidney showing renal corpuscle (center) and section of tubules. (×800.)

other contains loops of Henle, which pass into the medullary zone. The renal corpuscle usually found at the beginning of each nephron is somewhat smaller than the corresponding structure in mammals (Fig. 10-18). Apparently in the interest of conserving water, the glomerulus may sometimes be reduced to one capillary loop (Marshall and Smith, 1930). Both the avian and reptilian glomeruli have cores of nonvascular tissue of unknown significance (Vilter, 1935).

The Bowman's capsule surrounding the glomerulus is connected to the proximal tubule by a short neck. After making some convolutions and running back to the periphery, the tubule becomes smaller in diameter and connects with a very narrow tubule that corresponds to the Henle's loop of mammals. This segment is extremely important in the efficient reabsorption of water from the kidney filtrate. The next part of the tubule is the intermediate segment; it is thicker and joins the distal convoluted tubule. Eventually, this tubule joins a collecting duct. The tubules in the cortical and medullary types of nephrons vary in some minor particulars. The structure of the different segments of the nephron varies, and the structure of a particular segment in the bird nephron may also differ from the corresponding segment in the mammal. The cells lining the proximal tubule in both bird and mammal have brush borders (Fig. 10-18), but Henle's loop in birds has taller cells than in mammals.

Each lobule has in the center part of its cortical region a vein and one or more arteries. The avian kidney has a functional renal portal system, branches of which (interlobular veins) are found on the surface of the cortex. This portal system is absent in mammals. The branches of the renal artery pass to the glomeruli, and those of the renal portal vein carry blood directly to the tubules. The blood from the capillary plexus around the tubules is collected into central converging intralobular veins, which pass the blood eventually into the renal vein.

There is no bladder in birds (with the exception of the ostrich), and the ureters open directly into the urodeum, one of the three chambers of the cloaca. Investigators do not agree about the possibility of a urine-concentration function of the cloaca. Some believe that urine from the kidney is concentrated in the urodeum and coprodeum by cloacal resorption of water, but the problem is yet unresolved. The chief nitrogeneous waste product of birds, as in reptiles, is uric acid. The relative insolubility of this substance makes their urine well fitted for water conservation. Consequently, many birds can go long periods without water.

Most avian urine is only moderately hypertonic. In certain sea birds (e.g., cormorants) Schmidt-Nielsen and associates (1958) found that salt is excreted not in the urine, but as a strongly hypertonic salt solution by the nasal, or supraorbital, glands. These paired glands may be located on top of the skull, or elsewhere in the skull, and empty through ducts into the nasal cavity. The excess salt of ingested sea water can thus be taken care of in this way without burdening the kidneys.

MAMMALIA

Most mammals have compact bean-shaped kidneys with smooth surfaces and with a notch (hilus) in the medial region where the ureter, blood vessels, and nerves enter or leave. The kidney itself consists of a **cortex** in which the renal corpuscles are found and a **medulla** made up of collecting tubules and thousands of loops of Henle (Fig. 10-19). The medulla consists of a number of triangular masses called renal pyramids. The base of each pyramid is directed toward the broad area of the kidney, the cortex; the apex (papilla) of each one fits into a minor calyx of the renal pelvis. The tip of a papilla (which may be the common papilla of two or more pyramids) has a number of pitlike openings called the **foramina papillaria,** where the main excretory ducts of tubules open.

The pyramids of the mammalian kidney have a radially striated appearance because of the presence of many parallel tubules and blood vessels. The outer cortex zone of the kidney overlays the bases of the pyramids and also extends down between them, forming the **columns of Bertin.** Collecting tubules from the base of the pyramids invade the cortex and form the **pars radiata,** a series of lighter striations continuous with those of the medulla and hence called medullary rays, although they are in the cortex.

The kidney is a compound tubular gland that

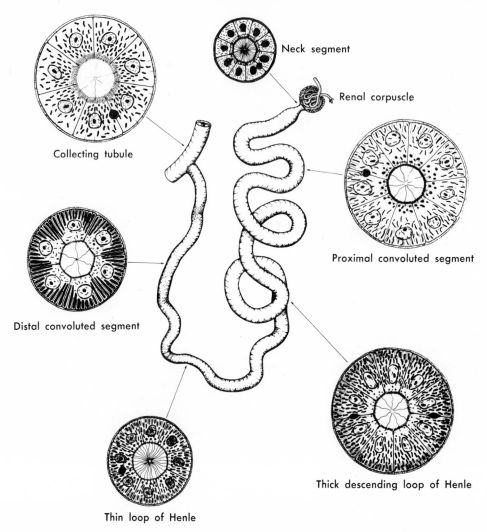

Collecting tubule

Neck segment

Renal corpuscle

Proximal convoluted segment

Distal convoluted segment

Thick descending loop of Henle

Thin loop of Henle

Fig. 10-19. Diagram of typical vertebrate nephron. Typical histologic sections (enlarged) of various divisions are shown as inserts. (Courtesy C. P. Hickman, Jr., Lexington, Va.)

secretes urine. Each single gland or functional unit is the nephron (Fig. 10-20). Thus the parenchyma of the kidney consists of many secretory nephrons, which are long, tortuous canals. These join excretory ducts, which convey urine into the minor calyces of the pelvis of the kidney. The nephron, or secretory part, is unbranched, with some portions straight and others convoluted. In large kidneys there may be more than a million nephrons. The excretory portions of the uriniferous tubules are the collecting ducts, which form a branching, treelike system. Each collecting tubule receives urine in the cortex from several nephrons and passes in a medullary ray down into the medulla (Fig. 10-21). In the medulla several collecting tubules join to form larger ducts, the **papillary ducts of Bellini,** which open on a papilla of the renal pyramid.

In the mammal the typical nephron (Figs. 10-12 and 10-19) consists of four segments that differ structurally and functionally: (1) the renal malpighian corpuscle containing the glomerulus, (2) the proximal convoluted tubule, (3) the intermediate loop of Henle, and (4) the distal convoluted tubule, which empties with many other nephrons into a collecting tubule.

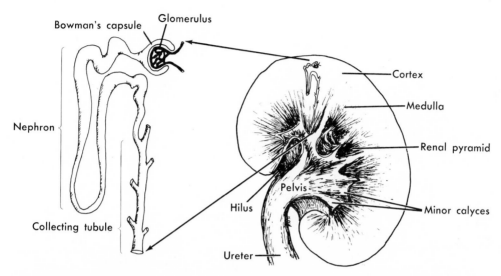

Fig. 10-20. Relationship of nephron and its collecting tubule to morphological structure of kidney. Renal corpuscle is found in cortex region, and collecting tubule opens near papilla on tip of pyramid.

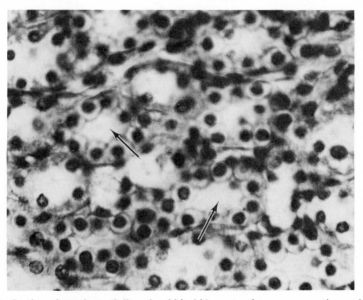

Fig. 10-21. Section through medulla of rabbit kidney to show cross sections of collecting tubules (mostly) as indicated by distinct cells in the wall (arrows). (×400.)

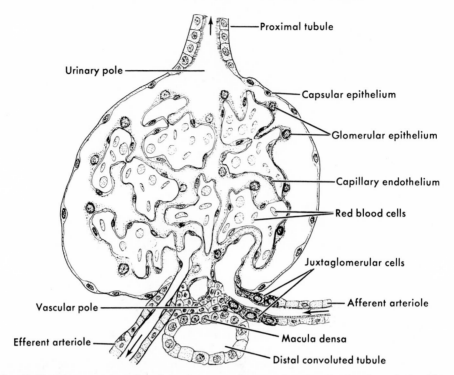

Fig. 10-22. Diagram of section through mammalian renal corpuscle showing relationship of Bowman's capsule to glomerulus and to its arteries. Note macula densa and juxtaglomerular cells. (Modified from Ham, 1969.)

Renal, or malpighian, corpuscle. The renal corpuscle is composed of an arterial vascular glomerulus that is enclosed in a double-walled, indented epithelial cup called the Bowman's capsule (Elias, 1957) (Fig. 10-22). It is a spherical body with a diameter of about 200 μm in man and thus makes up only a small extent of the entire nephron, which may reach a length of 50 mm or more. The glomerulus is a tuft of capillary-like blood vessels. Blood is supplied to the glomerulus by an afferent arteriole at the vascular pole of the corpuscle and leaves by the efferent arteriole at the same pole. On entering the renal capsule, the afferent arteriole divides into anastomosing loops **(rete mirabile).** These capillary loops reunite to form the efferent arteriole, which is only about half the diameter of the afferent arteriole. This difference in size is probably due to the thicker muscular walls of the afferent arteriole; the lumen is about the same in each.

Bowman's capsule consists of two layers of

simple squamous epithelium, formed from a single layer that has been pushed in by the invading blood vessels during development. The inner layer is the **visceral layer** (glomerular epithelium), which forms an intimate cover of the glomerulus; the outer **parietal layer** (capsular epithelium) makes up the wall of the corpuscle and is continuous with the rest of the tubule (Figs. 10-22 and 10-23). Only the parietal layer is easily seen, because of its nuclei bulging into the capsular space. It is a continuous layer of epithelial cells, ciliated in some lower vertebrates. The capsular space between the two epithelial layers receives the filtrate from the blood. It is continuous with the lumen of the proximal tubule at the urinary pole, nearly opposite the vascular pole where the blood vessels enter and depart.

An understanding of the glomerulus, especially of the visceral layer, has been one of the chief aims of kidney research involving work with the electron microscope. It is known that the capillaries of the rete mirabile consist of a single layer

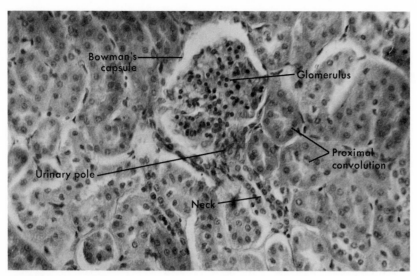

Fig. 10-23. Section through portion of renal cortex and renal corpuscle and sections of renal tubules of rabbit. (×200.)

of thin, flattened endothelial cells supported by a basement membrane, the **lamina densa.** In the very thin visceral layer the cells touch each other only at intervals, with spaces between them. These cells have morphologic specializations of tiny cytoplasmic end-processes, or **pedicels,** which contact the basement membrane adjacent to the capillary epithelium. These epithelial cells and their processes are often called **podocytes.** The capillary walls are also interrupted by slitlike pores **(fenestrae),** revealed by the electron microscope (Porter and Bonneville, 1963). Thus the basement membrane of the capillaries is the only intact barrier to the passage of blood constituents into the glomerular filtrate. This basement membrane also acts in preventing the capillaries from overstretching under high blood pressure.

The afferent arterioles lack true adventitia. Near the vascular pole of the capsule, the muscle cells of the media of these arterioles are largely replaced by special cells with rounded nuclei and granular cytoplasm. These cells are known as juxtaglomerular (JG) cells. They are absent in the efferent arterioles. JG cells have been found in all species of vertebrates that have mesonephric and metanephric kidneys. These cells are in close contact and apparent communication with the **macula densa,** a group of columnar cells in the

wall of the distal convoluted tubules of mammals (Fig. 10-22). There is some evidence that the JG cells secrete a hypertensive factor, **renin,** in response to a fall in blood pressure. The pressure of the filtrate passing through the distal convoluted tubule apparently affects the macula densa, which in turn affects the JG cells. There are many speculative ideas about this negative feedback mechanism that seems to be involved in the maintenance of blood pressure in the glomerulus.

Proximal convoluted tubule. This segment, the first part of the renal tubule, leads off directly from the urinary pole of the capsule of the renal corpuscle (Rhodin, 1958) (Figs. 10-12, 10-19, and 10-22). The first part of it is narrow and straight in many species, forming a short neck lined with low cuboidal cells that may be ciliated in lower vertebrates. This neck region is mostly absent in the tubule of man. The proximal convoluted tubule, as the name suggests, follows a looped and tortuous course. In microscopic sections, numbers of them can be seen concentrated in the periphery of the lobules around the renal corpuscles making up most of the cortex of the kidneys. It is the longest segment of the mammalian nephron.

The individual cells of the proximal convoluted tubule are cuboidal or low columnar, adapted for absorption, with a brush border of **microvilli.** This brush border may be very extensive, resulting in a tiny lumen that is often star-shaped. The cells are large and truncated and may be ciliated in some of the lower vertebrates. The nuclei are located near the bases of the cells and are large, spherical, and plate-staining.

Intermediate segment of the tubule. This is the most variable portion of the renal tubule of vertebrates (Figs. 10-12 and 10-19). A short, thin intermediate segment first appears in the kidneys of some bony fishes. In some amphibians it may be long and ciliated. It reaches its climax in mammals (and to some extent in birds) where it is thrown into a thin loop, the loop of Henle. This greatly elongated and narrow loop is adapted specifically for water recovery by means of a **countercurrent osmotic multiplier system.** Only animals that have the loop of Henle can excrete urine hypertonic to their blood.

The descending limb, a straight continuation of the proximal convoluted tubule, extends in a medullary ray directly into the medulla of the kidney. It is lined with simple cuboidal or squamous epithelium. The nuclei are flat and, in very thin segments, may bulge into the lumen. The general appearance is that of a large capillary. The ascending limb is a continuation of the descending limb. It has a thicker wall, lined with cuboidal epithelium. Its cells are more acidophilic than those of the thin segment. The thick segment is similar to the distal convoluted tubule into which it merges. As the tubule passes into the cortex, it becomes much thicker and contorted, forming the short distal convoluted tubule.

Distal convoluted tubule. At the junction of the ascending loop of Henle with the distal convoluting tubule is the macula densa, already noted. The distal convoluted tubule is usually considered as that part of the nephron from the macula densa to the collecting tubule. These segments are short, and in microscopic cross sections they differ from the proximal convoluted tubule in being fewer in number, smaller in diameter, and having low cuboidal cells without brush borders or striations but with distinguishable cell boundaries. Distal convoluted tubules have basement membranes around them.

Collecting tubules. Although these tubules are not considered as parts of the nephron, they have an intimate relationship with the nephrons and

play a part in renal physiology, especially in the countercurrent multiplier system of urine formation. Each distal convoluted tubule connects to a collecting tubule by way of a short junctional segment called an **arched collecting tubule.** Seven to ten of these arched tubules join the straight collecting tubule. Thus the nephrons far outnumber the collecting tubules.

The collecting tubules pass in a medullary ray down into the medulla, where they unite to form large ducts (the papillary ducts of Bellini). These large papillary ducts open onto the apex of a papilla of a calyx in the kidneys of higher vertebrates; in lower vertebrates they go directly into the wolffian duct. In cross section, the diameters of collecting tubules vary. In the smaller ones the lining cells are simple cuboidal or columnar; in the larger tubules the cells increase in height and are distinctly columnar (Fig. 10-19).

Ureter. The duct of the mammalian kidney is the ureter; its expanded proximal end, the renal pelvis (Fig. 10-20). The calyces that clasp the papillae are extensions of the renal pelvis. The ureters are paired tubes, located behind the peritoneum; they connect the renal pelvis with the urinary bladder, which is also retroperitoneal. All these parts have the same basic structure and, along with the urethra, serve as excretory passages to the exterior.

The lining mucosa of pelvis and ureter is made up of transitional epithelium supported by a lamina propria of dense fibroconnective tissue (Fig. 10-24). In the pelvis the epithelium consists of two to three layers of cells; in the ureter there are four to five layers. The cells vary from cuboidal to columnar, and the surface cells present a convex border to the lumen. A basal membrane is not visible under the light microscope.

A cross section of the ureter shows a stellate, irregular lumen. There seems to be no submucosa, but the muscularis is thick and consists of bundles of smooth muscle cells separated by connective tissue. The smooth muscle is arranged as an inner longitudinal and an outer circular coat (just the opposite of the arrangement found in the intestine). A third oblique coat of muscle may be present at the lower end of the ureter. In the pelvis the muscularis is arranged in a circular pattern around the papillae and may have a

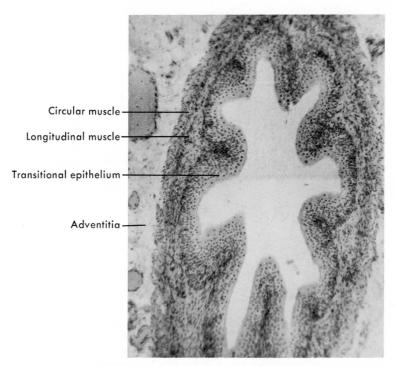

Circular muscle ———

Longitudinal muscle ———

Transitional epithelium ———

Adventitia ———

Fig. 10-24. Cross section of rabbit ureter. (×133.)

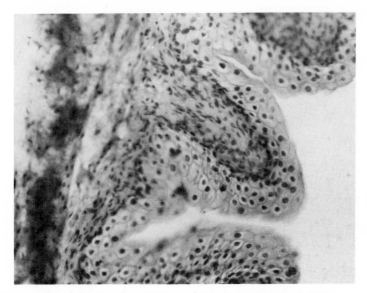

Fig. 10-25. Section of wall of rabbit urinary bladder. (×350.)

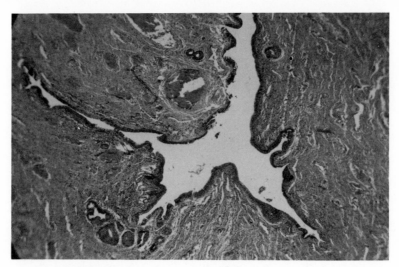

Fig. 10-26. Section through male urethra (human). Note blood vessels and numerous glands. (×8.)

sphincteric action on the papillae for squeezing urine from the ducts of Bellini. At the lower end of the ureter the circular muscle disappears, but the longitudinal layers continue down to the ureteric orifice.

A flap of mucous membrane prevents reflux from the bladder. External to the muscularis is a coat of fibroelastic connective tissue that is continuous with the surrounding connective tissue of the posterior abdominal wall along the ureter. The serosa is represented on the anterior surface of the renal pelvis and ureter, since this surface is covered by the peritoneum. The ureter has a supply of blood vessels and both motor and sensory nerve fibers.

Bladder. The wall of the bladder resembles that of the ureter except that its epithelium is transitional; it is made up of six or seven layers of cells when the bladder is empty, but fewer when it is distended with urine (Fig. 10-25). A few glands of mucus-secreting cells are present in the lamina propria, especially near the ureteric and internal urethral orifices. The lining mucous membrane is much folded in the contracted bladder. In the three-layered muscularis, the middle circular layer is developed as a sphincter around the internal urethral orifice and to some extent around the ureteric orifice. A fibroelastic connective tissue covers the outside (adventitia), with peritoneum covering the superior or anterior surface of the bladder.

Urethra. Accompanying their specialized external excretory and reproductive organs, mammals have an additional duct, the urethra (Fig. 10-26). The male urethra is much longer than the female one. It consists of three parts; the prostate urethra, the membranous urethra, and the cavernous urethra. The prostate urethra passes from the internal urethral orifice of the bladder through the prostate gland. Opening into it are the two ejaculatory ducts and ducts from the prostate gland. The membranous portion passes from the apex of the prostate to the penis. It extends through the urogenital diaphragm, from which it receives striated muscle forming the **external sphincter** of the bladder. The cavernous segment of the urethra extends through the penis, opening at the exterior **glans penis.**

The prostate urethra is lined with transitional epithelium; the other two portions have either stratified or pseudostratified columnar epithelium (Fig. 10-27). There is no distinct submucosa, but there is a muscularis of inner longitudinal coat and an outer circular coat. At the neck of the bladder, the circular coat thickens to form the **internal sphincter** of the bladder. A typical adventitia is lacking; the prostatic urethra is surrounded by the tissue of the prostate gland, the

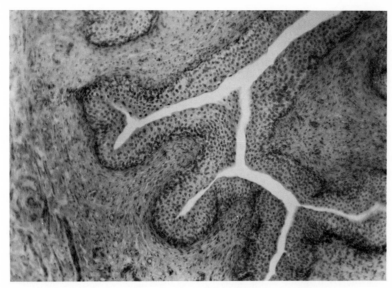

Fig. 10-27. Section through cavernous urethra of rabbit. Note stratified squamous epithelium. (×160.)

membranous by the muscular sphincter, and the cavernous urethra by erectile tissue and a dense outer sheath, the **tunica albuginea.** Pockets of mucous glands, the urethral glands of Littré, are present in the epithelium and lamina propria, especially in the cavernous urethra.

The female urethra is very short, extending from the bladder to its outlet in the **vestibule** (common cavity into which the urethra and vagina open). Its epithelial lining varies from transitional near the bladder to stratified squamous epithelium near its outlet. It has a striated muscle sphincter at its orifice, and some urethral glands. Specialized nerve endings **(pacinian corpuscles)** are also found.

Blood supply to kidneys. The renal artery and vein enter the kidney at the hilus and divide into a number of interlobar arteries and veins, which pass up between the pyramids to the boundary between cortex and medulla. There, as **arcuate** arteries and veins, they bend horizontally to form short arches. From each interlobar artery, a number of afferent glomerular arterioles arise, one to each renal corpuscle, as already described. Some branches of the interlobar artery also supply the capsule and underlying cortex.

After forming the capillary loops in the glo-

merulus, these capillaries recombine and leave (at the vascular pole) as the efferent glomerular arteriole. This artery then breaks up into a network of capillaries around the tubular portions of the nephron in the renal labyrinth and medullary rays. From glomerular efferent arterioles near the medulla, **arteriolae rectae spuriae** dip into and vascularize the medulla. The veins of the cortical region arise from the capillary networks and form the **stellate veins.** These combine to form the interlobular veins, which return along the corresponding artery to the arcuate veins. Blood is returned from the medulla by the **venae rectae** to the arcuate veins, which join the renal vein and then the post cava. There is no vascularization of the nephron tubules by a renal portal system as seen in the lower vertebrates.

References

Bentley, P. J. 1966. Adaptation of Amphibia to arid environment. Science **152:**619-623.

Edwards, J. G. 1928. Studies on aglomerular and glomerular kidneys. Am. J. Anat. **42:**75-107.

Elftman, H. 1950. The Sertoli cell cycle in the mouse. Anat. Rec. **106:**381-393.

Elias, H. 1957. The structure of the renal glomerulus. Anat. Rec. **127:**288-302.

Fänge, R., A. G. Johnels, and P. S. Enger. 1963. The

autonomic nervous system. Pages 124-136 *in* A. Brodal and R. Fänge, eds. The biology of *Myxine*. Universitetsförlaget, Oslo, Norway.

Gérard, P. 1933. Sur le système athrophagocytaire chez l'Ammocoete de *Lampetra planeri* (Bloch). Arch. Biol. Liège **44:**327-344.

Gérard, P. 1943. Sur le mésonéphros des Myxinides. Arch. Zool. Exp. Gen. Paris **83:**37-42.

Ham, A. W. 1969. Histology. J. B. Lippincott Co., Philadelphia.

Hickman, C. P., Jr., and B. F. Trump. 1969. The kidney. *In* W. S. Hoar and D. J. Randall, eds. Fish physiology, vol. 1. Academic Press, Inc., New York.

Holmgren, N. 1950. On the pronephros and the blood in *Myxine glutinosa*. Acta Zool. (Stockh.) **31:**233-348.

Jaffee, O. C. 1954. Morphogenesis of the pronephros of the leopard frog *(Rana pipiens)*. J. Morphol. **95:**109.

Jordan, H. E., and C. C. Speidel. 1930. Blood formation in Cyclostomes. Am. J. Anat. **46:**355-391.

Kempton, R. T. 1943. Studies on the elasmobranch kidney. I. The structure of renal tubule of the spiny dogfish *(Squalus acanthias)*. J. Morphol. **73:**247-264.

Kempton, R. T. 1962. Studies on the elasmobranch kidney. III. The kidney of the lesser electric ray, *Narcine brasiliensis*. J. Morphol. **111:**217-226.

Marshall, E. K., Jr., and H. W. Smith. 1930. The glomerular development of the vertebrate kidney in relation to habitat. Biol. Bull. **59:**135-152.

Porter, K. R., and M. A. Bonneville. 1963. An introduction to the fine structure of cells and tissues. Lea & Febiger, Philadelphia.

Rhodin, J. 1958. Anatomy of the kidney tubules. Int. Rev. Cytol. **7:**485-501.

Rugh, R. 1938. Structure and function of the peritoneal funnels of the frog, *Rana pipiens*. Proc. Soc. Exp. Biol. Med. **37:**717-721.

Schmidt-Nielsen, K., B. Jörgensen, and H. Osaki. 1958. Extrarenal salt excretion in birds. Am. J. Physiol. **193:**101-107.

Smith, H. 1953. The kidney. Sci. Am. **188:**40-49 (Jan.).

Smith, H. 1960. Evolution of chordate structure. Holt, Rinehart & Winston, Inc., New York.

Vilter, R. W. 1935. The morphology and development of the metanephric glomerulus in the pigeon. Anat. Rec. **63:**371-385.

Walker, A. M., and C. L. Hudson. 1937. The reabsorption of glucose from the renal tubule in Amphibia and the action of phlorizin upon it. Am. J. Physiol. **118:**140-143.

Willmer, E. N. 1960. Cytology and evolution. Academic Press, Inc., New York.

11 REPRODUCTIVE SYSTEMS

The chief function of the reproductive, or genital, system is the production of germ cells (gametes) as links in genetic succession. There are two kinds of gametes, spermatozoa and ova, or eggs. The sperm are produced by the male testis, and the eggs by the female ovary. Many variations are seen in the size, number, and shape of gametes among vertebrates. Sperm are flagellated, motile, and usually numerous; eggs are large, nonmotile cells, usually laden with yolk, and less numerous than spermatozoa.

Since sex is dimorphic in character (male and female), the mechanisms involved in the production of gametes in the two sexes are different. In both sexes organs have evolved for the formation of gametes, for their storage, for their transfer from male to female, and for mechanisms to ensure that they unite to form a union, the zygote. These are basic requirements that must be met, however varied the behavior of sexes may be with regard to mating, nesting, care of young, etc.

The organs are usually categorized as **primary sex organs** and **secondary,** or **accessory, sex organs** (Fig. 11-1). The primary sex organs are the gonads, **testes** in the male and **ovaries** in the female, which produce the gametes. The accessory sex organs are the ducts (and associated auxiliary glands) that carry the gametes from the gonads to the outside. In animals in which fertilization occurs internally, accessory sex organs open to the outside through specialized **external genitalia** —an **intromittent organ,** the **penis,** in the male and the **vulva** and related structures in the female.

EARLY DEVELOPMENT OF THE GENITAL SYSTEM

The gonads of vertebrates are made up of elements from two sources, structural and germinal elements. The structural elements arise from a pair of **genital ridges** that project into the coelom from the roof of the peritoneal cavity. The first indication of the development of the gonad is a thickening of the coelomic epithelium and a rapid proliferation of cells from its inner surface. In this position the ridges are definitely mesodermal in origin.

The germinal elements are the primordial germ cells. They are large cells that usually arise in extraembryonic endodermal regions. From their source of origin the primordial germ cells migrate to the genital ridges, either by amoeboid movement, transportation in the blood, or other means. Once in the genital ridges (the future gonads), the fate of the primordial germ cells is not clear (Everett, 1945). One theory rules out the possibility that the primordial germ cells are the progenitors of the definitive germ cells but suggests that they degenerate and are replaced by new cells from the gonads. Another theory states that the primordial cells are the sole progenitors. Still other theories propose that germ cells come from both sources. So the question is not closed. It is also possible that the development occurs differently in different groups of animals.

After the genital ridges are established, the gonads develop slowly and do not become functional until much later. The significant pattern of gonad development occurs at the surface of

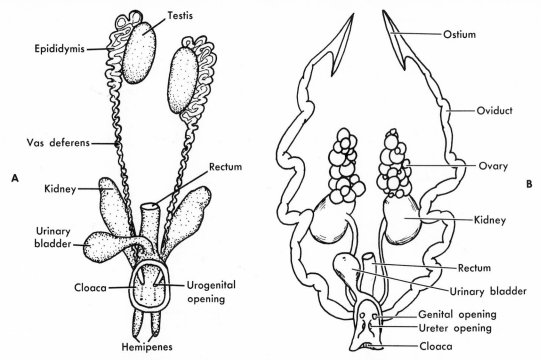

Fig. 11-1. A, Typical male urogenital organs. This type probably approaches that of the male lizard. The term typical must be used with reservations, since many lower vertebrates have modifications of this scheme. **B,** Typical female scheme of reproduction. Modifications of this plan are found among vertebrates.

the gonad where invagination and differentiation take place, according to sex and animal group. In the male testis the invagination forms hollow sex cords (seminiferous tubules), and the sperm are formed from their walls; in the female, the ovary forms a definite cortex (germinal epithelium) and inner medulla. Pockets of germinal epithelium, or secondary sex cords, pinch off and fall into the medulla to become hollow spheres, the graafian follicles. Ova develop from the walls of these follicles (Gruenwald, 1942).

Indifferent gonad. In the gonads of the young embryo, it is impossible to determine whether testes or ovaries will eventuate. By the stage when a cortex and a medulla region are recognizable, the gonad is covered by the simple cuboidal, **germinal epithelium,** which contains large, rounded cells, the presumptive germ cells. This epithelium thickens and sends cords of cells, the **primary sex cords,** into the connective tissue stroma of its interior. The primordial germ cells may accompany the sex cords, but as noted above,

their history is uncertain. Between the sex cords and the epithelium is the forerunner of the **tunica albuginea** (connective tissue). The inner ends of the primary sex cords are continuous with the **rete cords** of the rete blastema, a concentration of mesenchymal cells.

At this stage, a gonad is potentially hermaphroditic, or bisexual. If it is to be an ovary, then the cortical portion thickens and the sex cords break up into follicles; if the gonad becomes a testis, seminiferous tubules develop from the cords in the medulla, and the tunica albuginea becomes a fibrous connective tissue sheath around the testis. The rete cords hollow out and form the **rete tubules.**

The indifferent stage may last a long time before male or female sex is declared. In amphibians it may exist during the entire larval (tadpole) period. In the hagfish the indifferent stage may last through most of its life. Certain teleosts develop true hermaphroditic gonads and produce both eggs and sperm simultaneously throughout

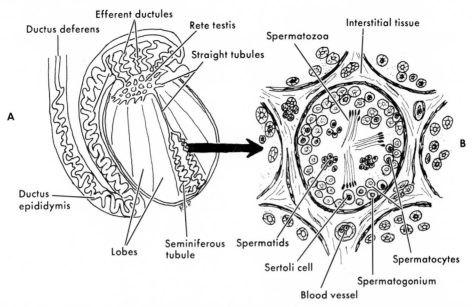

Fig. 11-2. Diagram of relationship between testis and seminiferous tubules. **A,** Longitudinal section of whole testis with one seminiferous tubule shown in one of the lobes. **B,** Cross section of seminiferous tubule showing stages of spermatogenesis.

their life. Many lines of evidence indicate that a multitude of factors are involved in the determination of sex. The chromosomal pattern is by no means the only influence. Sex reversals, abnormal hermaphroditism, and pseudohermaphroditism may ocur in most vertebrate species, although true or natural hermaphroditism is mostly restricted to certain cyclostomes and teleosts. In toads a small mass of cortical substance isolates early from the gential ridge and appears at the anterior end of each testis, as Bidder's organ. It is a remnant of the ovarian cortex, and if the testis is removed, the substance will develop into a functional ovary.

Duct system. After their formation, the gametes (or the embryo produced by fertilization) must be carried to the outside by some means. The vertebrate embryo begins its existence with a double set of sex duct primordia, parts of the excretory system for the prospective male and oviducts for the prospective female. Usually one set of these primordia in each individual will develop and the other degenerate. Males have made use of the kidney and kidney ducts for carrying their sex products; females have not. This bisexuality is the basis for the urogenital homologies

between the sexes, especially in the higher vertebrates.

MALE REPRODUCTIVE SYSTEM

The histologic pattern of testicular structure is generally similar in all adult vertebrates, although there are certainly some variations from the typical plan. The shape of the testis may be ovoid, cylindrical, tubular, or spherical. It is covered by coelomic epithelium underneath which is a stout fibrous capsule, the **tunica albuginea.** This capsule sends out a number of partitions, or septa, into the testis, dividing it into lobules (Fig. 11-2, *A*). Many of these septa reach to the capsule on the opposite side. The inwandering cords of cells become the seminiferous tubules. The rete cords at the ends of the seminiferous tubules form a network connecting the seminiferous tubules with several other tubules, the **ductuli efferentes.** These unite into a single tube, the convoluted **epididymis.**

Some teleosts have only one seminiferous tubule; there are more than 500 in man. Some tubules are short, but most are long and contorted. Each has a basement membrane and an inner germinal epithelium. The seminiferous tubules have two kinds of cells, spermatogenic cells and **Sertoli cells,** or nurse cells (Fig. 11-2, *B*). The Sertoli cells are rather tall, with their bases resting on the basement membrane. They are irregular in shape with small nuclei. The sex cells are massed between the Sertoli cells and frequently indent them.

Between the seminiferous tubules are spaces occupied by a connective tissue stroma containing blood vessels, nerves, and many types of cells including the interstitial **cells of Leydig** (Rasmussen, 1932). These large ovoid cells produce the important hormone testosterone.

Formation of spermatozoa

The main functions of the testis is the production of sex cells, the spermatozoa (Fig. 11-2, *B*), and the male sex hormone. Sperm formation is continuous in many vertebrates, including some mammals (e.g., man and bull), common fowl, some tropical birds, and some fishes. However, in the majority of animals, the process is seasonal.

Whatever the variations in timing, the mechanism of **spermatogenesis** is much the same among vertebrates (Figs. 11-2, *B,* and 11-3). There is first a proliferation of spermatogonia from the lining of the seminiferous tubule. The spermatogonia then go through periods of growth and

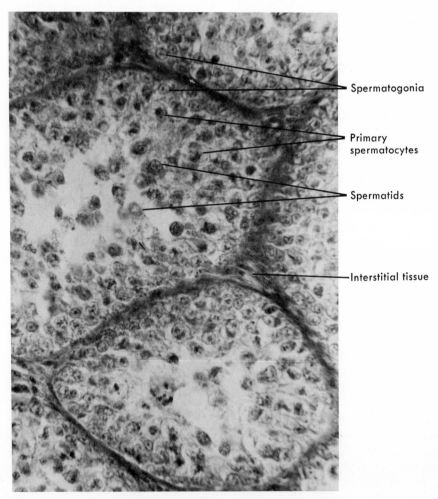

Fig. 11-3. Cross section of seminiferous tubules of pigeon, *Columba livia,* showing stages in development of the cells. Seminiferous tubules are separated by slight amounts of interstitial connective tissue. (×720.)

maturation, into primary and secondary spermatocytes and spermatids. The final stage is the transformation to spermatozoa. The identifying characteristics of each of the four types of spermatogenic cells is described below in the order of their development (refer to Figs. 11-2 and 11-3).

Spermatogonia. Spermatogonia form a layer of cells in touch with the basement membrane around the periphery of the seminiferous tubule. Spermatogonia are rather small cells with rounded nuclei, rich in chromatin. They increase in number by mitosis, and some undergo a period of growth and transform into primary spermatocytes. Other spermatogonia stay close to the basement membrane as undifferentiated cells (Duke, 1941).

Spermatocytes. The primary spermatocytes are the largest germ cells within the seminiferous tubule. They are found in the middle zone of the epithelium. Each cell is spherical and nuclei are often observed in some stage of division. Through meiosis, they give rise to the smaller secondary spermatocytes, each of which has half as many chromosomes as the primary. Cytoplasmic separation is not complete, and the two daughter secondary spermatocytes remain connected. They lie nearer the lumen. Since they divide quickly (by mitosis) to form spermatids, secondary spermatocytes are rarely seen in sections through the seminiferous tubules.

Spermatids. Spermatids lie close to the lumen of the seminiferous tubule and have spherical nuclei and scanty cytoplasm. They also have cytoplasmic continuity between clusters of spermatids. By a complicated series of transformations called spermiogenesis, the spermatids give rise to the mature spermatozoa.

Spermatozoa. **Spermiogenesis** is the process whereby the spermatid, now associated with the Sertoli cells, is gradually altered until it forms a complete spermatozoon. This process shows many variations in both the timing of the cycle and the type of mature spermatozoa produced among different vertebrates. However, the nuclear and cytoplasmic transformations are essentially alike in all of them.

In fishes and salamanders, for instance, the short seminiferous tubules (lobules) are arranged as compartments, with all the sperm cells at about the same stage of development in a given lobule.

In others forms, the seminiferous tubules are long, and a given section may show sperm cells in various stages of development. In animals with seasonal reproductive cycles, the testes between times of sexual activity may have an immature appearance when it is difficult to find sperm cells of any kind. In nonseasonal animals, the particular stage of development reached by the different generations of germ cells varies in different segments of the same tubule. This asynchronous condition may be due to the release of sperm in the region closest to the efferent tubules and may be influenced by the mating habits of the animal.

At the beginning of spermiogenesis the newly formed spermatid has a centrally located spherical nucleus, a Golgi apparatus, many mitochondria, and a pair of centrioles (Fig. 11-4, A). Spermiogenesis involves a striking differentiation of all these cellular structures (Clermont and Leblond, 1955). In the first phase (Golgi phase) several small granules appear in vesicles (from the Golgi apparatus), which fuse to form a large granule, the **acrosome,** within an acrosomal vesicle (Fig. 11-4, B). The membrane of the acrosomal vesicle then adheres to the outer layer of the nuclear membrane, and the acrosomic granule expands and forms a caplike covering over the nucleus (Fig. 11-4, C). In the meantime, the Golgi substance has migrated to the anterior end of the nucleus, where the fluid content of the acrosomal vesicle is resorbed. The acrosome now develops in size and shape and blends into the head cap characteristic for each species (Fawcett, 1958).

While the acrosome is forming at one pole of the nucleus, the centrioles move to the opposite side of the nucleus, and a slender flagellum grows out from one or both of them (Fig. 11-4, C). A sheath, the **caudal tube,** is laid around the submicroscopic filaments of the flagellum. The **distal centriole** migrates toward the cell surface and forms a ring, or annulus, around the longitudinal axial filaments. The mitochondria, at first scattered through the cytoplasm, migrate to the region between the proximal centriole and the annulus. They line up along the flagellum, and condense to form a spiral collarlike sheath that marks the **middle piece** of the future spermatozoon. The cytoplasmic surplus, not utilized in the

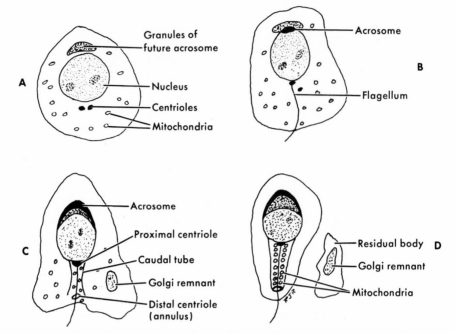

Fig. 11-4. Process of spermiogenesis. See text for description.

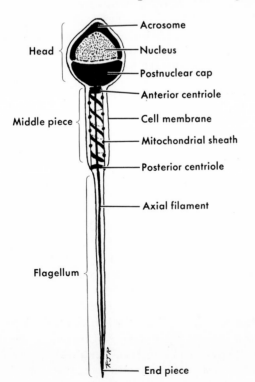

Fig. 11-5. Typical spermatozoon and its structure. Most vertebrate sperm have modifications of three basic divisions: head, middle piece, and tail (flagellum).

formation of the spermatozoon, is sloughed off as the **residual body** (Fig. 11-4, *D*), which is eventually absorbed by the cytoplasm of the Sertoli cell. A thin layer of cytoplasm covers the nucleus, the middle piece, and the tail piece (except the extreme end) of the spermatozoon. The tail piece has a structure similar to that of a cilium, with the 2 + 9 arrangement of longitudinal filaments.

When differentiation is completed, the spermatozoon (nonmotile at this stage) is released from the Sertoli cell and enters the lumen of the seminiferous tubule—pushed by the multiplication of other cells or by contraction of smooth muscle in the testicular capsule. Usually it is physiologically mature only at about the time it reaches the tail end of the epididymis. It then becomes actively motile by lashing its tail. The mature vertebrate spermatozoon consists of **head, middle piece,** and **tail** (Fig. 11-5). Each part contains important functional constituents for the role of the spermatozoon, which in fertilization contributes both nuclear and cytoplasmic components to the ovum. The head contains a condensed nucleus of the genetic material (DNA) and the acrosomic system. The acrosome may contain the enzyme

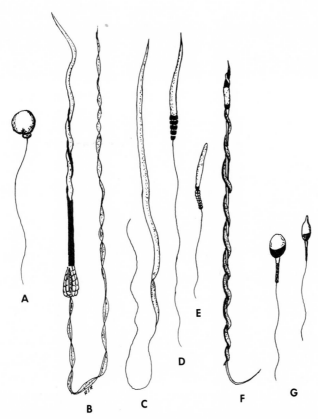

Fig. 11-6. Different types of sperm found among vertebrates. **A,** Teleost (salmon). **B,** Elasmobranch (skate). **C,** Anuran (frog). **D,** Reptile (limbless lizard). **E,** Bird (rail). **F,** Bird (robin). **G,** Mammal, front and side view (human). Variations of these are found in different species and groups. (Adapted from several sources.)

hyaluronidase, which facilitates the penetration of the spermatozoon through the follicle cells and the egg. The flagellar tail serves as an active propulsive organ, powered by the activity of the middle piece's mitochondrial sheath. In man the spermatozoon is about 60 μm long, of which the tail makes up about 50 μm. Sperm of other vertebrates may be much longer than this (Fig. 11-6).

FEMALE REPRODUCTIVE SYSTEM

As noted earlier, the indifferent gonad presents two structural regions, the outer cortex and the inner medulla. If the gonad is to become an **ovary,** the germinal epithelium of the cortex region thickens by proliferation of cells and elaborates **secondary sex cords.** The earlier-formed primary and rete cords of the indifferent gonad degenerate,

and the tunica albuginea becomes a rudimentary layer between the cortex and degenerate medulla. The primordial egg cells (Fig. 11-7) are found among the secondary cords, each enclosed by a single layer of epithelial cells provided by the cords. Such a structure is called a **primordial follicle;** the enclosing cells are the follicle, or nurse, cells.

The ovary shows many variations among the different groups of vertebrates. Although they arise from paired anlagen, ovaries may undergo many changes as they develop. In the cyclostomes, for example, the two ovaries fuse into one that surrounds the intestine. In the hagfishes the anterior part is an ovary and the posterior part a testis. Some fishes have only one ovary, as do most birds (the left), but many hawks have both. In alligators and a few lizards only the right ovary

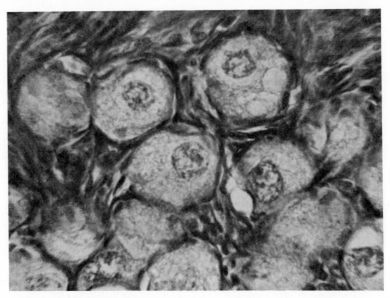

Fig. 11-7. Primordial egg cells in germinal epithelium of the cat. (×160.)

develops. Most mammals have paired ovaries, but the monotremes have only the left one.

Ovaries may be compact, elongated, or leaf-like. Within a single species they may vary in appearance and size according to the season of the year and to whether the sex cycle of the animal is seasonal or nonseasonal. The size and shape of the ovary may also be related to the size and number of eggs. The ovaries of teleosts are sac-shaped, blind at the anterior end. Some of them produce millions of small eggs. In contrast, the grape-bunch arrangement of the sharks produces relatively few, very large eggs. Amphibia have saccular ovaries. Each has a cortex and germinal epithelium, but the area of the medulla is represented by a lymph space. The same is true of reptiles and monotremes. The egg follicles of these forms have no cavities, but in mammals (except monotremes) many-layered follicles have cavities into which eggs project. In all cases the follicles move to the surface and discharge their eggs into the coelom.

Histologically, the vertebrate ovary is made up of **germ cells** and a variety of tissue cells (stroma) such as connective tissue cells, smooth muscle fibers, vascular elements, epithelial cells, and lymphatic cells. However, since the gross organization of these components varies, the ovaries and ova will be described separately, along with the process of egg development, for the different vertebrate groups.

CYCLOSTOMATA

In the cyclostomes, unlike the higher vertebrates, there is no combination of the gonad with the kidney to form either a vas deferens or an oviduct. The germ cells (Fig. 11-8) are freed into the peritoneal canals from which they enter either the urogenital sinus or abdominal pore, depending on the species, and then pass out into the water. This condition is somewhat like that in the annelids where the eggs or spermatozoa pass from the gonads into the abdominal cavity and out through genital pores unconnected with the excretory apparatus.

Petromyzon. The single testis of *Petromyzon* consists of numerous lobules. They are connected by a "mesorchium" of connective tissue to the great vessels and the kidneys and are covered by the serosa of the body cavity. Each lobule is a closed structure, its center filled with spermatozoa. Each lobule has its own connective-tissue membrana propria on which rests a row of "sperm mother cells," or spermatogonia. Between the lobules is connective tissue with numerous blood vessels. The ripe spermatozoa are about 100 μm long and show a head, middle piece, and tail.

The ovary is not as sharply divided into lobes

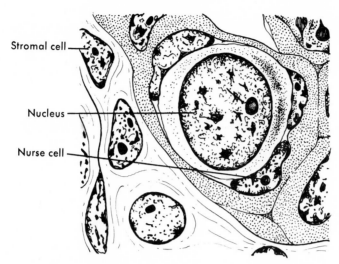

Stromal cell

Nucleus

Nurse cell

Fig. 11-8. Primordial sex cell of *Myxine* in the resting stage. Three nurse cells surround cell body. Large nucleus shows one conspicuous nucleolus and some chromatin bodies. Light area on left is stroma of testis. (Redrawn from Schreiner and Schreiner, 1905.)

as the testis. It has a mesovarium connecting it with the kidneys and great vessels, and it is surrounded by the serosa of the body cavity. Connective tissue and blood vessels penetrate to the interior of the organ and separate the individual eggs from one another with delicate septa. Each egg is surrounded by a double layer of follicle cells. Like a bird's egg, it has a blunt end and a more pointed end, or pole. The outer portion of cytoplasm of the egg itself generally appears radially striated. The inner portion contains great numbers of yolk nuclei which range in size from extremely minute structures up to some with a greatest diameter of 10 to 15 μm. The smallest ones are nearer the pointed pole, the largest nearer the blunt pole. In the region of the pointed pole is a nonvacuolated cytoplasm, or "pole plasm," whereas the rest of the cytoplasm is heavily vacuolated. Less vacuolated cytoplasm also forms a sort of cortex around the egg, and islands of it may be seen far in the interior.

Myxine. There are some differences between the genital systems of *Petromyzon* and its cousin *Myxine*. For example, the hagfish produces only 1 to 21 eggs, of relatively enormous size, compared to the 24,000 to 236,000 minute eggs formed by the ovary of the sea lamprey. But the most significant variation, as mentioned previously, is the presence of rudimentary hermaphro-

ditic features in *Myxine*. The discovery of a hermaphroditic species among the vertebrates was an event of signal importance in biology. Credit belongs largely to the great Norwegian scientist, and later Arctic explorer, Fridtjof Nansen. In two papers (1888 and 1889) he presented clear proof of hermaphroditism in the hagfish. Thus new significance was attached to all the evidences of hermaphroditism seen in the developmental stages of vertebrates, as well as to the occasional retention of hermaphroditic features in the adults of many species, including man himself.

The posterior testicular portion of the hermaphroditic gonad consists of many minute lobules, up to about 0.5 mm in diameter. It is surrounded by a single-layered epithelium, a part of the peritoneum. The structure of these lobules is very simple, indeed more simple than that of testicular lobules in many invertebrate animals. Schreiner (1955) reported the presence of some wandering cells in the interlobular connective tissue, but he found no "interstitial cells" comparable to the endocrine cells found in testes of higher vertebrates.

An outer theca around each follicle consists of two portions, outer fibrous and inner epithelial, the cells of the inner layer being a simple very flattened epithelium. Within the lobule there are two kinds of cells, the vegetative cells of Sertoli

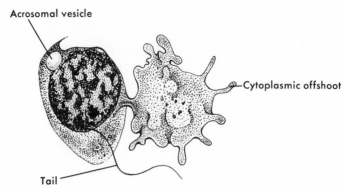

Fig. 11-9. Spermatid of *Myxine*. In upper left part of cell, the acrosomal vesicle has appeared. Developing tail is seen. Large cytoplasmic offshoot is a peculiar feature of spermiogenesis in this cyclostome. (Redrawn from Schreiner and Schreiner, 1908.)

and the germinating cells in the various stages of spermatozoa formation (Fig. 11-9).

Proliferation of oocytes takes place in the germinal epithelium. From there they appear to enter the mesenchymal tissue of the interior of the ovary, taking with them cells from the surface epithelium, which become the follicular cells. As development of the ovary proceeds, the first-formed eggs remain close to the insertion of the mesovarium, with the nucleus and animal pole facing it. There is thus a row of these oldest and largest eggs parallel to the alimentary tract at the boundary of the mesovarium. A mature follicle has four layers, in general similar to those found in higher vertebrates: an inner simple cuboidal epithelium, two layers of connective tissue, and an outer layer of peritoneal epithelium. As the egg becomes larger, chiefly by accumulation of yolk, the follicular epithelium becomes stratified and increases in thickness. The mature egg of *Myxine* has a tough, horny shell and tufts of anchoring filaments at each pole (Fig. 11-10).

CHONDRICHTHYES

Male reproductive system. The genital system of sharks is specialized for internal fertilization, a process unusual among aquatic vertebrates except for mammals that have left land existence and gone back to the sea. Internal fertilization requires a male copulatory apparatus. In the shark the pelvic fins of the male are fused on the ventral surface, and a rod (the **clasper**) projects backward from each of them. The clasper is really

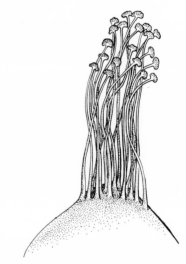

Fig. 11-10. Vegetative pole of an egg of *Myxine*. Anchoring filaments are seen. (Redrawn from Lyngnes, 1930.)

a rolled up extension of the fin. During copulation the claspers are inserted into the cloaca of the female, and the semen is injected or flushed into the female through the grooves of the claspers by means of a muscular siphon that leads into the clasper tube.

The testes are paired, and each hangs from the body wall by a mesorchium. A section through the testis of a shark reveals many lobules divided from each other by connective tissue cells with small oval nuclei (Fig. 11-11). During maturation these lobules become packed with potential germ cells in various stages of development:

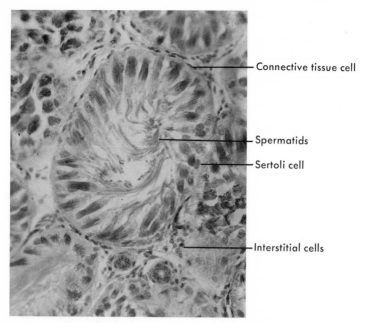

Connective tissue cell

Spermatids

Sertoli cell

Interstitial cells

Fig. 11-11. Section through seminiferous tubules of shark testis. Central lobule has advanced spermatids. All developing sperm cells in any particular lobule are at about same stage of development. (×340.)

spermatogonia, spermatocytes, spermatids, and spermatozoa. The center lobule in Fig. 11-11 has advanced spermatids. Note that all the developing germ cells in a given lobule tend to be in about the same stage of development.

From each testis delicate **vasa efferentia** carry sperm to the wolffian (opisthonephric) duct at the anterior, or reproductive portion, of the kidney. The vasa efferentia are lined with simple cuboidal epithelium encircled by a wall of connective tissue and smooth muscle. The wolffian duct (or vas deferens) is greatly coiled and provided with glands for aggregating the sperm into spermatophores. In all sharks and many amphibians the mesonephric duct is exclusively a genital duct; the urine from the caudal part of the opisthonephros is conveyed by one or more accessory ducts. Posteriorly, the duct widens to form a seminal vesicle, which opens into a urogenital sinus. The sinus opens to the outside through the tip of a urogenital papilla in the cloaca. These structures have already been described as part of the urinary system. In fact, the reproductive and excretory systems in elasmobranchs and most higher forms (Fig. 11-12) are so closely asso-

ciated that they are often considered together as the urogenital system.

Female reproductive system. The paired ovaries of the female are supported by a mesorchium. The oviducts (müllerian ducts) are large and are held in place by mesenteries. They have no direct contact with the ovaries. The ducts unite anteriorly forming an **ostium,** a mouth with long fimbriae that sweep the eggs into the müllerian ducts. Dorsal to the ovaries, in mature forms, the oviducts enlarge to form the shell gland but become narrow again before they widen to form the large uteri. The uteri open into the cloaca on each side of the urinary papilla.

A section through the shark ovary reveals follicles in different stages of development (Fig. 11-13). Shark eggs are large, and those about to break out of the ovary are located near the periphery. At ovulation the eggs weigh about 50 grams and are more than 1¼ inches in diameter. When the follicle ruptures at ovulation, a corpus luteum forms from the follicular granulosa at the site of the rupture. Degenerating (atretic) follicles are also common in the ovaries of dogfish.

The fertilized eggs develop in the uterus. The

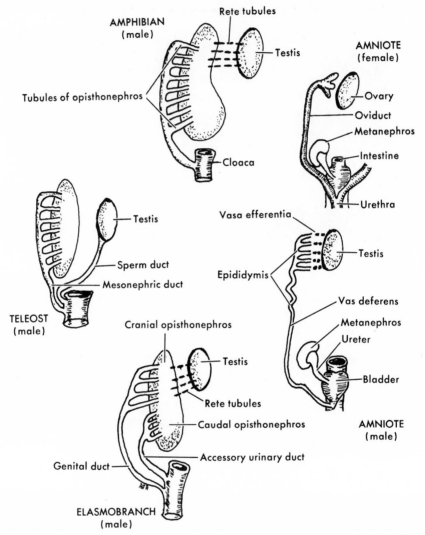

Fig. 11-12. Scheme of urogenital systems of vertebrates. Note evolution of functional ducts of kidneys. (Based on several sources.)

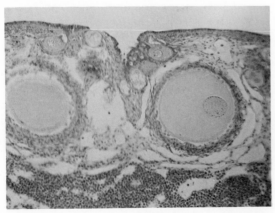

Fig. 11-13. Section through shark ovary showing two large follicles with eggs and several smaller follicles. (×120.)

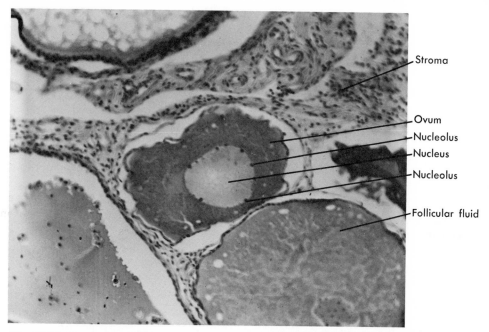

Fig. 11-14. Developing follicles in ovary of *Salmo gairdneri*. Follicle at center of field contains large ovum. Note many nucleoli against the inside of the nuclear membrane. (×125.)

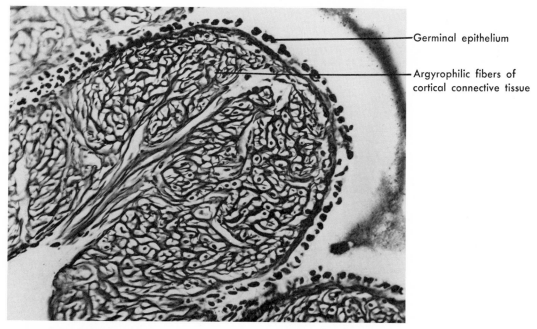

Fig. 11-15. Portion of stroma of trout ovary. Complex architecture of reticular fibers is seen. (Wild's silver stain; ×250.)

is ovoviviparous and main-
the uterus. There is no di-
~en the embryo and the
rks (e.g., *Mustelus,* the
~lk sac sends out villi that
~s forming a sort of pla-
~~ viviparous.

OSTEICHTHYES

The testes are elongated and somewhat lobulated. Their ducts, often an intense white, fuse to
form a single canal that opens between the rectum
and the urinary aperture. These male gonoducts,
carrying the spermatozoa, are formed in continuity with the testes and seem to have no connection whatever with the kidneys. The arrangement is very different from that seen in the shark
and the higher vertebrates. Therefore it is probably better **not** to use the terms "wolffian" and
"müllerian" for the male and female gonoducts
of bony fishes, since the homologies are, at best,
doubtful.

The ovaries correspond closely with the testes
in position and arrangement of the ducts. Each
ovary is an elongated sac, blind anteriorly. The
ova arise on the inner walls of the sac (Figs. 11-14
and 11-15) and are shed into the **ovisacs**—spaces
formed by the folding of the ovary so as to enclose a special pocket of the coelom. These ovisacs are prolonged posteriorly as funnellike tubes
to the exterior. The paired tubes meet and empty
by a single median pore between the anus and the
urinary papilla. There are many variations, but
in none do the posterior ducts have any relation
to the kidneys or urinary ducts. Note that in most
bony fishes the egg and sperm paths are comparable, which is not the case with other vertebrates.

In a few bony fishes (e.g., Salmonidae, the
family to which the brook trout belongs) the
gonoducts develop in connection with the gonads.
However, they undergo degeneration, and in the
adult they exist only as short funnels. The gametes are set free into the peritoneal cavity, enter
those peritoneal funnels, and leave by way of
genital pores behind the anus. This situation is
found in no other vertebrates, except the cyclostomes, and is reminiscent of invertebrate reproductive systems.

AMPHIBIA

Male reproductive system. The testes of the
frog are paired, ovoid bodies, yellowish white in
color, that lie alongside or ventral to the kidneys
near their anterior lateral border. A double membrane, the mesorchium, suspends the testes from
the dorsal side of the body cavity, where the
membrane becomes continuous with the general
coelomic lining. Each testis is a mass of seminiferous tubules with blood vessels, nerves, and
some connective tissue binding the tubules together. Around the whole testis is a connective
tissue membrane, the tunica albuginea, which is
covered by the peritoneum. The seminiferous tubules connect with the vasa efferentia, the efferent
ductules that pass to the anterior end of the kidney. These tubules join the uriniferous tubules in
the kidney and then pass to the wolffian (opisthonephric) duct and to the cloaca. In some forms
the lower ends of the wolffian ducts may be dilated as seminal vesicles for temporary storage
of sperm.

The seminiferous tubules consist of nonciliated,
cuboidal epithelial cells on which are found large
Sertoli cells and germ cells in various stages of development. In tubules of hibernating frogs (Fig.
11-16) it is difficult to find any maturation stages
between the primary spermatogonia and mature
sperm. But in a normal autumn frog, the tubules
are lined with mature sperm cells that have their
heads embedded in Sertoli cells and their tails
lying free in the lumen of the tubule (Burgos,
1955) (Fig. 11-17). Spermatocytes and spermatogonia may lie between the Sertoli cells.

Sertoli cells may be distinguished by their large
oval nuclei and prominent nucleoli. The seminiferous tubules are usually radially disposed,
converging toward the center of the testis where
they open into collecting tubules. The blind ends
of the tubules lie close to the tunica albuginea,
which makes incomplete radial compartments in
the testis.

Surrounding the seminiferous tubules are
masses of ovoid, granular, interstitial cells. This
tissue is especially abundant during hibernation
(Fig. 11-16) but becomes less so during the active breeding season when mature sperm are more
numerous. The characteristic interstitial cell is
the Leydig cell, which is probably the source of

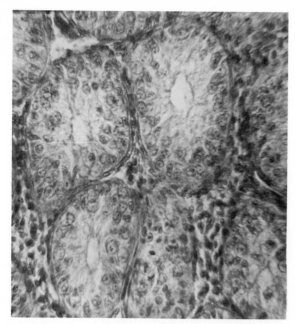

Fig. 11-16. Section through testis of hibernating frog, showing early stages of spermatogenesis in several seminiferous tubules. Large cells are primary spermatocytes. (×770.)

steroid hormones. Some smooth muscles are also found in the stroma of the testis.

Female reproductive system. The ovaries of the frog are saclike and lobulated. The walls are thin, and the appearance of the ovary depends on its seasonal activity. In the breeding season it is packed with developing eggs and is quite large; when the eggs are discharged, the ovary becomes reduced to a small wrinkled structure. Externally, the ovary is covered by a double peritoneal membrane, the mesovarium, which suspends the ovary from the dorsal body wall. The ovaries may be considered as mere folds of the peritoneum; they do not possess the solid stroma characteristic of higher vertebrates. Blood vessels and nerves run between the two membranes of the mesovarium.

Each developing egg (oocyte) is surrounded by a layer of cuboidal epithelial cells (follicle cells) that add yolk to the cytoplasm (Fig. 11-18). Outside the ova is the **theca interna** of connective tissue and smooth muscle. When the follicle ruptures, the mature egg breaks out of the **theca externa** and passes into the body cavity. From

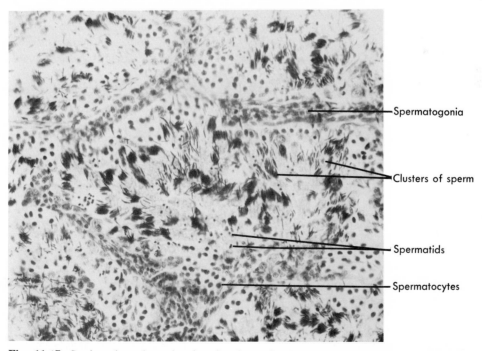

Spermatogonia

Clusters of sperm

Spermatids

Spermatocytes

Fig. 11-17. Section through testis of active frog, showing clusters of sperm released from Sertoli cells. Some Sertoli cells are also found among the sperm. Cross section of one whole seminiferous tubule and portions of others are shown. (×350.)

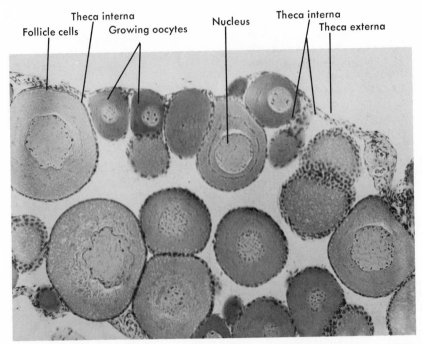

Fig. 11-18. Section through portion of ovary showing ova in various stages of development. Note that each maturing ovum is surrounded by layer of cuboidal epithelial cells (follicle cells) and by a delicate theca interna of connecting tissue with some smooth muscle cells. (×62.)

the body cavity the eggs are literally swept into the paired ostia, mouths of the oviducts, by the ciliary action of the mesothelial cells at the mouths. A section of an ovary shows eggs in various stages of development and many degenerating (atretic) eggs, as well. The eggs ripen under the influence of a hormone from the anterior lobe of the pituitary, the release of which is controlled by seasonal environmental factors.

The walls of the oviducts are relatively thick and highly glandular. The highly convoluted tubule is covered by peritoneum, beneath which is a connective tissue layer (adventitia). Internal to this layer is a circular smooth muscle layer, and lining the lumen of the oviduct is the glandular mucosa. The internal epithelium is made up of ciliated columnar cells with scattered goblet cells. Albumen glands, elongated and tubular, are found all along the length of the oviduct. These glands secrete the albumen that surrounds the egg. When this albumen takes up water, it forms the gelatinous layers that surround the eggs. The posterior ends of the oviducts are greatly dilated to form the ovisacs, in which eggs are stored until they are laid. These ovisacs lack glands and are lined by ciliated columnar epithelium.

A pair of fat bodies (corpora adiposa), yellowish tufts of flattened processes, are attached to the body wall or anterior end of each gonad (both testes or ovaries) (Fig. 11-19). They serve as storehouses of nutriment and undergo seasonal changes. Their development is closely associated with that of the reproductive organs.

REPTILIA

Male reproductive system. Reptiles, birds, and mammals possess a **metanephric** kidney that is involved solely with excretion; the **mesonephric** (wolffian) duct is restricted to serving as a **vas deferens** for the transport of sperm. (In the female both the mesonephros and the wolffian duct disappear.) The testes are two whitish, oval bodies just posterior to the right lobe of the liver. Each testis is held in place by a fold of the peritoneum, the mesorchium. The lizard testes undergo sea-

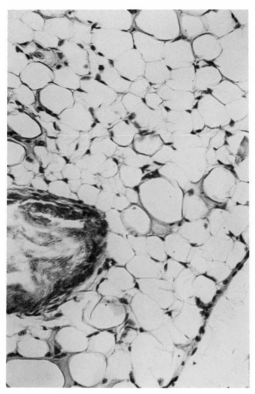

Fig. 11-19. Adipose tissue of frog, such as that found in fat bodies. Note that cells are distended to spherical shapes by fat deposits; nucleus and cytoplasm are shoved to periphery of cells. (×250.)

sonal variations. They are most active during the breeding season (usually early spring), when the seminiferous tubules are mostly filled with actively multiplying germ cells in all stages of development from spermatogonia to mature sperm (Fig. 11-20). Between breeding seasons, the seminiferous tubules are mostly cords of undifferentiated germinal cells.

Each is associated with an **epididymis,** which is a coiled sperm-carrying duct connected at one end with the seminiferous tubules of the testis and at the other end with the vas deferens. The cells of the epididymis secrete an acidophilic substance that is added to the semen. Acidophilic granules are prominent in the ciliated columnar lining of the reptilian epididymis. The vas deferens passes posteriorly to the cloaca, into which it empties along with the ureter from the kidney (metanephros). The vas deferens is lined with ciliated columnar epithelium with a basement membrane and lamina propria. The surrounding wall contains connective tissue and circularly and longitudinally arranged smooth muscle fibers.

Copulatory organs are characteristic of reptiles and mammals. Male snakes and lizards have paired saclike **hemipenes** lying in pockets beside the cloaca, which are everted for insertion into the female. Crocodilians and turtles have unpaired cloacal penes. A rudimentary penis, the **clitoris** may be present in the female. The surface of the reptilian penis bears a groove for the passage of sperm, as is also true in the lower mammals.

Female reproductive system. The paired ovaries are irregularly oval bodies with small elevations over their surfaces. These elevations mark the position of the underlying eggs, and the irregularity is caused by the presence of eggs of varying sizes. At an early stage of development each egg is surrounded by follicle cells; the more mature eggs have several layers around them. The increase in egg size is due mainly to the increasing amount of yolk. In addition, the egg is provided with a shell and enclosing membranes (amnion, chorion, and allantois). These membranes partly maintain a watery environment for the embryo and partly afford mechanisms for respiration and absorption of the yolk. Such an egg is called the amniotic egg and is characteristic of all reptiles, birds, and mammals, the amniotes.

A theca of connective tissue surrounds the follicle cells of the ovum. When the ovum is mature, this theca and the connective tissue of the ovary rupture and release the egg into the body cavity, where it is picked up by the oviduct. After the egg ovulates, the follicle regresses, with the formation of a corpus luteum (Gorbman, 1959). The egg is fertilized in the oviduct before the shell is added. Most reptiles are oviparous, but in some the shell glands at the posterior end of the paired müllerian oviducts serve as uteri. These forms are ovoviviparous; the egg remains in the uterus until it hatches.

AVES

Male reproductive system. The avian testes enlarge during the breeding season and then regress or shrink to pinhead size after breeding. The small

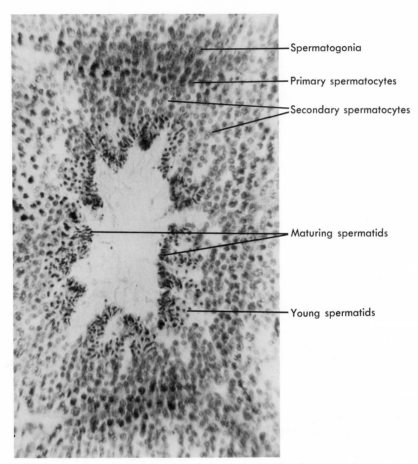

Fig. 11-20. Section through seminiferous tubule of *Anolis,* showing stages in spermatogenesis. Basement membrane or interstitial cells between adjacent seminiferous tubules are not shown. This section shows active phase of testis, the time of which varies with different species. (×640.)

size of the testes in some birds during the dormant season often makes sex determination very difficult.

The testes are two whitish organs near the cephalic end of the kidneys and ventral to them. They are composed of seminiferous tubules and interstitial tissue. At maturity each tubule consists of a multilayered epithelium in which the various stages of spermatogenesis are observable (Fig. 11-3), as is true in most forms. From the wall of the seminiferous tubule to the lumen are found spermatogonia, primary spermatocytes, secondary spermatocytes, spermatids, the nutritive Sertoli cells to which spermatids are attached, and the immature spermatozoa. Between the tubules is the connective tissue stroma composed of the interstitial cells of Leydig and vascular tissue.

The small epididymis and the much-coiled vasa deferentia conduct the sperm from the testis to the cloaca. Spermatozoa mature in the epididymis. In general, the mature spermatozoon has a long headpiece with pointed acrosome, a short midpiece, and a long tail (Fig. 11-6, *E*).

Copulation occurs when the cloacas of the male and female are brought close together and sperm is transferred to the female cloaca. An intromittent organ (penis) is found in ducks, ostriches, and some others (Benoit, 1950). It is a special erectile structure on the cloacal wall.

Female reproductive system. Potentially there are two ovaries, but in most birds only the left one develops. If this one is lost from disease or other causes, the right rudiment may develop into an ovary, or sometimes into a testis or ovotestis.

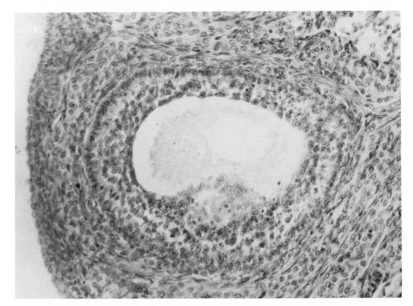

Fig. 11-21. An almost mature follicle in ovary of pigeon. (×43.)

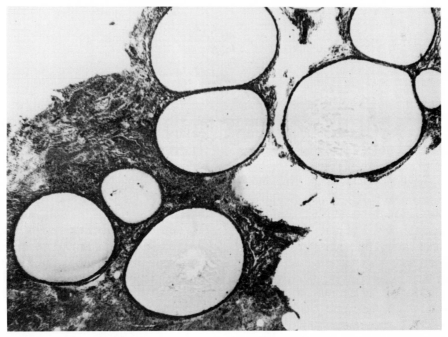

Fig. 11-22. Section through pigeon ovary, showing follicles in different stages of development. (×100.)

Hawks and owls normally have two functional ovaries. Like the testes, the ovaries greatly enlarge during the breeding season and regress dramatically after breeding.

The ovary consists of an inner medulla and an outer cortex, made up of follicles with ova. The avian ovary differs anatomically from that of other animals. The large follicles are not contained in the stroma but are borne on follicular stalks. There is no follicular fluid; there are a vitelline membrane, a membrane granulosa, and a theca folliculi around the egg (Fig. 11-21). Before maturity the ovary contains a mass of small immature ova, but before ovulation the individual follicles may reach a diameter of many millimeters, depending on the size of the bird (Fig. 11-22). The ovarian follicle is highly vascular. Its blood supply comes from the ovarian artery, which may branch directly from the dorsal aorta or from the left renolumbar artery.

If the ovary is single, the oviduct is single. It is a long, convoluted tube through which the egg is passed and in which the albumen, shell membranes, and shell are formed. The oviduct can be divided into five distinct areas. The most anterior end is the funnel-shaped **infundibulum,** which engulfs the ovum when it is released into the body cavity. The infundibulum is an arrangement of membranous fimbria around the ostium.

The second area is the **magnum,** which represents the longest single region of the oviduct and secretes most of the albumen (protein reserve) of the egg. Histologically, the magnum is very granular and contains both tubular and unicellular glands (goblet cells). The tubular glands secrete the thin albumen, and the goblet cells produce mucin. Forming the third region is the **isthmus** with few, small glandular folds. The inner and outer shell membranes are produced here.

The **uterus** is the fourth pouchlike division and has thick muscular walls. Little is known about the function of its tubular and unicellular glands, but they probably play a part in shell formation. In those birds that lay colored or speckled eggs, pigment is added to the shell in the uterus (Warren and Conrad, 1942). The most posterior area of the oviduct is the **vagina,** a more or less straight segment that opens into the urodeum of the cloaca and plays no part in the formation of the egg.

The wall of the oviduct, in general, is made up of an inner mucosa lining of ciliated columnar cells and the aforementioned glands, a submucosa of fibroelastic connective tissue much folded together with the mucosa, and a muscularis of circular and longitudinal smooth muscle. The whole may be surrounded by an adventitia coat of connective tissue and, at least partly, by a mesentery.

MAMMALIA

Male reproductive system. In mammals, the male genital system consists of the testes, where sperm are formed; the excretory ducts through which spermatozoa are discharged; auxiliary glands, which furnish part of the fluid secretions to the semen; and the penis, the intromittent organ used to inject spermatozoa into the female.

TESTES. In most mammals, the testes descend from the body cavity into a scrotum, a special sac formed from the coelom and integument. The paired cavities (one for each testis) are serosal extensions of the peritoneum. A serosal sac is called a tunica vaginalis. Its parietal layer lines the scrotal wall, and its visceral layer covers most of each testis and epididymis. In the wall of the sacs are the dartos muscles. The communicating passage between the coelom and scrotal sac is the inguinal canal. In some (e.g., the rabbit) this canal is very wide, and the testes can easily be pushed up into the body cavity. The lining of the scrotum is continuous with that of the abdominal cavity, and its muscles are also continuous with those of the abdominal wall.

Each testis is connected to the base of the scrotum by the **gubernaculum cord.** The testes are compound tubular glands with both exocrine and endocrine functions. They are enclosed by a tough capsule, the tunica albuginea, which has an innermost layer, the **tunica vasculosa.** Along the posterior margin of the testis the albuginea is thickened into the **mediastinum testis.** The mediastinum sends fibrous septa to the capsule, dividing the testis into compartments, or lobules, in which the seminiferous tubules are located.

The walls of the seminiferous tubules produce the spermatozoa. The process of spermatogenesis and the basic structure of the spermatozoon are similar among the vertebrates and have been described earlier in this chapter. Throughout spermatogenesis, and especially during spermiogenesis, the sex cells are associated with the Sertoli cells—tall, pillarlike supporting and nutritive cells. They usually have a cluster of sperm heads seemingly buried in their substance. However, electron microscope study shows that the heads are not truly within the cytoplasm of the Sertoli cell but that the cytoplasm is wrapped about them in a most intricate manner (Figs. 11-11 and 11-17). The Sertoli cells are relatively few, and they are spaced at regular intervals along the tubules. The ovoid nucleus with a prominent nucleolus has its long axis directed radially. These cells apparently serve a nutritive function during spermiogenesis and may produce a hormone that stimulates the next cycle of spermatogenesis (Lacy, 1967).

The interstitial tissue lying between the seminiferous tubules contains the **cells of Leydig** that produce the male hormone. These cells are usually found in compact groups. They are large eosinophilic cells with nuclei that contain coarse chromatin granules and a distinct nucleolus. A loose connective tissue stroma, made up of fibroelastic fibers with blood vessels and nerves, is also found in the interstitial tissue.

MALE DUCTS. In the amniotes (reptiles, birds, and mammals) the kidney is a **metanephros** drained by a true ureter. During embryonic life of males, the transient **mesonephros** has its anterior end linked to the seminiferous tubules of the testes by **vasa efferentia.** This relationship is very similar to the anterior or cranial opisthonephric condition found in sharks and amphibians. The mesonephric wolffian duct thus becomes the sperm-conducting duct, the **vas deferens.**

The seminiferous tubules end abruptly at the mediastinum by joining the **straight tubules,** which are lined with a single layer of Sertoli cells. The straight tubules enter anastomosing canals that form a network of irregular channels, the rete testis. They have a lining of low columnar or cuboidal epithelium, in which some of the cells may bear a single flagellum. The rete testis is connected to about a dozen vasa efferentia. These efferent ducts are lined with alternating regions of cuboidal and columnar cells, with pseudostratified regions in which a few basal cells are found. There is a basement membrane in this region. The tubules are surrounded by connective tissue and a thin layer of smooth muscle. Some of the taller cells bear cilia, the beating of which aids the as yet nonmotile spermatozoa on their way and into the epididymis.

The long, much-coiled ductus epididymis is lined with pseudostratified epithelium that rests on a basement membrane (Fig. 11-23). This duct, stretched along the posterior side of the testis, receives all the vasa efferentia. Its wall contains smooth muscle, and the free surface of the epithelium contains long, nonmotile **stereocilia.** These stereocilia do not contain the tubular filaments of true cilia but are composed of groups of long, slender microvilli-like structures. Spermatozoa are apparently motile by the time they reach the epididymis, since they get no help from these stereocilia, which may have a secretory function.

Contraction of the smooth muscle during ejaculation forces the contents of the epididymis into the ductus deferens, or vas deferens (Fig. 11-24). This duct passes through the inguinal canal and runs behind the peritoneum toward the urethra, where it ends in the dilated ampulla. The vas deferens has a thick muscular wall and a lamina propia of connective tissue. The lining epithelium is mostly pseudostratified, and some of the tall cells may bear stereocilia. There is a delicate basement membrane. Near the prostate gland the duct enlarges into the ampulla where it is joined by the seminal vesicles. In the ampulla the longitudinal folds of mucosa are exaggerated to form depressions and diverticula. The ampulla and the short duct of the seminal vesicle unite to form the ejaculatory duct. The epithelium of the latter is simple columnar or pseudostratified. Its wall is the dense fibrous connective tissue in the stroma of the prostate gland. It opens into the urethra by minute pores.

The only part of the male urethra that corresponds to the female urethra lies between the bladder and the openings of the paired ejaculatory ducts. The rest of the male urethra transports both semen and urine, as already noted. The male urethra consists of three parts: the **prostatic,** into which the prostate gland and the paired ejaculatory ducts open; the **membranous** part, lying between the prostate and the penis; and the **cavern-**

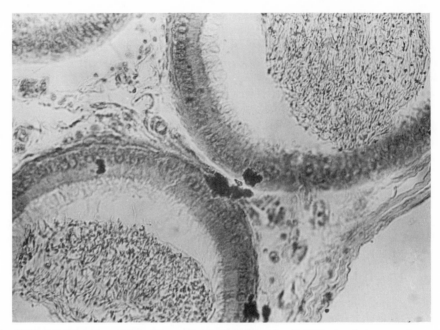

Fig. 11-23. Partial cross sections of two ducts of epididymis of rat. Both tubules contain aggregated masses of sperm, which acquire the power of motility and fertilizability as they pass through the ducts. Note stereocilia on lining of epithelium. (×200.)

ous, which runs through the penis and receives at its base the ducts of the paired bulbourethral glands. The histology of the urethra was described with the urinary system.

ACCESSORY SEX GLANDS. The functions of these glands include aiding in the process of fertilization, protecting the sperm, adding their contents to the seminal fluid, and coating the sperm duct against the harmful effects of an acid urine. The glands are typically mammalian structures, all secretory, represented by the seminal vesicles, the prostate gland, and the bulbourethral (Cowper's) gland.

Each seminal vesicle is a long folded glandular sac (Fig. 11-25), an outgrowth of the vas deferens at the ampulla. It secretes a sticky, mucoid, weakly alkaline fluid. Its epithelium may be partly simple columnar and partly pseudostratified. The seminal vesicles are surrounded by both circular and longitudinal layers of smooth muscle. In mammals these structures are purely secretory in function. They do not store sperm as in the lower forms.

The prostate gland surrounds the urethra at its origin from the bladder. This compound gland is made up of several tubuloalveolar glands grouped into four or five lobes (Fig 11-26). It is surrounded by a fibroelastic capsule; the glandular conponents are embedded in a dense stroma, which is continuous with the capsule and contains numerous smooth muscle strands. The secretory alveoli and tubules branch frequently and are irregular in size and form. The epithelium of the gland is folded and consists of simple cuboidal or columnar cells. The secretion is a thin, milky liquid. The gland frequently contains spherical or oval bodies called the prostatic concretions **(corpora amylacea),** which are condensations of secretion. The prostate gland pours its secretions into the urethra via numerous ducts.

The paired bulbourethral, or Cowper's, glands (Fig. 11-27) are tubuloalveolar glands with septa of connective tissue that separate each gland into lobules. They secrete a mucuslike substance that is added to the semen. These glands open into the urethra at the base of the penis.

The **glands of Littré** are slime-producing glands that open into the urethra. They are present in most parts of the urethra but are best developed in the cavernous urethra. They are branching

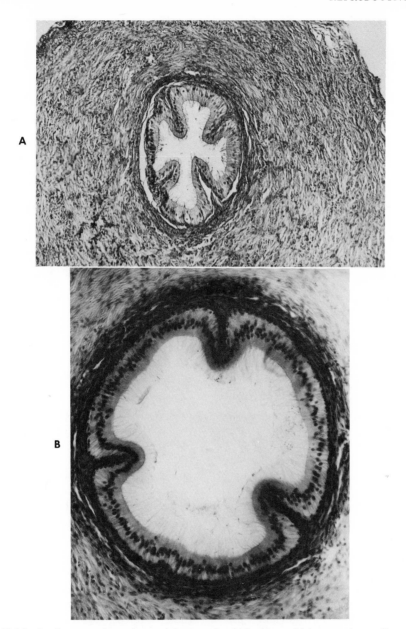

Fig. 11-24. A, Cross section of vas deferens of rabbit. Note thick muscular walls and epithelium-bearing stereocilia. **B,** Cross section of vas deferens of rat showing the arrangement of mucosa lining. Note stereocilia on epithelial cells. (**A,** ×100; **B,** ×215.)

mucous tubules that extend into the lamina propria or beyond. Their ducts contain intraepithelial pockets of mucous cells. They are lined with the same epithelium as the surface of the urethral mucous membrane (transitional, stratified columnar, etc.).

The sexual discharge from the penis is the seminal fluid or semen. It consists of the spermatozoa together with the fluid in which they are suspended. The fluid is a product of the accessory genital glands, with some secretions supplied by the genital ducts. In man, at least, the ejaculation of the semen occurs in a definite sequence. The mucus of the urethral and bulbourethral

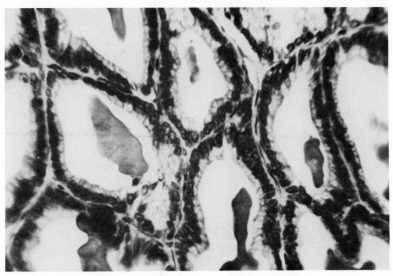

Fig. 11-25. Section through seminal vesicle of hamster. Mucosal lining forms an intricate system of folds, which in cross section appear as separate compartments. (×150.)

glands lubricate the urethra during the process of erection. During the actual ejaculation, the prostate acts first by discharging its alkaline fluid, which reduces the acidity of the urine-lined urethra. It also helps sperm mobility by diluting the thicker semen with its thin secretion. Next, the spermatozoa are forced out of the ductus epididymis and the vas deferens by the muscular contractions of their walls. The seminal vesicle acts last by pushing the sperm along with its thick secretion, which also furnishes fructose as a source of metabolic energy to the sperm.

PENIS. In the lowest mammals, the monotremes, the sperm pass down a groove in the penis. But in other mammals the groove is folded into a tube. The typical mammalian penis consists of three cylinders of erectile tissue: the paired **corpora cavernosa** dorsally and the single **corpus spongiosum** ventrally. The latter structure surrounds the urethra (Fig. 11-28). The corpus spongiosum extends beyond the end of the corpora cavernosa and expands into a cone-shaped glandular tip, the **glans penis** (Fig. 11-29). The glans is richly supplied with sensory endings and is covered with a loose skin, the **prepuce.** The glans is not surrounded by the connective tissue tunica albuginea that surrounds each of the three erectile bodies. In many mammals (bats, carnivores, insectivores, whales, etc.) the penis has

a bone **(os priapi** or **os baculum)** embedded in the connective tissue between the spongy bodies.

The histology of the penis is best studied in cross section. The major substance of the penis is formed by the three cylindrical bodies of erectile, or cavernous, tissue into which the blood rushes at erection. Each cavernous body is surrounded by a sheath of connective tissue (tunica albuginea). The three cavernous bodies are bound together by elastic connective tissue, the **fascia penis.** The cavernous bodies consist of a network of connective tissue and smooth muscle trabeculae covered with endothelium, which also lines the numerous blood spaces within.

Female reproductive system. The typical female reproductive system in mammals (Fig. 11-30) includes the follicular ovaries, the fallopian tubes, uterus, vagina, and auxillary glands. In addition, there are the external genitalia that make up the vulva.

OVARY. A cross section of the ovary reveals a **cortex** and a **medulla,** or zona vasculosa (Fig. 11-31, *A*). The cortex shows a riddled appearance because of the follicles of various sizes and degrees of development. Since not all the follicles and ova complete their development, **atretic follicles** may also be present. In atresia, fibroblasts grow into the follicle and replace it with connective tissue. **Corpora lutea** are also found. These

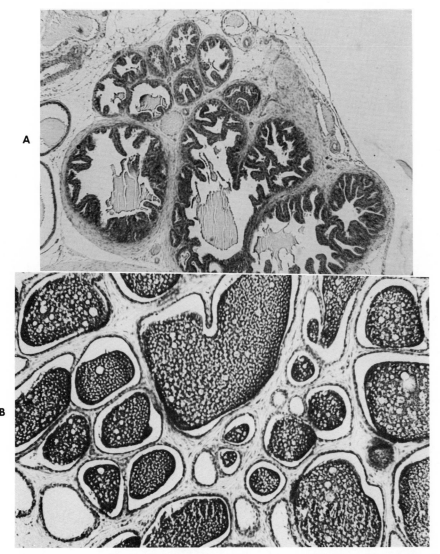

Fig. 11-26. A, Section through prostate gland of rat. Tubular spaces are separated by fibro-elastic stroma with smooth muscle strands. **B,** Section through prostate gland during secretory activity. The many tubuloalveolar glands are embedded in dense stroma of fibroelastic fibers and some muscles. (**A,** ×75; **B,** ×200.)

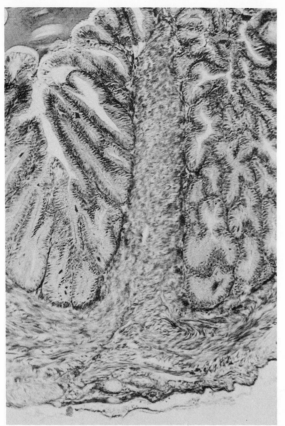

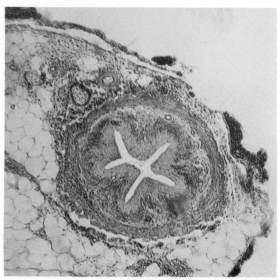

Fig. 11-27. Section through bulbourethral (Cowper's) gland, a compound tubuloalveolar gland. Septa of connective tissue separate gland into lobules, two of which are shown. Gland secretes a mucuslike substance, which is added to male discharge. (×600.)

Fig. 11-28. Cross section of the ureter of a rat. (×133.)

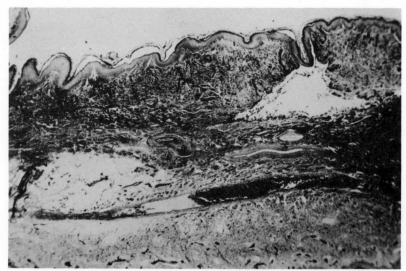

Fig. 11-29. Part of longitudinal section of prepuce (human), circular fold of skin over glans penis. Inner surface of prepuce adjacent to glans (at top) is moist and allows free movement. (×8.)

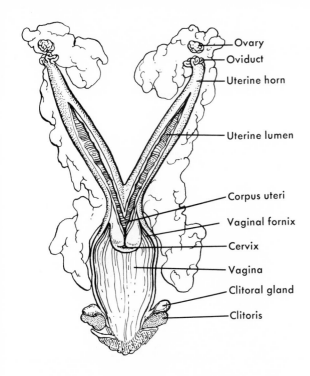

Ovary
Oviduct
Uterine horn

Uterine lumen

Corpus uteri

Vaginal fornix

Cervix

Vagina

Clitoral gland

Clitoris

Fig. 11-30. Ventral aspect of female genital system of mouse. Vagina, cervix, body of uterus, and uterine horns are shown as cut open along the midventral line. (Redrawn with modifications from a drawing by B. Bohen in Snell, 1941.)

are granular structures in which the granulosa cells of the ovulated follicle differentiate into large, pole-staining cells with vesicular nuclei (Fig. 11-31, *B*). In mammals, at least, the corpus luteum is the source of progesterone, the "hormone of pregnancy." The stroma of the cortex consists of spindle-shaped cells forming the **tunica albuginea,** a dense fibrous connective tissue layer under the germinal epithelium. The tunica albuginea also contains blood vessels.

The medulla is made up mostly of loose, fibroelastic connective tissue with a few strands of smooth muscle. At its attached **mesentery,** there is a point (the **hilus**) where blood vessels, lymphatics, and nerves enter and extend into the medulla. As noted earlier, this description of the mammalian ovary would not apply to the ovary of many lower vertebrates such as amphibians, in which there is no medulla; in those forms the follicles are distributed in the cortex around a large lymph space.

DEVELOPMENT OF THE EGG IN THE OVARY. The developmental history of the egg involves the events that take place in the follicle with its enclosed egg, or ovum, from the time of its inception to ovulation, the release of the egg. The first stage

is the transformation of the primordial germ cell into the **primary oocyte.** This young oocyte, surrounded by a single layer of cuboidal cortical cells (granulosa layer), becomes the primary follicle, large numbers of which can be seen in the peripheral regions of the cortex (Fig. 11-31, *B*). These various follicles are sometimes called **graafian follicles.** The granulosa continues to multiply and produce a multilayered covering around the oocyte.

The appearance of a transparent membrane, the **zona pellucida,** between the oocyte and the granulosa layers marks the secondary stage of development. Clefts now appear in the granulosa layer and ultimately coalesce to form a single **follicular cavity.** The cavity is filled from the first with an accumulation of **liquor folliculi** of unknown origin. This stage is called a tertiary follicle (Fig. 11-32). As the follicle grows, its cavity increases in size, and the oocyte, embedded in a portion of the granulosa, is attached to only one side of the granulosa proper.

The granulosa around the oocyte is now called the **cumulus oophorus.** At the same time, connective tissue has formed an additional sheath, the theca, outside the parietal granulosa. The

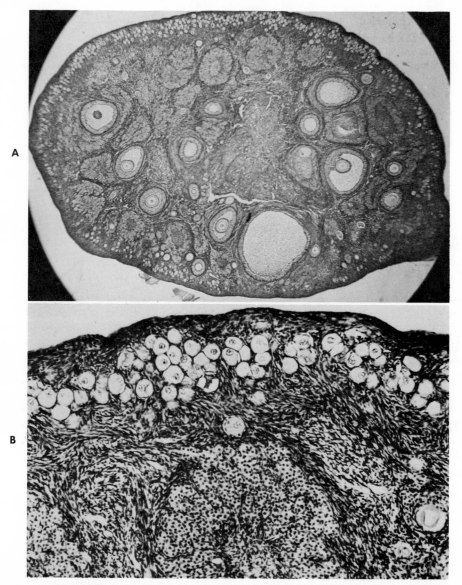

Fig. 11-31. Sections through ovary of cat. **A,** Primary follicles in periphery of cortex. **B,** Higher magnification of same region, showing corpus luteum in the lower center. (**A,** ×100; **B,** ×200.)

theca consists of the theca interna, containing cells and blood vessels, and the theca externa, an external, fibrous layer. Eventually, the cumulus becomes undercut with fluid-filled spaces, so that the oocyte is held in place by a slender stalk. Later still, the oocyte, with its pellucid membrane and remnants of the cumulus, breaks loose and floats into the follicular fluid. The cumulus cells around the oocyte form long, fibrous prolongations,

the **corona radiata.** When the liquid-filled follicle enlarges and bulges above the cortex, it finally bursts, releasing the egg. This shedding of the oocyte (now the ovum, or egg is **ovulation.**

Hormonal control of follicle development and ovulation occurs at nearly every step of the process, especially in higher vertebrates, but such details are beyond the scope of this book. In lower vertebrates, hormonal control is not as clearcut,

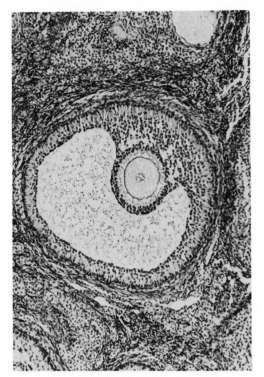

Fig. 11-32. Mature tertiary follicle of a cat. (×100.)

but there is evidence that pituitary hormones are vitally involved.

FEMALE DUCTS. In the mammals the regional differentiation of the oviducts has reached its climax (Fig. 11-33). There is usually a fusion of the caudal ends of the ducts, at least in the higher forms. In general, it may be said that mammals have stressed three major specialized regions of the oviduct. The anterior part is a narrow, slender **fallopian tube,** which may be convoluted. This leads into a broader, muscular **uterus,** which passes to the terminal **vagina.** The paired fallopian tubes never fuse; the uteri show variable degrees of fusion; and the caudal regions are invariably fused to form a single vagina.

The uteri show the greatest variation. In a partially fused uterus, the fused part is called the body and the separate parts are the uterine horns. In rabbits, for example, the uteri remain entirely separate (duplex), the lower ends joining to form the vagina. In other variations, the uterus may be unpaired externally but have two lumens on the inside and two uterine horns (bipartite), or there may be a single lumen in the uterine body

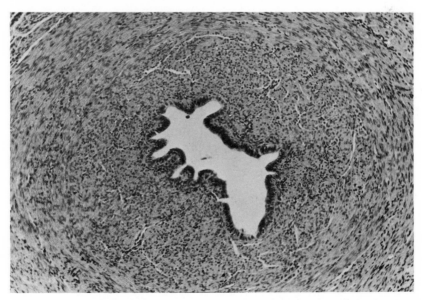

Fig. 11-33. Section through isthmus of oviduct (fallopian tube) of man. Tunica muscularis shows inner circular, or spiral, layer and an outer longitudinal layer. Complex longitudinal folds are cut in cross section. (×100.)

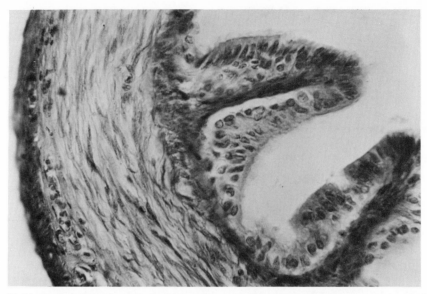

Fig. 11-34. Section through wall of rabbit oviduct. Wall consists of mucous membrane, middle muscular layer, and external serosa. Epithelium may be simple columnar or pseudostratified. There are variations in the wall in different regions. One kind of cell carries cilia, and the other kind is secretory. Rhythmic contractions of the inner circular and outer longitudinal muscles produce peristaltic waves that propel the egg toward the horn of the uterus. (×680.)

but two uterine horns (bicornuate). The ultimate condition, total fusion of the uteri (simplex) is reached in the primates and a few other placental animals. Many mammals have uteri that are transitional between the different types, especially between the bipartite and the bicornuate conditions. Wide variations exist within a single order of a class. Nearly all types are found in bats, for instance. There are also individual differences within a species (even in women).

The body of the uterus in many mammals narrows to form a cervix, the lower end of which projects into the vagina as the lips, or **portio vaginalis.** These lips surround the mouth of the uterus **(os uteri).** After mating, the sperm follow a course through the os uteri and the uterus to the fallopian tubes, where fertilization may occur.

Close to each ovary is a fimbriated funnel, the infundibulum, adapted for receiving the eggs when shed and passing them into the fallopian tube, the first segment of the oviduct. The fallopian tube is a ciliated, glandular, and muscular tube of narrow diameter. It is lined with simple columnar epithelium, and its mucosa is much

folded, forming a labyrinth. The ova are carried along primarily by ciliary and muscular action. Also, fluid is produced by the tube and probably aids in the propulsion of the ova (Hafez, 1963). Each fallopian tube is usually divided into three distinct regions: the infundibulum with its fringed folds; the ampulla, which is dilated and makes up most of the length of the tube; and the isthmus, the narrow part connected to the uterus (Fig. 11-34).

The wall of the uterus is made up of three coats: (1) a thin **serosa** on the outside, the peritoneal investment; (2) the **myometrium,** or muscular coat of smooth muscle cells arranged in circular, longitudinal, and oblique patterns, with the larger blood vessels of the uterus; and (3) the mucous membrane, or **endometrium** (Figs. 11-35 and 11-36). The endometrium has an epithelial lining and a connective tissue lamina propria that is continuous with the myometrium. The epithelium is of simple columnar cells and may be ciliated in some animals. Numerous simple tubular glands are found in the epithelium and lamina propria. The deeper glands reach almost to the myometrium.

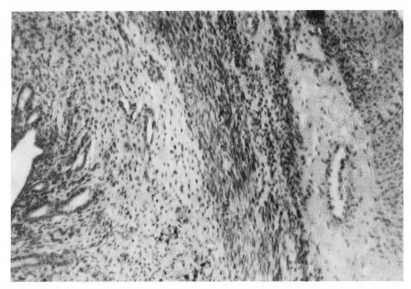

Fig. 11-35. Partial section through wall of cat uterus. Lumen and endometrium of uterus is at left. Note uterine glands of endometrium, which extends to myometrium. Smooth muscle cells of myometrium undergo extreme hypertrophy during the period of pregnancy. See text description. (×100.)

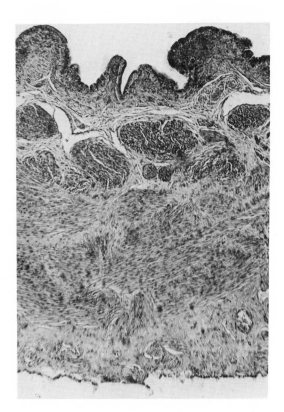

Fig. 11-36. Horn of rabbit uterus, upper end. Endometrium (at top) consists of epithelium and connective tissue lamina propria. Myometrium is thick, with circular, oblique, and longitudinal muscles. (×125.)

The endometrium undergoes many changes during the sex cycle of estrus or during pregnancy. Some histologists divide the endometrium into a **functional** layer, which is shed during estrus or menstrual cycles, and a **basilar** layer, which remains unchanged to regenerate another functional layer. In the resting uterus, when the endometrium is thinnest, the tubular glands are short and straight. As the follicles grow in the ovary and as the endometrium begins to thicken, the tubular glands lengthen and start to coil (Sharman, 1953). The blood supply increases, and the arterioles assume a spiral condition. Veins enlarge and capillaries increase. Near the time of ovulation, or estrus in most nonprimates, the endometrium reaches its greatest thickness. If fertilization occurs, the endometrium gets even thicker until implantation of the embryo. If there is no fertilization, then the endometrium regresses. This process is gradual in those animals that have an estrus period but is rapid in menstruation when the endometrial lining is sloughed off.

The narrow canal of the cervix has a mucosa of dense structure firmly attached to the fibrous connective tissue wall. It has branched, tubular glands that secrete a great deal of mucus. The cervical canal is lined mostly by the tall columnar cells of the uterus, but the portion that projects into the vagina has a stratified squamous epithelium, similar to that which lines the vagina.

VAGINA. The last part of the specialized oviduct is the vagina, a musculofibrous tube that receives the penis of the male. This structure is restricted to the mammals. However, the lower mammals, the monotremes, have no vagina; their paired uteri open directly into the cloaca, along with the urinary and intestinal tracts. The opossum, a marsupial, has three vaginae: the **median vagina,** formed by the fusion of the two oviducts, and two **lateral vaginae,** which are arched tubes that grow out at the anterior end, one on each side of the median vagina. The lateral ones either empty separately into the urogenital sinus or rejoin the median vagina. The median vagina serves temporarily as a birth canal and then grows shut after the young are delivered. Therefore, the chief function of the lateral vaginae is carrying the sperm to the egg. The penis of the male opossum is forked, and during copulation each tip enters a lateral vaginal canal.

The mammalian vagina is thin-walled and highly distensible. It is lined with stratified squamous epithelium and has a lamina propria containing elastic connective tissue and blood vessels. No glands are present in the mucous membrane, which is folded into two primary longitudinal ridges and numerous secondary ridges, or rugae. The muscle region contains both longitudinal and circular muscles, but they are not in layers. A fibrous adventitia on the outside of the muscularis connects the vagina with nearby structures. The upper part of the posterior wall is covered with peritoneum.

The vaginal epithelium undergoes various cyclic changes during the sexual cycle of animals. The famous vaginal smear of C. R. Stockard and G. N. Papanicolaou (1917) has revealed that the cells of the vaginal epithelium of the guinea pig become keratinized at the time of estrus, when the female is most receptive. This test has been used with other animals and has aided in the working out of many estrous cycles. The vaginal smear method is used in humans to detect uterine and cervical cancer. Although estrus as such does not occur in the human female, vaginal smears may indicate certain aspects of the sex cycle (Papanicolaou, 1933).

EXTERNAL FEMALE GENITALIA. Collectively, the external parts of the female reproductive system are called the **vulva.** The vulva may be considered the external opening of either the urogenital sinus or the vagina, if the urinary and reproductive tracts are wholly separated. In general, the lateral boundaries of the vulva are formed externally by two large folds of skin, the **labia majora,** within which are two delicate folds of skin, the **labia minora.** The labia minora skin merges with the mucous membrane of a shallow cavity, the **vestibule,** which lies between the labia minora and receives the openings of the urethra and the vagina. Protruding into the vestibule is the clitoris, the homologue of the male penis. The clitoris is rudimentary; it contains erectile bodies but lacks a urethra and corpus spongiosum.

The outer surfaces of the labia majora in the female mammal are covered with keratinized epidermis, provided with hair follicles, sweat glands, and sebaceous glands. The labia majora contain smooth muscle and fat tissue and may also be considered the homologues of the two

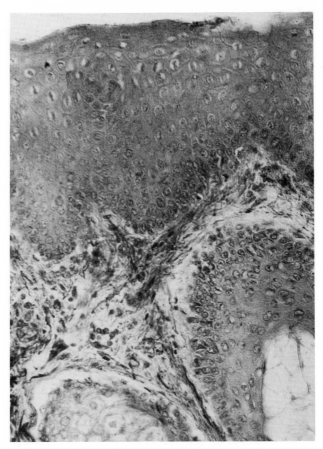

Fig. 11-37. Section of labium minus of mammal. The surface is of stratified squamous epithelium with a core of vascularized connective tissue. Parts of two glands of Bartholin are seen in lower area. (×200.)

halves of the male scrotum. The labia minora (Fig. 11-37) are covered with stratified squamous epithelium and have a core of spongy connective tissue and blood vessels, but they lack fatty tissue. Their surfaces are indented with high papillae that penetrate far into the epithelium. Sebaceous glands are found on both surfaces, but there is no hair. The inner surfaces of each fold have the color and appearance of a mucous membrane, even though they are skin.

References

Benoit, J. 1950. Traité de zoologie, edited by P. P. Grassé. Vol. XV. Oiseaux. Masson & Cie., Paris.

Burgos, M. H. 1955. Histochemistry of the testis in normal and experimentally treated frogs. J. Morphol. **96:**283-299.

Clermont, Y., and C. P. Leblond. 1955. Spermiogenesis in man, monkey, ram and other mammals as shown by the "periodic acid-Schiff" technique. Am. J. Anat. **96:**229-253.

Duke, K. I. 1941. The germ cells of the rabbit from sex differentiation to maturity. J. Morphol. **69:**51-68.

Everett, N. B. 1945. The present status of the germ-cell problem in vertebrates. Biol. Rev. **20:**45-70.

Fawcett, D. W. 1958. The structure of the mammalian spermatozoon. Int. Rev. Cytol. **7:**195-234.

Gorbman, A., ed. 1959. Comparative endocrinology. John Wiley & Sons, Inc., New York.

Gruenwald, P. 1942. The development of the sex cords in the gonads of man and mammals. Am. J. Anat. **70:**359-397.

Hafez, E. S. E. 1963. The uterotubal junction and the luminal fluid of the uterine tube in rabbits. Anat. Rec. **145:**7-12.

Lacy, D. 1967. The seminiferous tubule in mammals. Endeavour **26:**101-108.

Lyngnes, R. 1930. Beiträge zur Kenntnis von *Myxine glutinosa* L. I. Über die Entwicklung der Eihülle bei *Myxine glutinosa*. Z. Morphol. Okol. Tiere Berlin **19**:591-608.

Nansen, F. 1888. A protandric hermaphrodite *(Myxine glutinosa,* L.) amongst the vertebrates. Bergens Mus. Aarsber. (for 1887) **7**:3-34.

Nansen, F. 1889. Protandric hermaphroditism of *Myxine.* J. Roy. Microsk. Soc. **1**:188-189. (Abstr.)

Papanicolaou, G. N. 1933. The sexual cycle in the human female as revealed by vaginal smears. Am. J. Anat. **52**:519-531.

Rasmussen, A. T. 1932. Interstitial cells of the testis. Vol. 3, p. 1673, *in* E. V. Cowdry, ed. Special cytology, 2nd ed. Paul B. Hoeber, Inc., New York.

Schreiner, A., and K. E. Schreiner. 1905. Über die Entwicklung der männlicher Geschlechtszellen von *Myxine glutinosa* (L.). I. Vermehrungsperiode, Reifungstheilungen. II. Die Centriolen und ihre Vermehrungsweise. Arch Biol. **21**:183-357.

Schreiner A., and K. E. Schreiner. 1908. Über die Entwicklung der männlicher Geschlechtszellen von *Myxine glutinosa* (L.) III. Zur Spermienbildung der Myxinoiden. Arch. Zellforsch. Leipzig. **1**:152-231.

Schreiner, K. E. 1955. Studies on the gonad of *Myxine glutinosa* L. Univ. Bergen Arbok, 1955. Nat. Rek. **8**:1-40.

Sharman, A. 1953. Post-partum regeneration of the human endometrium. J. Anat. **87**:1-10.

Snell, G. D., ed. 1941. The laboratory mouse. McGraw-Hill Book Co., New York.

Stockard, C. R., and G. N. Papanicolaou. 1917. The existence of a typical oestrous cycle in the guinea-pig: with a study of its histological and psychological changes. Am. J. Anat. **22**:225-283.

Warren, D. C., and R. M. Conrad. 1942. Time of pigment deposition in brown-shelled hen eggs and in turkey eggs. Poultry Sci. **21**:515-520.

12 DIGESTIVE SYSTEMS

Oral cavity and pharynx

CYCLOSTOMATA

At the anterior end of *Petromyzon* is a large buccal funnel bordered by a fringe of papillae. On its inner surface are many horny epidermal, keratinous "teeth." At the bottom of the funnel is the mouth through which protrudes a tongue, also bearing horny teeth. Although the cyclostomes are "jawless" creatures, they have a cartilaginous ring surrounding the mouth opening.

The mouth leads into a buccal cavity. Posteriorly this cavity connects with not one but two tubes: a dorsal esophagus and a ventral pharynx. The latter actually is a blind pouch, having the seven gill slits on each side. There is a valvelike flap, the velum, between the buccal cavity and pharynx.

The olfactory apparatus warrants consideration as auxiliary to the digestive system, since it is of such importance in locating food. *Petromyzon* has an elaborately developed olfactory apparatus (Fig. 12-1). It is made up of the nasal tube leading from the surface of the body, the olfactory sac with its complicated folds of mucosa, and the nasopharyngeal tube. Vast numbers of receptor cells are found in the mucosal folds of the olfactory sac. They face each other in adjoining folds and do not occur on the peaks nor deep in the valleys of these folds (Kleerekoper and Van Erkel, 1960).

CHONDRICHTHYES

The buccal cavity of sharks is lined with the stratified epithelium of a mucous membrane. The basal cells of the epithelium are columnar. The only glands are the unicellular mucus-secreting ones found in the stratified epithelium. Under the mucous membrane is the connective tissue with elastic fibers and interstices filled with a jellylike matrix.

The mucous membrane of the mouth is thrown into ridges and papillae. The lips are merely hard ridges of compact tissue. There is some degree of connection between the mouth and external nostrils. The tongue, primitive and largely immobile as in all fishes, is a simple fold of tissue supported by the hyoid cartilage. Taste buds in elasmobranchs are located in both the mouth and pharynx.

Although teeth show a variety of different forms in elasmobranchs, they are usually triangular and pointed in sharks. In bottom-dwelling species, the teeth are mostly flattened plates for crushing the shells of the mollusks on which they chiefly feed. The teeth of sharks are similar in origin to the scales of their integument. The selachian type contains an inner core of osteodentin, a primitive form of dentin, in which calcified trabeculae are laid down within the pulp.

The mesenchyme is differentiated into an inner zone that forms the pulp elements and an outer zone that forms the odontoblasts for calcifying the dentin. This arrangement produces canals for carrying connective tissue, blood vessels, and nerves. The osteodentin is covered superficially by a hard dermal layer. The enamel organ of higher vertebrates is functionless in sharks, so that there is no outer layer of enamel. A second type of shark tooth, bradyodont, with highly vascular dentin has also been described by Radinsky (1961).

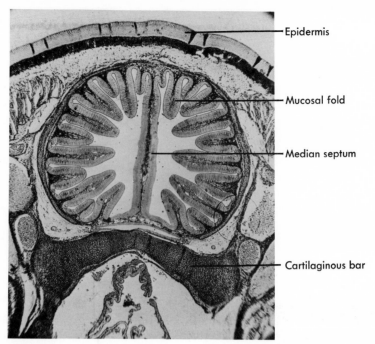

Fig. 12-1. Cross section through olfactory sac of *Petromyzon*. (From Kleerehoper, H., and G. A. van Erkel. 1960. Can. J. Zool. **38:**209-223. Reproduced by permission of the National Research Council of Canada.)

The teeth are not embedded in the jaws but are attached to fibrous bands of connective tissue. The finely serrated teeth may occur in a single row, with several linear series of teeth in developmental stages behind the functional row. New teeth can thus simply move forward and replace worn-out ones. In some sharks (e.g., *Mustelus*) several rows may function at the same time. Teeth are formed at the dental ridge, the anterior wall of a crescent-shaped groove of stratified epithelium along the inner margin of the jaw.

Pharynx. The pharynx is a continuation of the buccal cavity, lined by stratified epithelium. The limits of the pharynx are defined by the paired pharyngeal pouches, or slits. These internal gill slits are protected by a series of projecting gill rakers, rods attached to the branchial cartilages at their anterior edges. The gill rakers may also serve to prevent the escape of prey, since they project part way into the pharyngeal cavity. A pair of spiracles also opens into the pharynx. The epithelium of the pharynx has numerous mucous glands, the mucus of which assists in swallowing food.

OSTEICHTHYES

The oral cavity of fishes in general is lined by an epithelium that is much like a continuation of the epidermis (Figs. 12-2 and 12-3). The lining is of stratified squamous cells and includes many unicellular glands. In general, salivary glands are lacking. No tonsils are found. The epithelium over the tongue is similar to that of the rest of the oral cavity. Although it contains some sensory endings or taste buds, such organs are also found elsewhere in the oral cavity and pharynx and indeed may be distributed widely over the **external surface** of the body (Fig. 12-4).

In the teleost fishes the mouth is relatively large. Both jaws have numerous small, recurved, conical teeth. These are borne on several bones: the maxillae, premaxillae, palatines, vomer, dentaries, and basihyal. Their function clearly is to prevent the escape of the captured animals. The teeth are of no use for rending or chewing.

Histologically, the free portion of the tooth of *Esox lucius* is covered over by a thin layer of enamel. This layer is thickest in the apical region, becoming thinner and finally dwindling

Image labels: Epidermis, Mucosal fold, Median septum, Cartilaginous bar to nothing at the base. The main mass of the tooth consists of dentin (called by the Germans "tooth bone"). The dentin appears to be a direct extension of the bone of the tooth socket. Both consist of bundles of calcified connective tissue fibers running generally parallel to the surface of the tooth.

The outermost layer of dentin, just under the enamel, is compact. The deeper-lying dentin shows many canals, running longitudinally through its substance. The outer compact layer and the portion with narrow canals together form the **vitrodentin.** Under them is the **trabecular dentin,** where the canals are larger and show numerous cross anastomoses. The canals contain blood vessels and are called haversian canals because they are similar to those in the true bone of the tooth socket.

The trabecular dentin is traversed throughout by a system of fine canals that correspond to the canaliculi of bone. They serve to transport nourishment and oxygen and excretory products, including carbon dioxide, away from the dentin-forming cells, the odontoblasts.

The mucous membrane of the mouth has a stratified epithelium with numerous goblet cells and flattened superficial cells. The roof of the mouth presents some longitudinal folds. The

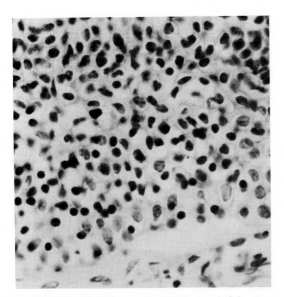

Fig. 12-2. Dorsal surface of tongue of goldfish, *Carassius auratus.* Lack of cornification and absence of papillae are seen. Numerous lymphocytes with deeply staining nuclei occur in the stratified cuboidal epithelium. (×400.)

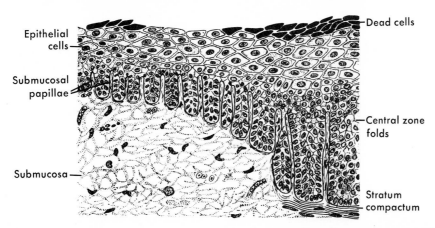

Fig. 12-3. Buccal mucosa from roof of mouth of stone-roller, *Campostoma anomalum* (Rafinesque). In this transverse section, midline region is to right. Epithelium is thickest there. Submucosal papillae show high degree of regularity. Desquamating dead cells occur at surface; conspicuous stratum compactum is found beneath epithelium of central area. (Redrawn from Rogick, 1931.)

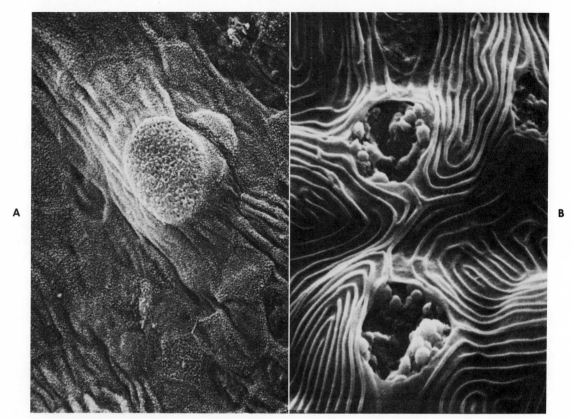

Fig. 12-4. A, Taste bud on barbel of adult catfish. **B,** Two taste buds on lip of common guppy. (Scanning electron micrographs; **A,** ×1500; **B,** ×8500.) (Courtesy Dr. P. P. C. Graziadei, Department of Biological Science, Florida State University, Tallahassee, Fla.)

lamina propria consists of fibrous connective tissue, and no distinct submucosa is present.

The tongue is free only in its anterior third. It has a bony-cartilaginous support from the branchial skeleton. In the free portion the skeletal tissue is replaced by cell-rich connective tissue. The mucosa over the tongue is very similar to that lining the mouth in general, but in the caudal portion a number of small lingual teeth are found. The mouth cavity ends caudally at the beginning of the pharynx, which is bounded on either side by the branchial arches.

AMPHIBIA

The epithelium lining the amphibian mouth is stratified in four or five layers. Unlike other vertebrates, however, the amphibians usually have cuboidal cells on the surface, and these cells are ciliated in many species including frogs, toads, and most of the salamanders. Two types of cells are commonly found in the oral mucosa: ordinary epithelial cells and goblet cells, which are unicellular, mucus-secreting glands. The basal cells usually are fairly tall columns, with several layers of polyhedral cells above them.

The oral mucosa serves as an **accessory respiratory surface** in many amphibians. Capillaries penetrate the epithelium. In the anurans they pass all the way up to the superficial ciliated layer. In the urodeles they are confined generally to the basal layers. From the penetrating capillaries rounded diverticula arise, increasing the extent of the vascular bed and serving to slow the rate of blood flow to permit a larger degree of gaseous interchange.

The ciliation of the superficial layer is probably related to the respiratory function, since it helps to prevent mucus and other material from

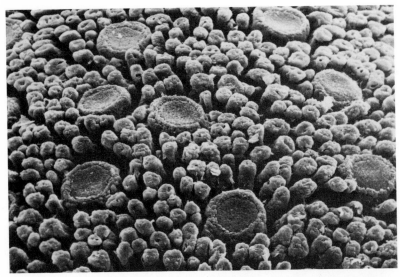

Fig. 12-5. Dorsal surface of frog tongue showing both filiform and fungiform papillae. (Scanning electron micrograph; ×180.) (Courtesy Dr. P. P. C. Graziadei, Tallahassee, Fla.)

accumulating on the oral mucosa. Aquatic larvae and gill-bearing adults do not have a ciliated oral epithelium. These cells have a **cuticular border** similar to that seen on the surface cells of the epidermis of other amphibians and of many fishes. Even in the fresh condition, the border shows vertical striae. It is easily seen in stained sections.

The goblet cells are scattered through the layers. Some of them are narrow and so elongated that they extend through the entire thickness of the epithelium. Others are very large and globular. Their nuclei often are darkened and pressed toward the base of the cell by the mucus. There are many similarities between the buccal epithelium and the epithelium of the skin, the epidermis. Differences are seen in the absence of the "club cells," the presence of true goblet cells, and the somewhat thinner epithelium of the mouth.

Conspicuous are the many vacuoles among the epithelial cells. They contain other cells or cell debris. The leukocytes that migrate into the oral epithelium in amphibians consist of both polymorphonuclear and mononuclear types. However, farther down the tract they usually are exclusively mononuclear.

Taste buds occur in various parts of the lining of the oral cavity of amphibians. In *Necturus* they are most abundant on the posterior part of the tongue. Papillae on the tongue vary in shape,

many being filiform and others rounded or "fungiform" (Fig. 12-5). Glands in the connective tissue beneath the epithelium in various parts of the oral cavity contain both serous and mucous alveoli.

The identification in amphibians of tissue that may be called "tonsillar" is a question of general interest because of the role of tonsils in the immune mechanisms of the animal body (Chapter 8). In fact, frogs and toads show two **palatine fossae,** invaginations of epithelium surrounded by dense masses of lymphoid tissue. There also are many smaller epithelial invaginations surrounded by lymphocytes. In the urodeles the epithelium in regions corresponding to the palatine fossae of anurans shows outgrowths whose cores contain lymphoid tissue. In *Proteus* and *Salamandra* these protruding masses are definitely tonsilloid.

Teeth are generally numerous in the amphibians. In some cases, however, they are lacking, as in most toads and in the lower jaw of frogs. The teeth consist of dentin covered with a small amount of enamel. They are bifid and conical shaped. In both urodeles and anurans the teeth are fused to the jaw.

REPTILIA

The reptiles show a tremendous variety in gross and microscopic details of structure of the

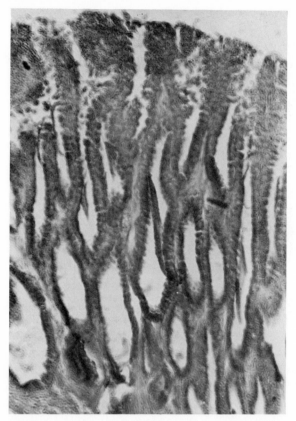

Fig. 12-6. Section through tongue of chameleon showing glandular structure. Invaginations of surface epithelium bear large numbers of mucous cells. Alveoli of true glands can be seen to penetrate into muscular layer. (×100.)

oral cavity and pharynx. The epithelial lining in different species, and in different parts of the oral mucosa of an individual, may vary all the way from simple columnar ciliated with goblet cells to stratified squamous type with heavily cornified superficial layers. Differences in individuals, as in various specimens of the turtle *Emys europea,* appear to be dependent on age. Young animals have a simple ciliated columnar epithelium; older ones the stratified squamous type.

Papillae are common in the lining of the oral cavity of reptiles. Individual branched tubular glands often are widely distributed, but in some of the lizards they are grouped as definite organs or **salivary glands.** Mucus-secreting cells are found over the surface of the tongue, and in many lizards

there are also abundant glands scattered through the tongue musculature (Fig. 12-6).

The tongues of reptiles are structures that are much varied in their gross and microscopic structure and in the functions they serve. In the snakes and lizards the tongue generally has a remarkable motility. In turtles and crocodiles its motility is slight.

The tongue of *Sphenodon,* a primitive kind of lizard from New Zealand, has many long filiform papillae on its anterior part, mingled with some fusiform ones that contain taste buds. The summits of the papillae are covered with a stratified cuboidal epithelium, whereas the sides and bases have a simple columnar epithelium. In some specialized lizards such as the arboreal chameleons food is seized by the long, active tongue and is drawn back into the mouth. The tip of the tongue is adhesive. Mere contact of this organ with an insect is enough to spell its fate!

The serpent tongue has a stratified epithelium with a well-developed stratum corneum. It has also a **sheath,** essentially an extension of the wall of the mouth, lined also by stratified epithelium, into which the very mobile tongue can be retracted.

The question of the origin and distribution of tonsils is again an important one in relation to the development of immune mechanisms. The epithelium of the posterior part of the buccal cavity and of the pharynx often shows considerable numbers of leukocytes, in particular the small lymphocytes. Pharyngeal tonsils definitely occur in the common lizard *Lacerta agilis,* and in *Crocodilus niloticus.* They show as narrow crypts around which are grouped elongated nodules of lymphoid tissue. The epithelium over the nodules seems to lack a basement membrane, and there is a mingling of lymphoreticular and epithelial tissue.

The reptiles, like the amphibians, show a range in numbers of teeth, from none to many. The teeth usually are ankylosed to the jaw and consist of orthodentin with a tip of enamel. The teeth of lizards are rounded or pointed cones. The largest of living lizards, *Varanus,* the monitor, has about thirty small ankylosed teeth in each jaw.

The poisonous snakes have certain teeth (poison fangs) specialized by elongation and by the presence of a groove to transmit the venom. These teeth, like the others, are replaceable. In the cobra the poison fangs are fused to the maxilla. The groove is on the anterior surface of the fang and is converted into a tunnel by the meeting of its edges. The tunnel is lined with enamel. There is no direct connection between the poison (parotid) gland and the poison fang, but a flap of mucous membrane helps to convey the venom to the tooth.

Crocodiles and alligators have sharp conical teeth. Here there is a vertical succession of teeth, new teeth erupting into sockets occupied by the preceding ones. As in other reptiles however, the teeth have no distinct roots, only broad, widely open bases covered by cementum.

AVES

In the reptiles a variety of types of epithelial lining of the oral cavity were found. In all the birds the lining is a **stratified squamous** epithelium. Many species show a high degree of cornification of this epithelium. The stratum corneum in the pigeon *Columba livia* not only covers the upper surface of the tongue and the roof of the mouth but also extends along the roof of the pharynx down to the upper larynx and the beginning of the esophagus.

The beaks of birds are composed of keratinous material. They often are exquisitely adapted to food-acquiring habits. The beaks of eagles and hawks are strong, sharp, "cruel" instruments, hooked for greater efficiency. Those of swans, ducks, and geese have flattened edges with numerous sensory endings, serving as an efficient means to sieve their food from the mud and sand. Birds of the shore, who pick and choose as they walk, have fine, long beaks almost like delicate forceps for picking up their insect prey.

The tongue of the bird has papillae on its dorsal surface. Its glands are largely if not entirely mucus-secreting. Usually intrinsic muscles are lacking, although adipose tissue is common. Many birds have a bone, the **os entoglossum,** in the midline of the tongue, which serves to give the organ rigidity. A curious modification is the horny plate found at the tip of the tongue of some birds, for example, the duck and goose. On either side of it are cornified "hairs." The taste buds of the

pigeon are found mostly on the dorsal portion of the tongue, scattered through the epithelium, rather than in papillae as in mammals.

Deep below the epithelium of the oral cavity is a connective tissue lamina propria, which contains diffuse lymphoid tissue and occasional nodules that may show clear germinal centers of immunologic importance. None of the lower vertebrates—fishes, amphibians, reptiles—seems to have developed these centers. Many birds have lymphoid masses in the pharynx, and some have also an esophageal tonsil.

It is a matter for regret that we cannot know more about the internal anatomy of extinct animals. Modern birds carry out the process of mastication by means of the gizzard, which is a specialized portion of the stomach. Fossil remains from such as the primitive *Ichthyornis* and *Archeopteryx* show that these "early birds" actually possessed **teeth.** Whether they had a gizzard at the same time, we do not know. Their teeth were reptilian in character.

Avian salivary glands may be simple, branched, or compound. Salivary glands are adapted to the type of food. Many aquatic birds whose food is slippery have a few glands. Most granivorous and insectivorous birds have a complete set of salivary glands. The chief function of the salivary glands is lubrication. Their histology in birds usually shows only mucous cells.

MAMMALIA

The basic pattern of the digestive system is essentially the same in all vertebrates, with some special modifications imposed by varied diets. Most of the differences between the alimentary systems of different vertebrates are in the buccal cavities. In mammals evolution of the buccal cavity has apparently been stressed more than in most other animals.

The oral epithelium in man and most mammals is a stratified squamous, **nonkeratinizing** type. But a number of others, especially those whose food requires a great amount of abrasion, show significant keratinization in various places such as the pharyngeal roof, the upper surface of the tongue, and the "dental pads" in Ruminantia, the cud-chewers.

In the mouse the skin of the outer portion of

the lips contains deeply embedded hair follicles. Where the skin passes over into mucous membrane there is a rather sharp line of demarcation. Here the hairs disappear and the stratified squamous epithelium becomes much thicker. The surface layers are well cornified, and, indeed, the epithelial lining of the entire oral cavity is cornified in the mouse. The underlying connective tissue, or lamina propria, of the mouth is of fibrous nature, with some wandering cells. It forms low, broad papillae. Large sebaceous glands at the angles of the mouth open through short ducts onto the surface of the lips.

The hard palate forming, as in man, the anterior part of the roof of the mouth, has rows of ridges from front to back. Of these, the first three are transverse, whereas the last are U-shaped with the concavity directed forward. The mucosa here is firmly attached to the surface of the bone.

The soft palate forms the posterior part of the roof of the mouth. Above the oral mucosa in this region there is no bone, but there are voluntary, striated muscles. The oral surface of the soft palate and its posterior margin have stratified squamous epithelium; the nasal surface has pseudostratified columnar ciliated epithelium. Mucous glands occur in the mucosa of the oral surface and empty their secretions into the mouth cavity by way of short ducts.

Tongue. The tongue is divided grossly into two portions. The posterior portion, representing about two thirds of the organ, is attached to the floor and sides of the mouth, while the anterior third lies free in the oral cavity and can be moved about readily. All the free surfaces of the tongue are covered by stratified squamous epithelium. Usually the epithelium is thicker on the dorsal side, and papillae of several different shapes are common here. Some mammals, including the cats, show cornification of the superficial layers (Fig. 12-7). In others such as man, rat, guinea pig, and rabbit it is known that even the most superficial epithelial cells retain their nuclei, except at times on the very summits of the papillae. The ventral surface is smooth, but the dorsal surface shows numerous projections, the papillae (Fig. 12-8).

There are four types of papillae. Most numerous are the threadlike filiform papillae; they are

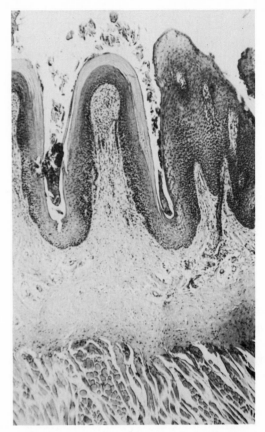

Fig. 12-7. Dorsal surface of tongue of leopard, *Panthera pardus.* Note highly cornified lingual papillae. (Hematoxylin-eosin; ×35.)

cone-shaped projections consisting entirely of epithelial cells (Fig. 12-9). They have no taste buds and seem to serve a mechanical function.

The fungiform type (Fig. 12-10) are lower, broader structures in which the connective tissue (lamina propria) as well as the epithelial layer take part. In the mouse each fungiform papilla has a single taste bud at its center, whereas in corresponding human papillae a number of taste buds are found.

The third type of papilla is the circumvallate, a large, flattened papilla surrounded, as the name implies, by a moat, or ring-shaped fossa (Fig. 12-11). These are usually arranged in V formation at the root of the tongue. In the mouse, however, there is only a single circumvallate papilla, located close to the base of the tongue in the midline. Taste buds are found in the epithelium of the

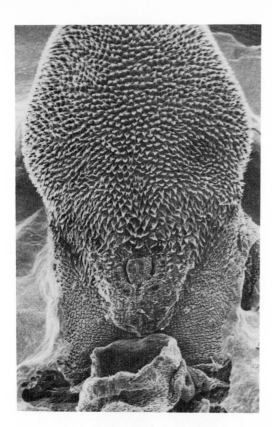

Fig. 12-8. Dorsum of rat's tongue, viewed from behind, showing wealth of papillae. (Scanning electron micrograph; ×35.) (Courtesy Dr. P. P. C. Graziadei, Tallahassee, Fla.)

Fig. 12-9. Filiform papillae on dorsal side of tongue of mouse. Papillae consist almost entirely of epithelial cells and their product, cornified material that forms the pointed apices. (Wilder's reticulum stain; ×250.)

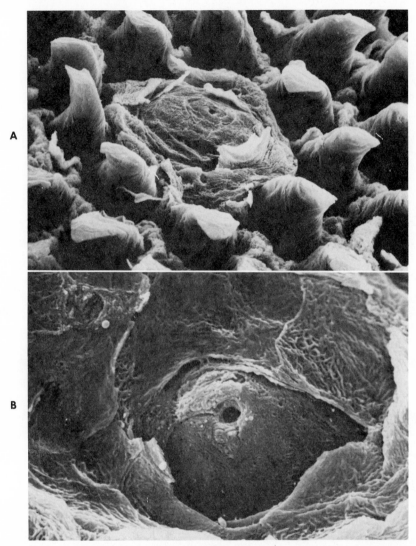

Fig. 12-10. A, Fungiform papilla of rat tongue surrounded by filiform papillae. Note pore center of fungiform papilla. **B,** Detail of fungiform papilla of rat tongue. (Scanning electron micrographs; **A,** ×1000; **B,** ×4800.) (Courtesy Dr. P. P. C. Graziadei, Tallahassee, Fla.)

periphery of the papilla, abutting on the moat about it.

Foliate papillae are well developed in the rabbit, poorly so in man. Papillae of the circumvallate type are seen in monotremes in the location where foliate papillae occur in higher mammals. It seems probable that foliate papillae represent circumvallate papillae that have become arranged in linear order.

The lamina propria of the tongue consists of a thin layer of connective tissue that is every-

where underlaid by striated skeletal muscle. The muscle fibers run in three dimensions, some coursing longitudinally, some transversely, and some vertically. These fibers actually are of two kinds: (1) intrinsic, originating and inserting within the tongue, and (2) extrinsic, originating outside of the tongue from various skeletal attachments and inserting within the tongue.

A group of serous glands, with short ducts opening into the moat of the single circumvallate papilla, is found near the base of the mouse

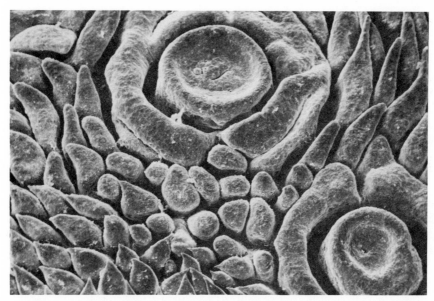

Fig. 12-11. Two circumvallate papillae among filiform papillae on tongue of young dog. (Scanning electron micrograph; ×75.) (Courtesy Dr. P. P. C. Graziadei, Tallahassee, Fla.)

tongue. Mucous glands located on either side of the tongue open by ducts directly onto its surface.

Tonsils. The tonsils, which in man form a circular guard near the upper end of the alimentary tract, vary considerably in different mammals. In man, palatine, pharyngeal, and lingual tonsils are found. The rabbit has no pharyngeal or lingual tonsils, the mouse and rat have no palatines, and lingual tonsils are not found in cat, dog, or sheep. The lymphoid nodules that help to form the tonsils are generally relatively large ovoid ones. They have conspicuous clear centers, usually showing more of the character of "reaction" than of "germinal" centers—i.e., much phagocytosis, little mitotic activity.

Teeth. The dentition of the mouse is very unlike that of man. It normally has four incisors and twelve molars, a total of sixteen teeth. The mouse incisors, like those of other rodents, grow throughout life from persistent pulps. The apical foramen of the root remains open. The crown of the tooth is worn away constantly and replaced by formation of new tooth tissue. The enamel forms a thicker layer on the anterior convex surface of the incisors than on the lingual surface, so that as the tooth wears away, there is always a sharp edge.

The molars, on the other hand, have all their surfaces covered by enamel, as is true of all human teeth.

Salivary glands. In mammals generally, there are three pairs of major salivary glands: the parotid, the submandibular (or submaxillary), and the sublingual. The following descriptions refer specifically to the mouse, but the glands of man and other higher mammals show similar characteristics.

The parotid glands, one on each side near the external ears, consist of a number of lobes. The chief duct of each gland is formed by the union of branches from the lobes and opens into the mouth cavity opposite the upper posterior molar teeth. The secretion of the parotid glands has a high concentration of the important enzyme amylase, which begins the digestion of carbohydrate in the mouth.

The substance of the lobes is divided into lobules by thin septa of connective tissue carrying blood vessels, lymphatics, and nerves. Each lobule consists of a large number of small saclike structures, the alveoli. In the parotid gland all the cells of all the alveoli are of the serous type; that is, they secrete a watery, nonmucoid fluid. The gland cells are pyramid-shaped, with the somewhat rounded apices surrounding a small central lumen. Nuclei, basally located, are round or oval and show chromatin granules and nucleoli. Apical to the nuclei are the secretion, or "zymogen," granules. Basal to the nuclei the cytoplasm is basophilic and somewhat striated in its aspect.

The exit duct arising from each alveolus is known as the intercalated duct. It is formed by very low cuboidal epithelial cells. Intercalated ducts empty into the intralobular ducts, which are lined by columnar cells. These are distinguished by the markedly striated aspect of their cytoplasm, giving the name "striated ducts" to the tubules that they compose. The intralobular ducts, in turn, open into larger ducts until the main duct of the gland is formed.

The submandibular glands are large structures that slightly overlap at the midventral line beneath the lower jaw. Their division into lobes and lobules is similar to that of the parotid. The main duct of each lobe of the gland opens separately on the floor of the mouth.

The alveoli are composed of serous cells of a type different from those of the parotid gland and are called "special serous cells." Like the cells of the parotid, these cells are pyramidal in shape, but their nuclei stain more deeply and their cytoplasm lacks the basophilic substance. These cells, because of their dark and shrunken-appearing nuclei and the coarsely reticular aspect of their cytoplasm, are often difficult to distinguish from mucous cells in ordinary preparations. But their contents do not stain with any of the reagents used to demonstrate mucin.

Close to the lateral surfaces of the submandibular glands lie the sublingual glands. Each usually consists of one large lobe divided into several lobules by septa of connective tissue. The duct of the gland runs a course parallel to that of the submandibular gland but opens separately on the floor of the mouth. The alveoli consist almost entirely of mucous cells. The smallest ducts, arising from the alveoli, are lined by a low cuboidal epithelium. They open into striated ducts within the lobules similar to the intralobular ducts in the other salivary glands. These, in turn, empty into larger interlobular ducts that run, together with blood vessels, lymphatics, and nerves, in the interlobular septa. The main duct is lined by a pseudostratified epithelium.

In addition to the secreting epithelial cells, the alveoli of all the salivary glands have certain basal cells just inside the basement membrane. These appear to be of myoepithelial nature, and apparently are able by their contraction to help in emptying the secretory product into the ducts.

Pharynx. The pharynx, a kind of "crossroads" of the digestive and respiratory systems, lies between the oral cavity and the esophagus. Thus the oral cavity and the internal nares open into it; the eustachian tubes, passages to the middle ear, open from it; and it is continuous posteriorly with the esophagus dorsally and the larynx ventrally. Almost the entire lining of the pharynx is stratified squamous epithelium, although over a very small area near the internal nares the epithelium is pseudostratified and ciliated. Outside the epithelium is a lamina propria of dense connective tissue, resting directly on the striated fibers of the constrictor muscles of the pharynx. Among these fibers occur groups of mucous glands opening into the lumen by means of short ducts lined with stratified squamous epithelium.

The histologic structure of the pharynx of the mouse differs from that of man in two rather striking features. The mouse has keratinized epithelium but no lymphatic tissue in its pharynx. These two features may well be functionally related. For example, it seems probable that tonsils, so conspicuous in man, either are not needed or could not function with the heavily cornified overlying epithelium.

Esophagus and stomach
CYCLOSTOMATA

The esophagus proceeds posteriorly as a straight tube. There is no real stomach, although some have regarded slight enlargements at anterior and posterior end of the intestine as "stomach" and "rectum," respectively. Histologically, there are some important modifications in the wall of the gut at various levels. The lining mucosa of the pharynx has longitudinal folds, with tall cylindrical, nonciliated epithelial cells. In the esophagus cilia are present in younger specimens but apparently absent in older ones.

CHONDRICHTHYES

The narrow esophagus of the shark (Fig. 12-12) has thick, muscular walls. It is only a passage-

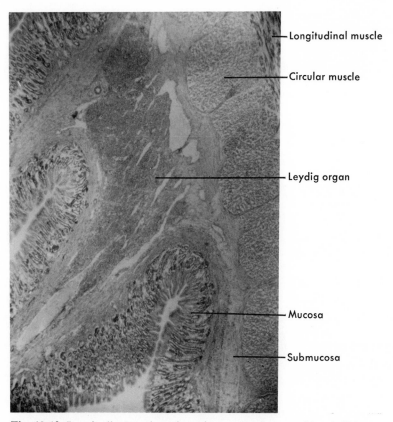

— Longitudinal muscle

— Circular muscle

— Leydig organ

— Mucosa

— Submucosa

Fig. 12-12. Longitudinal section of esophagus, *Squalus acanthias*. (×53.)

way for food; no digestion occurs in it. The lining is stratified epithelium with numerous mucus-secreting cells and often small papillae. The Leydig organ of lymphoid structure extends all the way through the esophagus into the stomach.

The esophagus leads into a large stomach, which is bent into the shape of a U, so that its exit as well as its entrance face anteriorly. This divides the stomach into two parts: a descending cardiac limb and an ascending pyloric limb. This bending of the stomach is made possible because the anterior end is anchored by the transverse septum and the posterior end by the falciform and hepatoenteric ligaments (parts of the ventral mesentery). The stomach may have first evolved to serve as storage for food, but its mucous glands have become modified to produce acid for digestion and prevention of bacterial decay.

The gastric glands of sharks, indeed of all vertebrates, open at the bottoms of gastric pits, called

"foveolae," or crypts. They appear to be differentiated into fundic and pyloric glands. Fundic glands (Fig. 12-13) are scattered over most of the body of the stomach, but the pyloric glands are generally restricted to the pyloric end. In each gastric gland there appears to be only one type of cell, even though they secrete two very different products, pepsin and hydrochloric acid. Neck cells, found in higher forms at the junction of the gland and the crypt, seem to be absent in sharks.

A transverse section of the stomach wall (Fig. 12-14, *A*) reveals a surface of simple columnar epithelium with mucus in the apical ends of the cells; a lamina propria of scanty connective tissue; a muscularis mucosae of small bundles of longitudinal smooth muscle fibers; a submucosa of longitudinal tissue, blood vessels, nerves, and lymphoid cells; and a muscularis externa of an outer longitudinal layer and an inner circular layer. In part of the stomach's curvature, oblique muscle

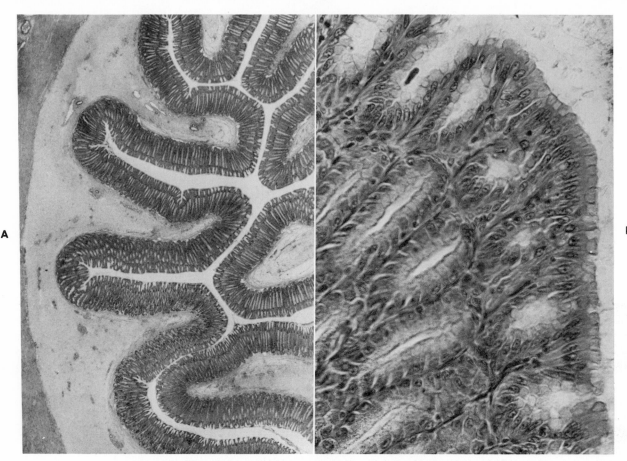

Fig. 12-13. Sections (partially oblique) through fundic region of shark's stomach, showing gastric glands. (**A,** ×16; **B,** ×300.)

fibers may also be found in the wall. The longitudinal section in Fig. 12-14, *B,* shows the extensive folding of the mucous membrane.

OSTEICHTHYES

Esophagus. In the bony fishes, the esophagus lining shows a number of longitudinal folds, so that when the organ is empty, the lumen is very small (Fig. 12-15). A cross section through the esophagus shows a mucosa, submucosa, muscular layers, and an outer serosa. The mucosa and submucosa are complicated not only by the primary longitudinal folds but also by secondary and even tertiary foldings. The epithelium is stratified, consisting chiefly of mucus-secreting cells. The basal layer is generally free of mucus and made up of cylindrical and cuboidal cells. The superficial layer is usually flattened cells without mucus.

Taste buds are found in the esophagus epithelium of some fishes. Multicellular glands are usually lacking, although a few may be seen near the junction with the stomach. A muscularis mucosae of longitudinally oriented fibers separates the mucosa from the submucosa. In the latter there are granular cells and collections of lymphomyeloid tissue, similar to the organ of Leydig noted previously in the shark.

The muscular coat through most of the length of the esophagus is composed of bundles of skeletal muscle. The inner layer is longitudinally oriented, whereas the outer layer is circular. Bundles of longitudinal muscle also are found in the primary folds. Nearing the stomach, the skeletal muscle gradually changes over to the smooth type, beginning with the inner portion and proceeding outward.

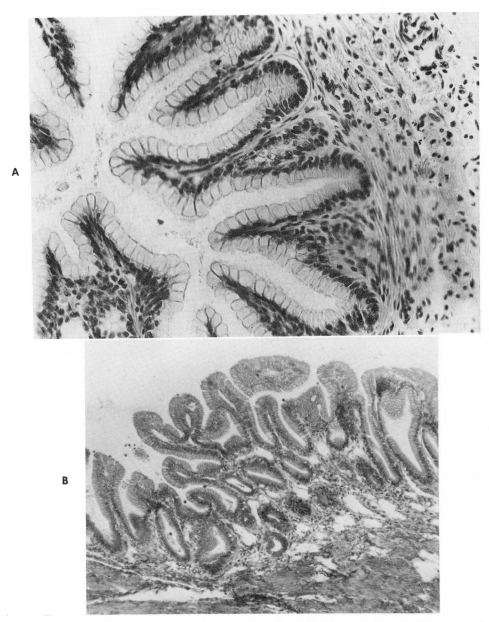

Fig. 12-14. Sections through pyloric stomach wall of *Squalus acanthias.* **A,** Transverse. **B,** Longitudinal, showing extensive folding of mucous membrane. See text for description. (**A,** ×800; **B,** ×750.)

Stomach. As in the sharks, the stomach is U-shaped, consisting of a wide cardiac and a narrow pyloric part. Since fish have no salivary glands, the process of digestion begins in the stomach. The wall of the stomach of the brook trout, *Salvelinus,* is rather thin. It is interesting to note that in the lakes of Ireland there is a form of trout, known locally as the gillaroo,

which lives largely on shellfish and which has a remarkably muscular and thick-walled stomach.

The histologic structure of the stomach of *Salvelinus* bears a rather striking resemblance to that of higher vertebrates, including man. The inner lining is marked by folds, or rugae, of the mucosa and submucosa. These are direct continuations of the folds in the esophagus, but here

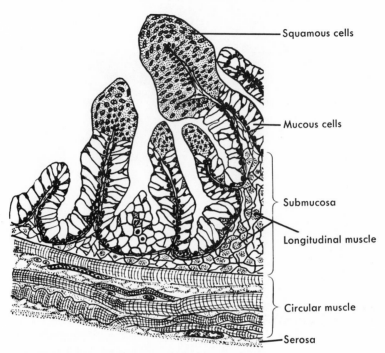

- Squamous cells
- Mucous cells
- Submucosa
- Longitudinal muscle
- Circular muscle
- Serosa

Fig. 12-15. Portion of cross section of esophagus of stone-roller, *Campostoma anomalum* (Rafinesque). (Redrawn from Rogick, 1931.)

they show branches and anastomoses, forming a longitudinally oriented network. The mucosa contains numerous gastric pits. The surface of the mucosa is covered by a special type of mucus-secreting, simple columnar epithelium; this same type lines a portion of the pits. In these cells the nucleus is usually located in the middle third of the cell. The distal third is filled with a clear mucus that is different from that found in goblet cells or other mucus-secreting cells generally. It tends to be acidophilic (i.e., to have an alkaline reaction) rather than basophilic. Also it is not cast off in one mass as often occurs with the mucus of goblet cells but is apparently discharged in a continuous fashion. The function of this mucus secretion is probably protection of the epithelial cells against the hydrochloric acid produced by the gastric glands.

In the pits the epithelium changes to lower and broader cells with more basally located nuclei; the mucus-secreting function gradually diminishes. Each primary pit divides into several secondary ones.

The gastric glands open into the secondary pits, at least two glands for each pit. The glands themselves are elongated tubular structures, occupying all of the rest of the thickness of the mucosa, with a cellular lamina propria lying between them. The narrow lumen of each gland is surrounded by cuboidal, somewhat dome-shaped cells with projecting apical ends. The nuclei are spheroidal, and the cytoplasm contains secretory granules. It is believed that these granules are the enzyme pepsinogen, which is capable of digesting proteins. Pepsin, derived from pepsinogen, acts only in an acid medium, which is also produced by these cells—in the form of hydrochloric acid.

In the portion of the stomach near the esophagus and through much of the main body of the stomach are the **fundic glands;** near the intestinal, or pyloric, end of the stomach are the **pyloric glands.** A constricted upper, or "neck," portion of the gland of either type, just below the foveola, often shows a few special "mucous neck cells."

The lamina propria between the glands is highly cellular. Especially prominent here are peculiarly motile cells with large cytoplasmic granules. The bony fishes apparently are the lowest vertebrates in which such cells are found. They are so abundant in the stomachs of some as

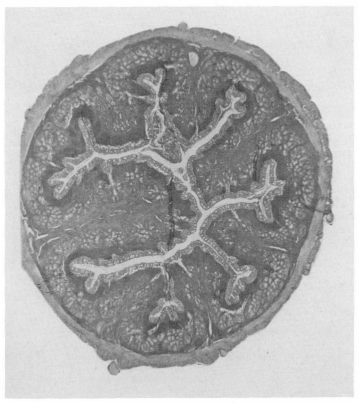

Fig. 12-16. Cross section of frog's esophagus. Note folds of inner surface, goblet cells in lining epithelium, and prominent muscularis coat. (×50.)

to form aggregations that may be mistaken for lymph nodules. Some investigators have called them mast cells, but the granules in certain species are eosinophilic rather than basophilic as are mast cells in higher forms.

The farther caudad one goes in the stomach, the more widely the gastric glands are separated from one another, and the greater the amount of interglandular connective tissue. The pepsinogen-secreting cells are gradually replaced by taller mucus-secreting ones until this replacement is complete by the time the pyloric sphincter is reached.

A conspicuous layer in the wall of the stomach of many fishes is the **stratum compactum.** It is apparently the deep portion of the lamina propria, but its appearance is in marked contrast to the more superficial, highly cellular part. It is a homogeneous-appearing fibrous layer containing very few cells. Beneath it is a muscularis mucosae with its fibers oriented in a longitudinal direction.

There is a submucosa of fibrous connective tissue, staining much less deeply than the mucosa.

The submucosa contains blood vessels, branches of which penetrate far into the mucosal folds, and nerve fibers and cells that form a "submucous plexus."

The muscular coat of the fish stomach usually consists of an inner circular and an outer longitudinal layer, with an intermuscular plexus of nerve cells and fibers between them. The muscle is exclusively smooth muscle. Delicate bundles arise from the inner circular layer, penetrate the submucosa, and form a network that penetrates to the boundary of the lamina propria. This is the muscularis mucosae. Outside the muscle layers is the serosa, made up of a thin layer of connective tissue and the peritoneal epithelium.

AMPHIBIA

Esophagus. The esophageal wall of amphibians consists of four major layers: mucosa, submucosa, muscularis, and adventitia. The mucosa is represented primarily by the epithelium, with a narrow strip of denser connective tissue serving as a lamina propria. The inner surface of the organ

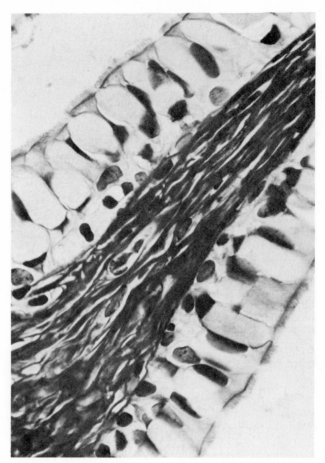

Fig. 12-17. Cross section of longitudinal fold of esophagus of *Ambystoma mexicanum*. Note ciliated pseudostratified columnar epithelium and numerous goblet cells. Core of fold is primarily collagenous connective tissue. (×250.)

is thrown into folds (Fig. 12-16) involving the looser connective tissue of the submucosa as well as the mucosa. The epithelium is a simple one-layer or two-layer tissue. The cells may be cuboidal or columnar (Fig. 12-17). In fact, size and shape of the epithelial cells vary according to location on the folds. Those in the depressions may be only half as tall as those on the crests.

Regardless of topography, three kinds of cells are found: ciliated columnar or cuboidal, goblet, and spindle-shaped. The ciliated cells are the most numerous. Unlike many columnar cells, these usually have the nucleus somewhat closer to the apical end. Goblet cells are generally numerous, but vary in numbers in different specimens. The mucus mass may often be seen protruding at the surface. The "spindle" cells, which do not reach the lumen, help to give the epithelium an appearance of stratification; it is actually of the type generally called pseudostratified.

Glands are present in the distal portion of the esophagus in some genera (e.g., *Proteus, Bufo,* and *Rana*) and absent in others (e.g., *Salamandra* and *Pipa*). These are large, simple alveolar glands. They contain both mucous and serous cells, the former generally occupying the neck region of the gland. These esophageal glands have been shown to secrete pepsin. However, the question of homology with stomach glands of higher vertebrates is still open.

The lymphoid tissue appears to be an important part of the esophagus in the Anura. Many nod-

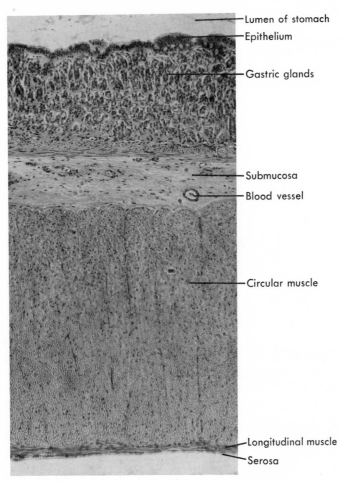

Lumen of stomach
Epithelium
Gastric glands
Submucosa
Blood vessel
Circular muscle
Longitudinal muscle
Serosa

Fig. 12-18. Longitudinal section through frog's stomach, showing different layers in wall. Note extensive region of glandular mucosa with sections of glands. (×130.)

ules normally occur in the lamina propria. A peculiar arrangement is seen in *Bufo* where each nodule shows a **central cavity.** The epithelium dips down to line these cavities of the individual nodules.

Beneath the connective tissue of the mucosa are some scattered bundles of smooth muscle tissue, the muscularis mucosae. This layer is found near the lower end of the esophagus, but appears to be lacking in some species. The submucosa consists of a loose connective tissue with few cells compared to the lamina propria. As in the fishes, it contains blood vessels, a nerve plexus, and lymphatics.

The **muscularis externa** lies outside the submucosa. It has a relatively thick inner subdivision of circular smooth muscle. The outer, longitudinal muscle is thin and, in the upper part of the esophagus, consists of separate bundles between which are masses of connective tissue. The outermost portion of the wall is the **adventitia,** formed by loose connective tissue that binds the esophagus to surrounding organs.

Phylogenetically there are perhaps not too many significant points on which to compare the esophagus of the fishes and the amphibians. The need for cilia would seem to be greater in animals that live at least partly out of the water; they are conspicuous on the epithelial lining of amphibians. In both fishes and amphibians the muscle coat forces the food along toward the stomach. In fishes, in which striated skeletal muscle extends far along the esophagus, "swallowing" may be a more prolonged voluntary act.

Stomach. There is no sharp demarcation of the stomach from the esophagus, but the termination of the longitudinal folds generally marks the ending of the esophagus. In contrast to what is seen in man and other mammals, the histologic transition is a gradual one. Ciliated cells are found here and there, well down into the more characteristic gastric epithelium, and some of the gastric epithelial cells extend up into the caudal end of the esophagus.

A study of a longitudinal section of the frog stomach (Fig. 12-18) reveals several layers: a thick layer of epithelium, connective tissue, and glands; a layer of longitudinal and circular muscles, the muscularis mucosae; connective tissue and blood vessels, the submucosa; two layers of muscles, an inner circular layer and an outer longitudinal layer, together called the muscularis externa; and a thin outer layer of flattered cells, the serosa, formed from the peritoneum.

The gastric glands and mucosae of the frog have been recently reexamined by Norris (1960). With the help of modern histologic techniques, he established three distinct mucosal zones in the stomach: forestomach, fundic portion of stomach, and pyloric region. Six special cell types were recognized in these zones: zymogenic cells, acidophilic cells, argentaffin cells, and three types of mucous cells. Norris introduced the term forestomach to that short transition zone where the many longitudinal rugae of the pale esophageal mucosa are replaced by the brownish gastric mucosa with fewer rugae. This zone is also characterized by having simple ciliated columnar surface epithelium with numerous goblet cells. A few glands are present here, as in the esophagus; they are made up of zymogenic (enzyme-producing) cells. There is a definite muscularis mucosae, which is absent in the esophagus. At the margin between the forestomach and the fundic stomach the muscularis externa (outer longitudinal and inner circular coats) becomes much thicker. The fundic portion of the stomach has a lining of simple columnar mucous epithelium. (There appear to be no goblet cells or ciliated epithelium.) The gastric glands are simple or branched tubular glands that, as in the fishes and higher vertebrates,

empty into gastric pits. These glands are packed in fairly close to each other, being longest in the fundic region and shortest at the esophageal end.

Two prominent cell types are distinguished in the epithelial lining of the glands—neck mucous cells and an acidophilic type (Fig. 12-19). The mucous cells have their nuclei located near the bases and a vacuolated appearance in their apical cytoplasms. The acidophilic cell, making up the major part of the glandular epithelium, has an ovoid nucleus near the base of the cell and a finely granular eosinophilic cytoplasm. This type secretes hydrochloric acid and some pepsinogen, and therefore it may correspond to both the parietal and chief (zymogenic) cells of the mammalian stomach. These acidophilic cells may represent a type from which the parietal and chief cells have not yet differentiated.

A third type of cell, the argentaffin cell, is filled with argyrophilic granules and may be observed after silver staining between epithelial cells in the upper part of the fundic glands. These cells are known to secrete serotonin, in some animals a potent vasoconstrictor substance.

The pits are deeper and the gland tubules shorter in the pyloric regions of the stomach. Acidophilic cells are scarce or entirely absent, but the surface epithelium is about the same as in the fundic region. Pyloric glands are mostly lined by large mucous cells, and scattered argentaffin cells may also be revealed by silver staining.

To summarize the distribution of the various cell types according to Norris (1960), the zymogenic cell is restricted to the glands of the esophagus and forestomach; the acidophilic cell is found in the fundic glands; the argentaffin cell is in all four zones; and the mucous cells are variously distributed in the different zones, the so-called neck mucous gland being found in all four zones.

Beneath the glands is a muscularis mucosae, a thin layer of smooth muscle. In the fundic part it consists of longitudinal fibers, whereas in the pyloric region an inner circular subdivision is added.

The muscularis externa, or muscle coat, consists of inner circular and outer longitudinal layers. In the pyloric region the circular portion is particularly well developed, but also in this region, the individual cells are only about half as

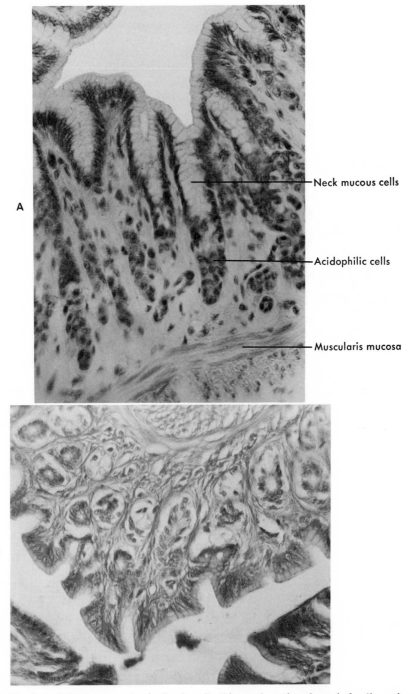

Fig. 12-19. Fundus of frog's stomach. **A,** Section. **B,** Diagonal section through fundic region, showing "nests" of glands containing mostly acidophilic cells. (**A,** ×480; **B,** ×300.)

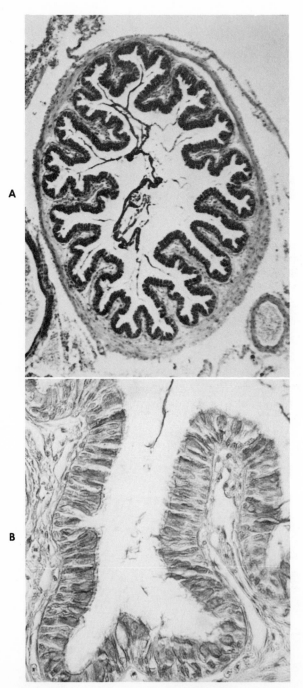

A

B

Fig. 12-20. Cross sections of the esophagus of the lizard, *Anolis carolinensis*. Note lining of double layer of epithelial cells and rugalike extensions of mucosa layer. (**A,** ×35; **B,** ×400.)

long as in other parts of the alimentary tract. An interesting feature of the upper portion of the stomach is the presence of bundles of smooth muscle running between inner circular and outer longitudinal subdivisions. These bundles may represent the middle subdivision of the muscle coat in mammals, the part called the oblique muscle.

REPTILIA

Esophagus. The microscopic structure of the esophagus of many reptiles (e.g., the lizard) shows similarity to that of the amphibians, with a double-layered, columnar or cuboidal, ciliated epithelium (Fig. 12-20). However, in those turtles that spend more of their life back in the water, the epithelium is a cornified, stratified squamous one. Prominent pointed papillae line the esophagus in these reptiles (Fig. 12-21).

Goblet cells are present, but distinct esophageal glands are absent. As in the amphibians, lymphoid nodules may be scattered just under the epithelial layer in the lamina propria. In some species such as the "slow worm" there is a tremendous amount of lymphocyte migration into the epithelium, reminding one of the condition in the mammalian tonsils.

An indistinct muscularis mucosae of smooth muscle may usually be seen, at least in the posterior part of the esophagus. This layer seems to

Fig. 12-21. Esophageal lining of large marine leatherback turtle. Forest of tall pointed papillae is seen. Penny in field gives an idea of their dimensions. (From Bleakney, J. S. 1967. Marine turtles as introductory material to basic biology. Turtox News 45(3):82-86.)

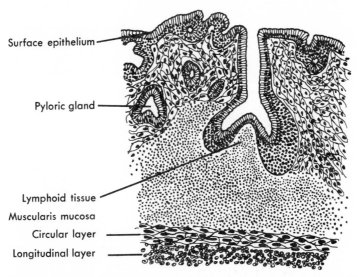

Fig. 12-22. Pyloric stomach of turtle *Emys europea*. Large mass of lymphoid tissue lies at bottom of crypt and causes elevation of epithelium. (Redrawn from Oppel, 1897; after Ballmer, 1949.)

be missing altogether in the turtle esophagus. The submucosa, as in other animals, is made up of connective tissue, blood vessels, and nerves. It pushes the mucosa into the lumen in folds. In the smooth muscle coat the longitudinal layer is poorly developed. It often is lacking anteriorly, beginning as a series of bundles and becoming a complete layer only near the esophagogastric junction. The outer tunica adventitia is composed of fibroelastic tissue, much as in the amphibians.

Stomach. There is a generally close resemblance of the surface epithelium and that lining the gastric pits of the reptile stomach to these elements in fishes and amphibians. The cells are tall columnar with clear mucus in the apical portions. There has been no further specialization of cells of the fundic glands. Other than the mucous neck cells, the gland cells are all of one type, each being a producer of both pepsinogen, a forerunner of pepsin, and hydrochloric acid.

As in the other cold-blooded vertebrates, the mucus-producing pyloric glands are shorter and less branched than the fundic glands. Lymphoid tissue may be reticular or of solitary nodules (Fig. 12-22). There is usually a distinct muscularis mucosae of both circular and longitudinal fibers. The areolar connective tissue of the submucosa is similar to that of the esophagus. In

contrast to the esophagus, both circular and longitudinal layers of the muscularis coat are usually well developed. The outermost layer of the stomach is the thin serosa.

AVES

Esophagus. The histology of the bird esophagus is similar to that of most vertebrates. The mucosa of the esophagus consists of a stratified squamous epithelium, which may be strongly cornified in some cases, for example, the pigeon. In contrast to mammals, the esophageal glands are found only in the mucosa, not in the submucosa. In fact, some birds (e.g., the swallow *Hirundo*) have glands that lie entirely within the epithelium. These tubular glands are entirely mucogenic and thus have lubricating function only. Their enormous number in the pigeon (Fig. 12-23) indicates heavy secretion of mucus.

The mucosa also shows accumulations of lymphoid tissue, both nodular and diffuse, often in close association with glands and their ducts. In the duck *Anas* such tissue is so well developed as to form an actual "esophageal tonsil." Stratified squamous epithelium lines the crypts of the tonsil, and it is infiltrated by hosts of lymphocytes. Farther posteriorly the crypts seem to be replaced by mucous glands, but the epithelium of these

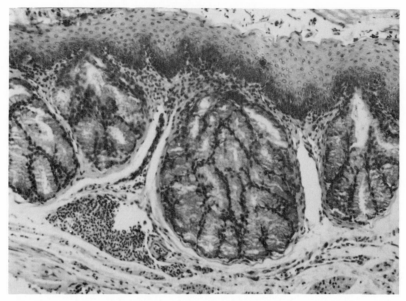

Fig. 12-23. Cross section of pigeon esophagus. Deeply folded tubular glands are entirely in lamina propria and are found in clusters separated by connective tissue partitions. Note extensive folding of mucous membrane and stratified squamous epithelium. (×100.)

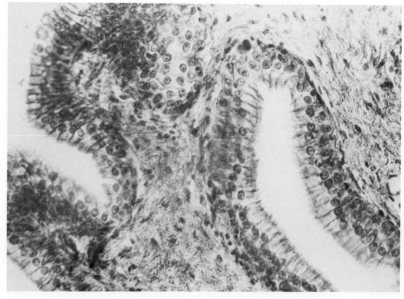

Fig. 12-24. Section through crop of pigeon. Note lining of stratified epithelium. (×200.)

glands is also heavily infiltrated by lymphocytes.

The muscularis mucosae of the bird esophagus varies in different species with regard to the number of layers present (one or two) and the orientation of the muscle fibers and bundles. Beneath the muscularis mucosae is a submucosa and then, as in most other vertebrates, inner circular and outer longitudinal layers of the muscle coat proper plus a covering adventitia.

In gramnivorous, or grain-eating, birds especially, a large **storage sac** is found, the crop. Although this resembles a stomach by its size and shape, it apparently represents the dilated middle or lower portion of the esophagus, as seen by the elements of its wall (Fig. 12-24). Food is moved from the crop to the stomach by reflex action, depending on the fullness of the stomach. The crop may be a temporary expansion of the esophagus (false crop) or a permanent specialized diverticulum (true crop).

The crop of the pigeon has both serous and mucous elements and may play a part in the predigestion of food, especially starches. In the pigeon the crop is specialized to the extent that it produces "milk" which is fed to the young for several days after they are hatched. The milk-producing glands are under the control of **prolactin,** a hormone of the anterior pituitary gland that is curiously similar to the hormone that stimulates the **mammary** glands to activity in mammals (Dumont, 1965).

Shortly before a brood of young doves is hatched, the wall of the crop of the mother increases in thickness, the vascularization increases, and the epithelium hypertrophies. The milk is produced by a desquamation of fat-laden epithelial cells. Consequently, it is particularly rich in fat (35%).

Interestingly enough, **both males and females** produce crop milk, but that of the female has some aspects peculiar to it, including elements similar to the colostrum corpuscles of mammalian milk. Although the functions are similar, crop milk and true milk differ in composition; neither casein (a protein) nor lactose (a sugar) is found in this nutritive fluid of the avian parents.

Stomach. Beyond the crop is the stomach, which is itself subdivided into two parts: (1) a glandular proventriculus, which most resembles the stomach found in the other classes of vertebrates, and (2) a thick-walled, muscular gizzard, evidently derived from the pyloric part of the stomach but highly specialized as a "masticatory" organ for the **edentulous** birds. In many birds that eat hard vegetable matter, the gizzard has a keratinous lining formed by numerous glands. In addition, it contains stones that have been swallowed and forms a veritable "mill" in this part of the alimentary canal. The gizzard in carnivorous birds is much less developed than in the grain-eaters.

A curious condition is seen in the herring gull, *Larus argentatus.* As a fish-eater, this bird shows relatively little specialization of the gizzard, but when it takes to invading the fields and becomes a confirmed grain-eater, the pyloric part of the stomach develops a thick muscular wall and a horny lining, thus transforming into a true gizzard, The great British anatomist and surgeon, John Hunter, first pointed out this interesting fact.

The character of the mucosa of the avian true stomach, or proventriculus, is distinctive when compared with vertebrates of other classes. The glands here are of a **composite** nature, a large number of individual tubules emptying into a large central chamber, which in turn opens into the main lumen of the stomach. The cells lining the central chamber and the duct resemble the "mucous neck" and surface cells that we have seen in lower vertebrates. In the carnivorous species, the composite glands appear less complex than in the herbivorous and insectivorous ones.

The cells of the proventricular glands are all of one type and show little difference from those of reptiles. They are functionally homologous with both the chief and parietal cells of mammals, producing both hydrochloric acid and pepsin. The muscularis mucosae of the proventriculus has inner circular and outer longitudinal subdivisions. They lie rather far apart from one another in the birds, separated by connective tissue.

Two features of the gizzard, or ventriculus, are outstanding: (1) the tough, noncellular lining, which consists of a keratinoid substance called koilin, and (2) the thick, powerful smooth muscle coat of the wall (Fig. 12-25). Although the glands of the mucosa are specialized in producing the peculiar lining material, they are morphologically comparable with gastric glands in general and

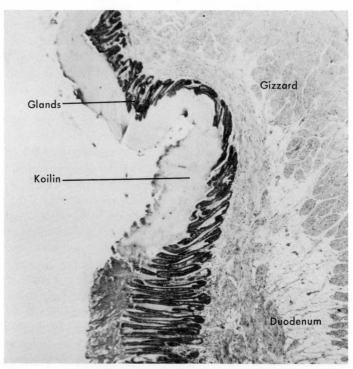

Fig. 12-25. Junction between gizzard and duodenum of *Columba*. Gizzard is characterized by strong musculature and noncellular lining of koilin. This lining appears more or less homogeneous but has horizontal laminations and striations. It is secreted by glands in the gizzard. (×35.)

probably contribute some proteolytic enzyme and acid to the contents of this part of the alimentary tract. The transition between the mucosa of the proventriculus and ventriculus is a gradual one, in contrast to the sudden change from esophagus or crop to the proventriculus. The nature of the glands alters little by little.

The noncellular lining material shows vertical and horizontal striations, the former coinciding with the individual tubular glands, the latter perhaps being related to varying content of desquamated cells, bile pigment, and other elements. The lamina propria is a highly cellular lymphoreticular tissue, usually with many cells that appear to be "tissue eosinophils," local residents rather than recently arrived from the circulating blood.

There is a positive correlation between the degree of development of the "cuticle," or koilin lining, of the ventriculus and the development of the stratum compactum beneath the lamina propria. On the other hand, the better developed

the lining, in general, the less developed is the muscularis mucosae of this part of the alimentary tract. The submucosa of both the true stomach and the gizzard consists of areolar connective tissue with blood vessels and nerves, but no glands.

The muscle coat of the true stomach consists of the usual inner circular and outer longitudinal layers. In the gizzard, however, the circular division shows a tremendous development. An interesting feature is a special arrangement producing a strong point of origin for the muscles of the gizzard wall. There is a concentration of collagenous fibers forming an iridescent mass on one side of the gizzard, from which muscle fibers fan out along the wall. The muscular coat of the gizzard tends to be thickest in gramnivorous birds, less so in carnivorous species, and least in frugivorous ones, the fruit-eaters.

The proventriculus functions in the secretion of gastric juice and as a storage organ. Many birds of prey are capable of dissolving even the skeletons

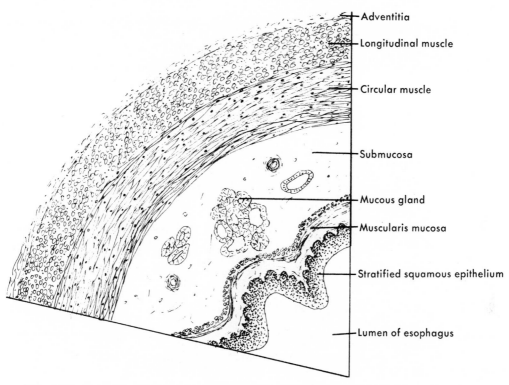

Fig. 12-26. Diagram of cross section of rabbit's esophagus. Contours of layers are very irregular.

of the victims in the stomach contents. Some, however, especially the owls, form pellets of fur, feather, and bones in the proventriculus and eject them through the mouth. The ventriculus, or gizzard, is chiefly a mechanical device (with the aid of small pebbles) for grinding up the food, taking the place of teeth. However, as a result of the secretions of the ventricular glands, some digestion apparently takes place here.

MAMMALIA

Esophagus. The mammalian esophagus (Fig. 12-26) is a simple tube containing some mucous glands for lubricating the passage of food. It is lined with layers of stratified squamous epithelium, surprisingly similar to those of the epidermis, and has both striated and smooth muscle in its walls. Longitudinal folds of the mucosa and submucosa reduce the lumen to a very small opening when food is not passing through it.

There is usually a thin stratum germinativum, a well-developed stratum granulosum with coarse basophilic granules in its cells, and a stratum corneum on the surface that forms about half the total thickness of the epithelium (Fig. 12-27). The degree of cornification is greatest in mammals that have a diet of coarse food such as grains and tough, fibrous plants and is least in flesh-and insect-eating mammals. The **stratum granulosum,** much as in thick epidermis, is usually prominent. However, in mammals in which the cornification is fairly marked but in which the cells of the cornified layer still show remains of nuclei, there is no stratum granulosum. The ox, *Bos,* is an example of a mammal with this condition.

A number of mammals show cornified papillae in the esophagus, similar to those of the tongue but lacking taste buds. These are prominent in the Bovidae, a large family of ungulates that has more than a hundred species including cattle, bison, goats, and sheep.

It is a mammalian characteristic that glands of

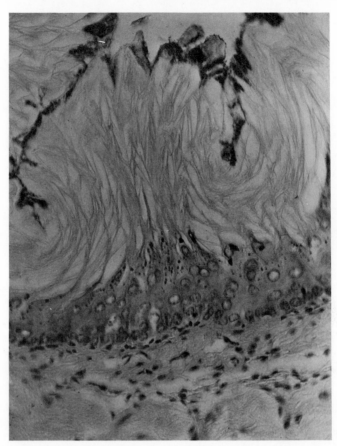

Fig. 12-27. Esophagus of mouse. There is a thick stratum corneum in correlation with the generally coarse, abrasive food of this mammal. (Wilder's reticular stain; ×250.)

the esophagus occur predominantly in the submucosa. An interesting evolutionary question is why and how these glands have retreated to this deeper position in these descendants of the reptiles whereas in the birds they have remained in the mucosa.

Human esophageal glands are of mucous character. Although this is true in most mammals, there are species in which the glands are mixed, often with peripheral demilunes of serous cells on the mucous alveoli. Some others (e.g., the mouse) have no glands in this portion of their digestive tracts. When the esophageal glands continue on into the cardiac part of the stomach, as they do in some kinds of animals, their ducts travel back to empty into the lumen of the esophagus. Under this arrangement their secretion is not mixed directly with the gastric one.

Different species of mammals vary remarkably in the degree of development of lymphoid tissue. In some it is almost or entirely lacking from the wall of the esophagus, but in others it is abundant. The pig, *Sus,* has so many lymphoid nodules and so much infiltration of epithelium by migrating lymphocytes, along with the presence of cryptal invaginations, that one seems justified again in speaking of "esophageal tonsils" as was done in the case of the duck.

A muscularis mucosae is a constant feature of the mammalian esophagus, at least somewhere along its course. In many mammals such as the mouse, it is lacking from the upper portion. As in the lower vertebrates and birds, the submucosa is areolar tissue. The elastic connective tissue elements in mammals, however, are especially well developed around the glands and in relation to the muscularis mucosae.

There is much variation in the muscle coat of

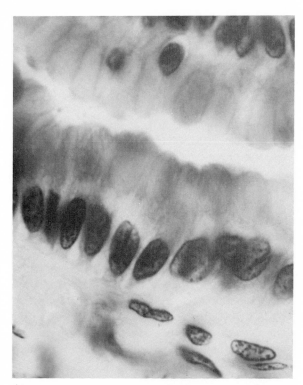

Fig. 12-28. Simple columnar epithelium lining gastric crypt in stomach of dog. (×1495.)

the mammalian esophagus. In all species except the monotremes there is an extension of striated skeletal muscle fibers along the esophagus. In the dog, rabbit, and pig they reach almost to the stomach; in the cat they go only about half the distance. Descending the esophagus, it is the inner portion of the muscle wall that first begins to show smooth muscle fibers. The striated muscle remains longest in the outermost layer.

The subdivisions of the muscle coat are not actually circular and longitudinal, although in section they appear very much this way. Rather, they are arranged as tight and loose **spirals.**

The outer layer of the esophagus in all mammals is an adventitia of fibrous connective tissue, often containing fat and serving chiefly for anchoring the esophagus in place.

Stomach. The stomach (Fig. 12-28) is an important part of the alimentary tube in mammals. It is found in its simple fundamental form among the insectivores, the carnivores, and some of the primates, including man. Many others, herbivores especially, have compound-type stomachs. In some mammals, the stomach is bilocular, consisting of two clearly distinguishable divisions. More frequently, however, the division into two or more portions is a matter of internal structure and type of epithelium.

As mentioned in the previous discussions, simple columnar epithelium serves as both the surface and crypt lining of the stomachs of nearly all cold-blooded animals. However, many different mammals have an epithelium of the **stratified squamous** variety lining a greater or lesser part of their stomach. These include monotremes, marsupials, edentates, ungulates, many rodents, and some primates. Usually, there are no glands in this part of the stomach.

This unusual stomach lining is very similar to the lining of the esophagus and is reminiscent of the avian crop. In fact, there may be a similarity in function. However, it will be recalled that the crop is actually the lowermost part of the **esophagus,** whereas these areas are integral parts of the true stomach. The development of such a lining, well suited for receiving rough, solid foods, arose in various orders of mammals apparently independent of crop evolution. It seems to represent the phenomenon of "convergence," well known among evolutionists, by which adaptation to similar conditions leads to development of similar structure in groups that are taxonomically and genetically unlike.

HUMAN. In the simple mammalian stomach such as that found in man, the cardiac (esophageal) end is much wider than the pylorus. The cardiac stomach itself is the transitional region between the esophagus and the stomach. Like the esophagus, it contains mostly mucous glands. Part of the cardia bulges to form the fundus. This portion of the stomach is lined with simple columnar epithelium, folded to form deep pits into the bases of which open the ducts of long, tubular glands.

One should recall that in the lower vertebrates and birds there is only one kind of cell in the gastric glands proper. The same cells secrete both pepsin and hydrochloric acid. Only in the mammals are there two distinct types differentiated—chief cells and parietal cells. During embryologic development of the mammalian stomach, the glands originally consist of one type of cell, resembling that found in lower vertebrates.

The most numerous cells of the fundic glands are the serous chief cells, also called zymogenic because of their secretion of the enzyme pepsin. These are columnar cells of medium height with oval, centrally located nuclei and granular cytoplasm.

A second type of cell, the mucous neck cell, is found in the upper portion of the tubules, in the region of the slight constriction or "neck" where tubule and pit are joined. These cells are taller and stain lighter than the chief cells. They secrete mucus, which is mixed with the enzyme-containing serous fluid from the chief cells. The relative importance of the mucus from the neck cells, as compared with the abundant secretion formed by cells lining the foveolae and covering the surface, is not understood as yet.

Among the zymogenic and mucous neck cells are other cells, generally large and conspicuous by their lighter, often brightly acidophilic cytoplasm, which appear as though wedged in between the other cells. These are the parietal cells. Their nu-clei are large and round; frequently there are two nuclei in one cell.

Fundic glands occupy a relatively large portion of the thickness of the mucosa; hence the pits in their part of the stomach are shallow. The pits in the lower stomach are longer, however, because the pyloric glands themselves are shorter and deep in the mucosa. The glands of the pyloric region (Figs. 12-29 and 12-30) are lined by tall columnar cells with basal nuclei. They secrete mucus only. Transition between the regions of fundic and pyloric glands is a gradual one.

In general, the stomach mucosa is thick, and the wall has the characteristic submucosa and muscular layers. A prominent muscularis mucosae can usually be seen in cross sections.

MOUSE. The histologic structure of the stomach of the mouse is considerably different from that of man. Internally, the stomach consists of two distinct parts (Fig. 12-31), as discussed previously. The mucous membrane on the left side has a stratified squamous epithelium with a cornified

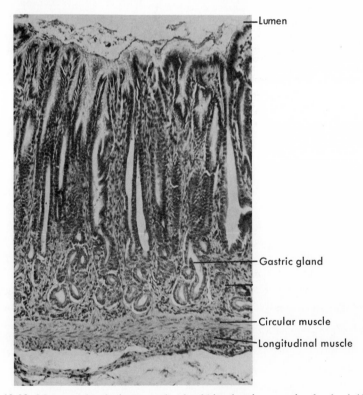

Fig. 12-29. Mucosa of pyloric stomach of rabbit, showing gastric glands. (×100.)

surface layer and no glands (Fig. 12-32). This nonglandular part of the stomach is thin-walled and appears slightly translucent and grayish internally, whereas the rest of the stomach (to the right) is thick-walled, white, and opaque.

The mucosa of the nonglandular region has many papillae, constructed from the connective tissue lamina propria and the cornified epithelium. At the boundary between glandular and nonglandular parts, the nonglandular mucous membrane forms a prominent ridge. Near the entrance of the esophagus into the stomach, about midway

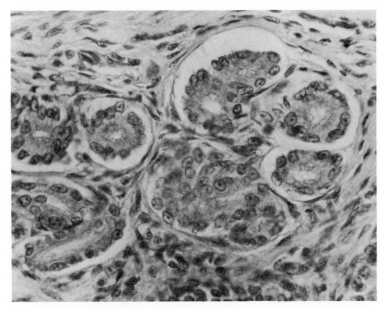

Fig. 12-30. Pocket of gastric glands in pyloric stomach of cat. In preparation, these glands have been cut off from their ducts. (×600.)

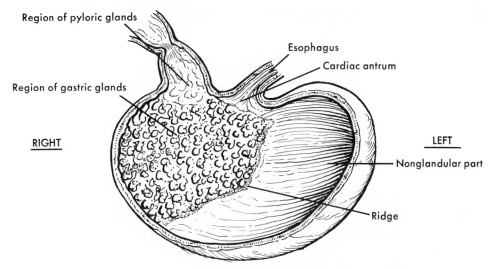

Region of pyloric glands

Esophagus

Cardiac antrum

Region of gastric glands

RIGHT

LEFT

Nonglandular part

Ridge

Fig. 12-31. Outline drawing of mouse stomach. Nonglandular part has a lining of stratified epithelium; the glandular part, one of simple epithelium. (Redrawn from a figure by B. Bohen in Snell, 1941.)

Fig. 12-32. Region of the ridge in mouse stomach. Ridge and the stomach to its left are lined by stratified squamous epithelium; gastric glands on right are lined with simple columnar epithelium. Many large, light staining parietal cells are seen in lower regions of glands. (Hematoxylin-eosin; ×250.)

along the lesser curvature, this ridge is particularly large. It lies just to the right of the entrance and forms the edge of a sort of corridor leading to the large cavity of the nonglandular side. Evidently the ridge functions to direct the food for storage prior to entering the glandular part of the stomach.

In the glandular part of the stomach closest to the nonglandular part near the boundary, there is a narrow transitional zone where the gland tubules themselves are lined by epithelium much like that generally found lining the pits. These were considered to be cardiac glands by Bensley (1902). To the right of this zone, the glandular stomach is divided roughly into two main parts. The gastric or fundic glands are found in the first part, and the pyloric glands in the second, closest to the small intestine.

The connective tissue of the mucosa constitutes a lamina propria. The cell population is more varied here than in the esophagus and nonglandular part of the stomach. Numerous lymphocytes, plasma cells, and eosinophil leukocytes, as well as fibroblasts, are seen. There is a muscularis mu-

cosae, consisting chiefly of longitudinally arranged smooth muscle, in both parts of the stomach. On the glandular side, delicate strands of this muscle course between the glands. They apparently give motility to the glandular mucosa and help discharge the secretion.

Beneath the muscularis mucosae is a submucosa of loose connective tissue with blood vessels and lymphatics. Beneath it, in turn, is a muscle coat consisting of three layers: inner oblique, middle circular, and outer longitudinal. At the pylorus the circular layer is thickened to form the **pyloric sphincter.** The muscle coat of the nonglandular side of the stomach is relatively thin.

The outermost layer of the stomach is the serosa, connective tissue covered by the flattened mesothelial cells; this tissue constitutes a part of the visceral peritoneum of the abdominal cavity.

ADAPTATIONS OF THE MAMMALIAN STOMACH. The lining of the entire stomach in monotremes is stratified squamous epithelium. Although the order Monotremata represents the lowest or most primitive of the mammals, it is most unlikely that this type of lining is due to a primitive condition.

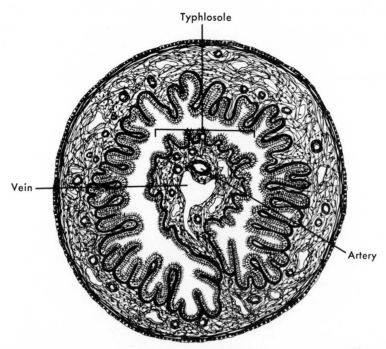

Typhlosole

Vein

Artery

Fig. 12-33. Cross section of small intestine (midgut) of lamprey. Conspicuous artery and vein are seen in typhlosole, which involves both mucosa and submucosa. (Redrawn from Krause, 1923.)

Rather, it appears to be an adaptation to the type of diet, which consists generally of large quantities of food of low nutritive quality that is macerated in the stomach before being passed on to a place of enzymatic digestion.

In the next lowest order, the Marsupialia, the relation of stomach structure to diet is conspicuous. Insectivorous marsupials such as *Didelphis* (opossum) have simple columnar epithelium with mucous cells on the surface and in the foveolae (crypts). Gastric glands are of the type found in the glandular stomach generally, and the fundic glands have both chief and parietal cells. On the other hand, herbivorous marsupials such as the Macropodidae (kangaroos and wallabies) have a lining of stratified squamous epithelium for the upper part of the stomach and a glandular lower portion.

Similar to the insectivorous marsupials, mammals of the order Insectivora have a simple stomach with a simple columnar epithelium and gastric glands.

It is interesting to note that the important cellulose-splitting function in the great group of even-toed ungulates (order Artiodactyla) is carried out by symbiotic microorganisms in a highly specialized stomach, whereas in the odd-toed ungulates (order Perissodactyla) it occurs in the **cecum,** as is the case in lagomorphs and some other groups. The involvement of symbionts in digestion in the nonglandular portions of stomachs has been suggested for other mammals, too (Young, 1962).

The intestine

CYCLOSTOMATA

In this group, with no stomach as such, the intestine accounts for the major portion of the alimentary canal. It is a tubular structure that shows no winding or coiling. At the transition of the esophagus to the intestine, there is a follicle or mass of hemopoietic-lymphoid tissue.

The intestine of the lamprey shows a prominent longitudinal fold, which follows a slightly spiral course and almost fills the lumen of the empty intestine. This is the "typhlosole," which the student may recall is the name applied to the prominent, but straight, fold in the intestine of the earthworm. In both cases, whether or not the structures should be thought of as homologous, they serve to increase the digestive and absorptive surface area of the intestine.

In the dorsal portion of the fold is a large artery, and ventrally a large vein (Fig. 12-33). This spiral fold has so many vessel spaces within it that it has been described as consisting largely of cavernous tissue. There are many smaller longitudinal folds projecting into the lumen, and the typhlosole itself bears secondary folds.

The lining epithelium of the intestine generally bears short cilia (Fig. 12-34). In places, however, especially on the side toward the liver, there are nonciliated cells with large, acidophilic secretory granules. Mucous cells are **not** found, and there are no intestinal glands as such.

Between the mucosa and smooth muscle coat is a plexus of nerve cells and fibers, which seems to correspond to Meissner's plexus of the alimentary tract of mammals.

The ciliated cells apparently serve to propel the food through the intestine. Peristaltic action is weak, and the muscle coat is correspondingly poorly developed. The ciliated cells also have the function of absorption. This type of intestine probably represents a truly primitive condition such as existed in the ancestors of the vertebrates and their earliest derivatives.

CHONDRICHTHYES

The intestine of the elasmobranch fishes is divided into a small and large intestine. The small intestine, which actually has a larger diameter than the large intestine, consists of a short duodenum (Fig. 12-35) into which the bile and pancreatic ducts open and a long, wide, and straight ileum provided with a typhlosole-type spiral valve (Fig. 12-36). This conspicuous spiral valve greatly increases the internal surface and thus substitutes functionally for the long coiling of the intestine seen in higher forms.

At the junction of the large intestine and the rectum, a small rectal gland opens into the gut. This gland, found only in this group of fishes, is now known to regulate the salt content of the sharks' body fluids (Burger and Hess, 1960a and 1960b). Bulger (1963, 1965) has given a description of the fine structure of this unique gland.

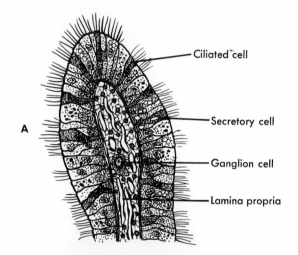

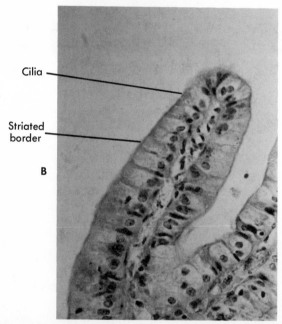

Fig. 12-34. A, Detail of fold of mucosa of small intestine of lamprey. There are many ciliated cells here; hagfish lacks such cells. B, Photomicrograph of mucosa fold of small intestine of lamprey. Cilia are seen on some cells; others show only a striated border. (B, Hematoxylin-eosin; ×400.)

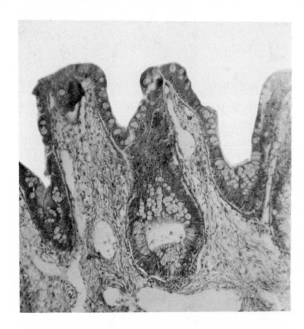

Fig. 12-35. Section through duodenum of shark *(Squalus acanthias),* showing numerous mucous glands. (×150.)

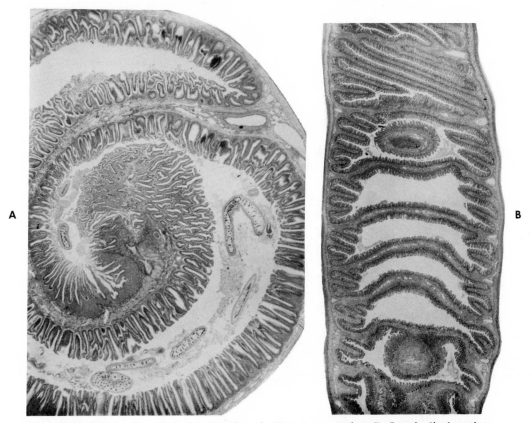

Fig. 12-36. Spiral valve of shark's intestine. **A,** Transverse section. **B,** Longitudinal section. (**A,** ×77; **B,** ×75.)

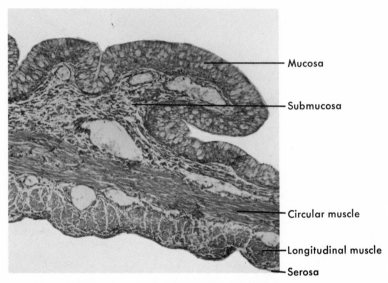

Mucosa

Submucosa

Circular muscle

Longitudinal muscle

Serosa

Fig. 12-37. Cross section of large intestine of *Squalus acanthias.* (×150.)

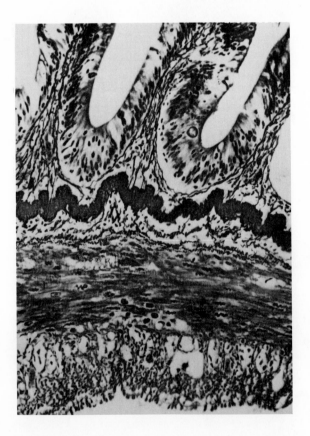

Fig. 12-38. Small intestine of *Salmo gairdneri.* Beneath folds is layer of dense connective tissue (dark), the stratum compactum. External to connective tissue is layer of smooth muscle; outside of this is serosa of connective tissue and peritoneal epithelium. (Wilder's reticulum stain; ×125.)

It consists of many branched tubules intermingled with large quantities of lymphoid tissue. The rectum opens into the cloaca, which also receives the openings of the urinary and genital ducts.

The cells of the epithelial lining of the intestine have striated borders. Goblet cells are also present. The intestinal epithelium is not provided with cilia (Fig. 12-37), and so peristalsis has apparently become adequate for the job of food propulsion. Thus, as might be expected, the muscular coat of the intestine is well developed.

OSTEICHTHYES

The trout intestine (Fig. 12-38) is a rather uncomplicated tube that first passes forward as the duodenum and then becomes bent on itself and passes backward, without further convolution, to the anus. The posterior portion of the large intestine has its mucous membrane raised into ring-shaped ridges, which probably represent the spiral valve that is such a conspicuous structure in the intestine of the elasmobranchs. The name "large intestine" for the posterior portion of the intestine of fishes is somewhat arbitrary, since in many cases, its diameter is about the same as that of the "small intestine."

In the majority of bony fishes, tubular pouches (the pyloric ceca) open into the duodenum (Fig. 12-39). These structures vary in number according to species, from two or three in flat-fishes to nearly 200 in the mackerel. Pyloric ceca are not found in higher vertebrates.

The small intestine is lined by a simple columnar epithelium. Here the absorbing cells are interspersed with goblet cells, the mucus-secreting unicellular glands. The epithelium of the pyloric ceca

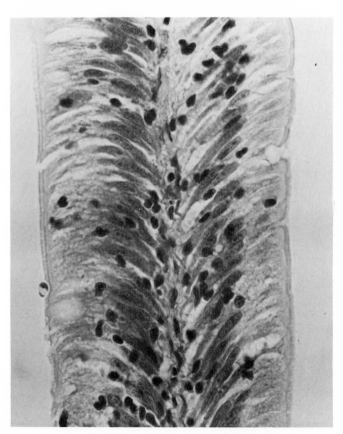

Fig. 12-39. Portion of mucosal fold of pyloric cecum of *Salmo gairdneri,* the rainbow trout. Tall columnar epithelium shows striated border, and there is active migration of lymphocytes. (×400.)

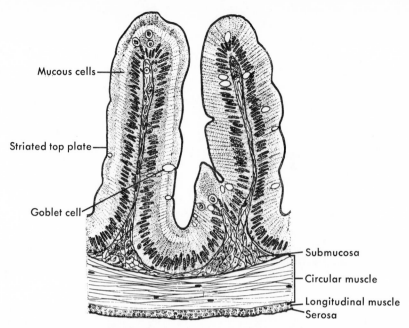

Mucous cells

Striated top plate

Goblet cell

Submucosa

Circular muscle

Longitudinal muscle

Serosa

Fig. 12-40. Cross section of the intestine of the stone-roller fish, *Campostoma anomalum* (Rafinesque). Two irregular "zigzag" folds of mucosa are shown. Layers of wall are indicated. Mucous cells may be early stages in development of goblet cells. (Redrawn from Rogick, 1931.)

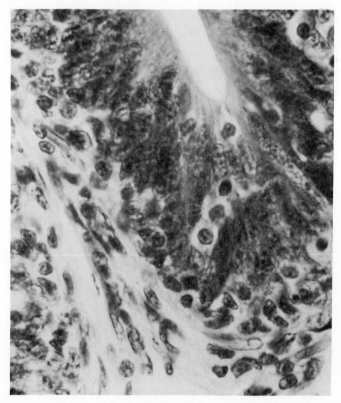

Fig. 12-41. Intestinal mucosa of shark, *Scylliorhinus canicula*. Active migration of lymphocytes into the epithelium is seen here. In places they form "chains" of cells. (×400.)

shows a general similarity to that of the small intestine. However, the absorptive cells of the ceca are somewhat taller and more slender, and the goblet cells are fewer. There are no submucosal glands such as the glands of Brunner of higher vertebrates, nor are there equivalents of the mucosal glands or crypts of Lieberkühn. A few species of fish have structures formed by the intestinal mucosa resembling the villi of the mammalian intestine, but folds of grosser nature are the rule (Fig. 12-40).

Leukocytes, especially small lymphocytes, often can be seen migrating into the surface epithelium (Fig. 12-41). In the conger eel we have found them in vacuoles apparently within the epithelial cells (Fig. 12-42). There are usually several per vacuole, and they appear to be degenerating. In some species such as the African lungfish, *Protopterus,* or the brook trout, *Salvelinus fontinalis,* wandering pigment cells as well as lymphocytes are common (Fig. 12-43).

Histologically, the large and small intestines seem to differ less in bony fishes than in the higher vertebrates. The layers of the wall in both include the mucosa (epithelium and lamina propria, with a prominent stratum compactum of dense connective tissue); submucosa; muscularis with inner circular and outer longitudinal layers; and the serosa.

The length of the intestine relative to the length of the body and other parts of the alimentary tract is greater than in Chondrichthyes. In many species some coiling occurs. As discussed previously, folds of mucosa and submucosa are seen, but in most species other than the elasmobranchs there is no spiral valve.

The alimentary tract of fishes terminates in a chamber called the cloaca, which serves as a **common receptacle** for urine, reproductive cells (ripe gametes), and the undigested residue of food. The epithelium is stratified cuboidal and resembles that of the skin. Skeletal muscle re-

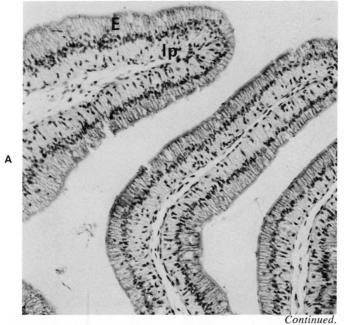

Continued.

Fig. 12-42. Intestine of conger eel, *Conger conger.* **A,** Small intestine. Tall epithelium covers long, branching folds of mucosa. *E,* Epithelium; *lp,* lamina propria. **B,** Detail of intestinal fold. Note delicacy of connective tissue of lamina propria. Epithelial cytoplasm highly vacuolated. A vacuole with degenerating lymphocytes is seen (arrow). **C,** High-power view of basal portion of vacuolated intestinal epithelium. Large vacuole (arrow) contains three degenerating lymphocytes. Many migratory cells are seen in the area. *LP,* lamina propria. (**A,** ×25; **B,** ×400; **C,** ×1000.)

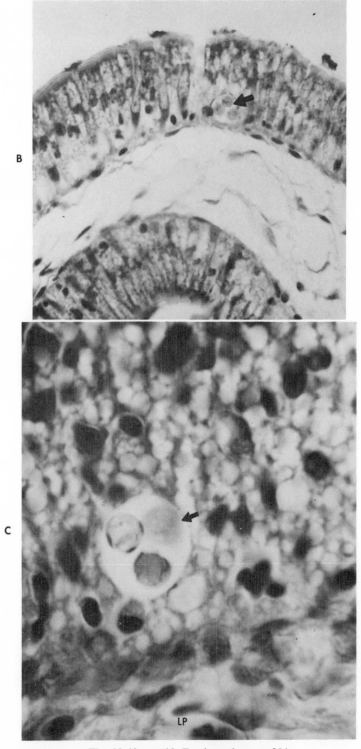

Fig. 12-42, cont'd. For legend see p. 281.

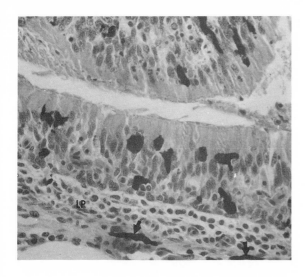

Fig. 12-43. Small intestine of African lungfish, *Protopterus*. Epithelium characteristically contains many large cells with pigment. Lamina propria *(LP)* is highly cellular and contains few pigmented cells (arrows). (×125.)

places the smooth muscle of the wall in the region of the vent.

AMPHIBIA

The small intestine of the amphibians is relatively short. It is marked off from the stomach by a pyloric sphincter, and after a small amount of coiling, it joins with the large intestine. In frogs there is a valvelike structure between the two intestines, but this is lacking in most other types of amphibians. It is interesting to note that the small intestine of the tadpole, which is omnivorous, is far more coiled than that of the adult, which feeds almost exclusively on insects.

Both mucosa and submucosa are raised in longitudinal folds, as in the spiral valve of the shark or the valvulae conniventes of mammals. There are no folds of the mucosa per se that would be comparable to the villi found in higher forms.

The intestinal epithelium consists of two kinds of cells, the simple columnar, general absorptive cells (Fig. 12-44) and the unicellular mucous gland, or goblet cell. On the outer borders of the absorptive cells are striations that the electron microscope reveals to be tiny microvilli. These probably function by increasing the absorptive surface. Assigning such a role to the striated borders seems logical because they are broader in the proximal parts of the intestine and much narrower distally, perhaps indicating the functional difference in degree of absorptive activity.

As in the esophagus, the epithelial cells usually vary in height and shape according to topographic location. On the summits of the folds they are tall and slender; in the depressions, broad and short. Along the sides of the folds they are intermediate and set obliquely on the basement membrane. Goblet cells can be seen wedged between the bases of absorptive cells along the entire length of the intestine. In *Necturus,* they become more numerous distally until, just above the cloaca, the epithelium consists almost entirely of goblet cells.

In the epithelium, at all levels of the alimentary tract that have been described, there are often vacuoles containing white blood cells (leukocytes). These cells exhibit peculiarly varied aspects. Sometimes they are very large, with nuclei of irregular shape and with rather clear cytoplasm. Such large cells frequently contain smaller leukocytes, or what appear to be the breakdown products of such leukocytes. Wandering cells containing granules of yellow pigment are also frequently seen in the epithelium of the intestine. From their relationship to epithelial cells, it seems that they may be migrating to the surface, perhaps to discharge their contents into the lumen.

Peculiar "cell nests" are conspicuous features in many of the urodeles (Fig. 12-45). These are masses that lie beneath the general level of the epithelium but are usually connected to it by a narrow bridge of cells. The nests present a rather bewildering aspect. They contain a variety of cells: some apparently ordinary epithelial cells,

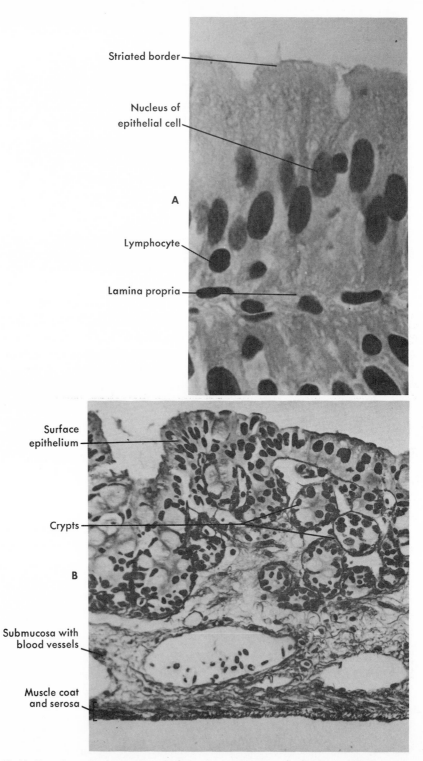

Striated border

Nucleus of
epithelial cell

A

Lymphocyte

Lamina propria

Surface
epithelium

Crypts

B

Submucosa with
blood vessels

Muscle coat
and serosa

Fig. 12-44. Intestine of salamander, *Ambystoma mexicanum*. **A,** Surface epithelium of small intestine. Epithelium consists of tall columnar cells. A few lymphocytes are seen among them. **B,** Large intestine. Surface epithelium consists of columnar cells. "Cell nests" or intestinal glands are seen beneath them. (**A,** Iron hematoxylin and orange G, ×1000; **B,** ×100.)

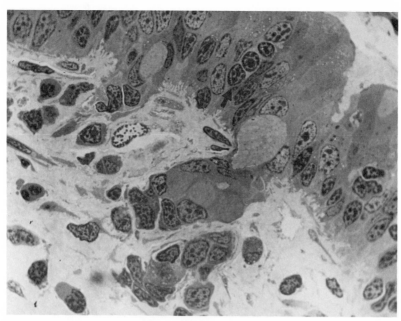

Fig. 12-45. Mucosa of small intestine of *Ambystoma mexicanum*. Typical "cell nest" lies below general level of epithelium. Phenomena of mucus secretion and migration of lymphocytes into epithelium are conspicuous. In one place (arrow), three lymphocytes in a series are seen. (1 μm section of material embedded in epon and stained with toluidine blue and azure II; ×400.)

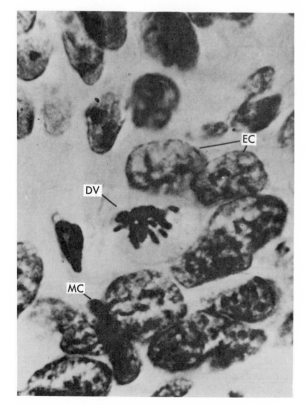

Fig. 12-46. Mitotic figure in cell nest of small intestine of *Triturus viridescens*. *DV*, Dividing cell; *EC*, epithelial cells; *MC*, mesodermal cell. (Pappenheim's stain; ×1000.) (From Andrew, W. 1963. Ann. N. Y. Acad. Sci. **106:** 502-517.)

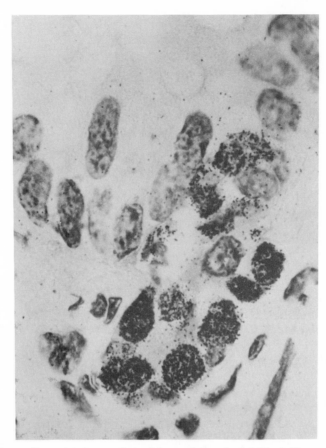

Fig. 12-47. Autoradiograph of a portion of base of intestinal fold of *Necturus maculosa* 12 hours after injection of tritiated thymidine into the animal. Many labelled nuclei are seen in cell nest, and some are in surface epithelium. (Hematoxylin-eosin; ×620.) (From Patten, S. F., Jr. 1960. Exp. Cell Res. **20:**638-641. Copyright by Academic Press, Inc.)

some with droplets of mucus, some lymphocytic, others with clear cytoplasm, and still others in mitotic division (Fig. 12-46). This histologic appearance suggests a passing of these cells, or at least some of them, up into the overlying epithelium.

Indeed, the early incorporation of radioactive thymidine into the cell nests observed by Patten (1960) (Fig. 12-47) points to these cell nests as sites of proliferation of new cells. Furthermore, the presence of mucus-secreting cells would not indicate a **glandular function** for the nest but rather a type of **premature differentiation** of a cell with the potency to form a typical goblet cell. The frog does not ordinarily show prominent cell nests. However, the regeneration of epithelium after its

degeneration during hibernation occurs from small nests, or **nidi,** that resemble those seen regularly in the urodeles.

There are no intestinal glands as such, although the pitlike entrances to the nests of epithelial cells have been called "glands." These pits probably correspond to the crypts of Lieberkühn found in mammals, where they apparently serve to increase the surface area of the intestinal mucosa. The crypts of Lieberkühn (in mammals) also seem to have a regenerative function.

A cross section of the amphibian intestine reveals about the same coats generally characteristic of the alimentary tract (Fig. 12-48). The mucosa is made up of the epithelium described above and a lamina propria of richly cellular tissue (Fig. 12-

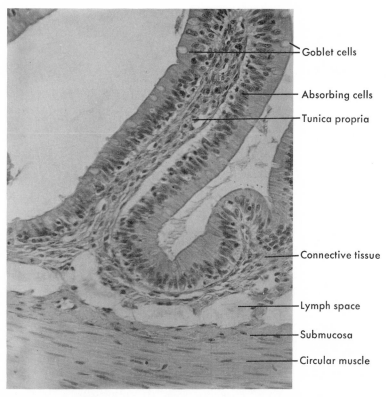

Goblet cells

Absorbing cells

Tunica propria

Connective tissue

Lymph space

Submucosa

Circular muscle

Fig. 12-48. Section of fold in small intestine of frog. Note large lymph spaces at base of mucosa and tunica propria. (×330.)

49). Outside of it is the muscularis mucosae. In some amphibians (e.g., the frog) the muscularis mucosae may consist only of scattered cells among irregular lymph spaces. The submucosa consists of connective and lymphoid tissue, with blood vessels. It is covered by a thick layer of circular muscles, then a thinner layer of longitudinal muscles. Finally there is a thin serosa coat of peritoneum.

The cloaca is the common chamber into which empty the ureters and sperm ducts in the male and the ureters and oviducts in the female. We have mentioned the increasing numbers of goblet cells in the distal end of the intestine. These are also abundant in the first part of the cloaca, but at about its middle they become few in number or lacking. Farther caudad, the epithelium becomes stratified. In the males of some species the cloaca shows tracts of ciliated cells, both ventrally and dorsally. They lie immediately behind the entrance of the ureters and sperm ducts and have the ob-

vious function of propelling sperm cells out of the male cloaca.

REPTILIA

The intestines of reptiles show a variety of topographic patterns. Numerous longitudinal folds are seen in the mucosa and submucosa. In the turtle, for example, high folds occur. There are still no villi in the reptile intestine.

The small intestine is lined with a single layer of tall, nonciliated columnar cells. The epithelial cells bear marked striated borders—microvilli; these microvilli are very long in some cases, such as *Lacerta*. Goblet cells are scattered along the epithelial layer and become very numerous toward the large intestine, a condition that also prevails in mammals.

The lamina propria, as in amphibians, is a highly cellular tissue. The muscularis mucosae usually consists of bundles of smooth muscle, some circular and some longitudinally arranged.

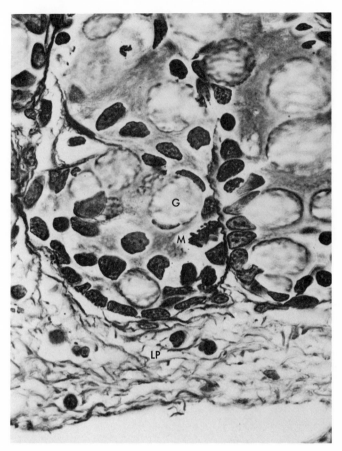

Fig. 12-49. Large intestine of *Ambystoma mexicanum.* Groups of cells seen here lie beneath surface epithelium and seem more clearly glandular than cell nests. *G,* Mucous cell; *LP,* lamina propria; *M,* mitotic figure. (Wilder's reticular stain; ×400.)

The reptilian intestine shows considerable variation among different genera and species with regard to the prominence of the submucosa and the details of arrangement of the muscle coat.

The large intestine has about the same plan as the small intestine, except that it has many tubular glands extending to the muscularis mucosae. These glands are lined with simple columnar cells among which are many goblet cells. Diffused lymphoid tissue is abundant in the submucosal layers (Fig. 12-50), as it is in some amphibians. Longitudinal much-branched submucosal folds give the appearance of villi in cross sections.

AVES

In birds the small intestine is the principal site of digestion. It consists of a looped duodenum (Fig. 12-25) and a more or less elongated ileum (Fig. 12-51, *A*). The mucosa (Fig. 12-51, *B*) shows columnar and goblet cells and contains crypts of Lieberkühn. There are no Brunner's glands (mucus-secreting glands) in the submucosa. True villi are found for the first time in avian intestine (Fig. 12-51, *C*). In fact, they often occur through its entire length, from stomach to cloaca. Villi are fingerlike projections of the mucous membrane, covered by epithelium and having a core of lamina propria. Besides the lamina propria of reticular tissue, isolated patches of lymphoid tissue (Fig. 12-51, *C*) are found throughout the length of the small intestine.

Between the small and large intestine there is usually a pair of ceca (Fig. 12-51, *A*). The ceca are blind pouches with linings similar to that of the intestine proper, but there are much more

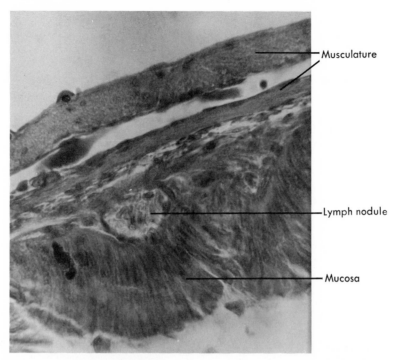

Fig. 12-50. Cross section of large intestine of *Anolis,* showing a lymph nodule. Villi are lacking. Note that thicker longitudinal muscular layer is separated from inner circular layer by a lamina of connective tissue. (×900.)

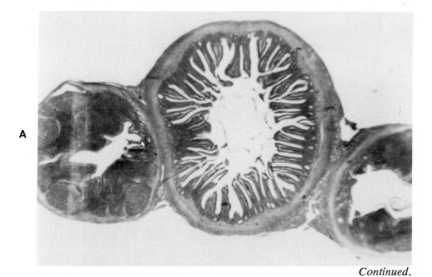

Continued.

Fig. 12-51. A, Cross section through the small intestine and paired ceca of the pigeon. **B,** Portion of the mucosa lining of the small intestine of *Columba.* Arrows indicate goblet cells. **C,** Section of small intestine of *Columba* showing duodenal villi and lymphatic nodule (bottom center). (**A,** ×53; **B,** ×400; **C,** ×100.)

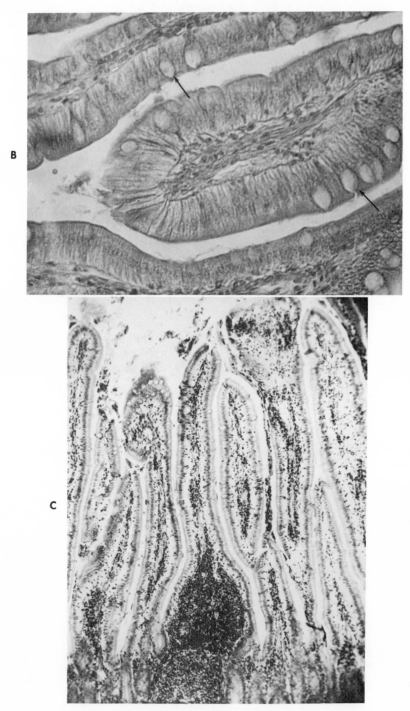

Fig. 12-51, cont'd. For legend see p. 289.

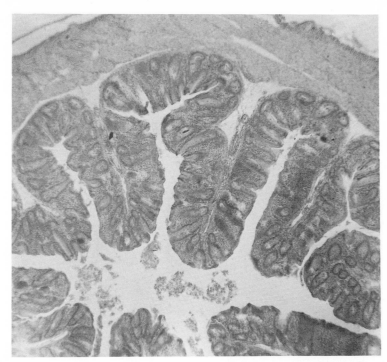

Fig. 12-52. Portion of large intestine of chick. Mucosa is much folded, but no villi are present. (×200.)

lymphoid tissue, broader villi, and a thinner muscle coat. The ceca may be vestigial or absent in some birds. Little is known about their function.

The short large intestine (Fig. 12-52) is a continuation of the small intestine. Histologically, it resembles the small intestine and is primarily concerned with the temporary storage of fecal material. There usually are no villi, but the large intestine has folds that resemble villi, and many goblet cells.

At the junction of the large intestine and cloaca is the bursa of Fabricius, a dorsal diverticulum that becomes lymphoid in adult birds (Ackerman and Knouff, 1959). In fact, recent evidence (Glick, 1970) indicates that the bursa in the young chick is involved in antibody-mediated immunity in many sites of the body.

MAMMALIA

The majority of the mammals show a clear distinction between the small intestine and large intestine as measured by their girth. The histologic distinction is even more striking. The mucosal surface of the small intestine shows millions of villi. The large intestine lacks villi altogether. In it the crypts of Lieberkühn appear like closely set wells with only small areas of flat surfaces between their mouths.

Small intestine. The minute, fingerlike **villi** are the only features "in relief" on the inner surface of the mouse intestine. However, in man and many other mammals large, circular folds **(plicae circulares)** occur. They are also called the **valves of Kerkring.** These folds include both submucosa and mucosa. They are readily visible to the naked eye.

The relationship between the villi and the plicae is somewhat like that between individual trees and the hills they cover; a veritable forest of villi climbs the slopes of the folds and fills the valleys as well. Distention of the small intestine does not erase the plicae circulares, as it does the longitudinal folds or rugae of the stomach.

On the surface of the villi and the intestinal lining between them are tall columnar epithelial

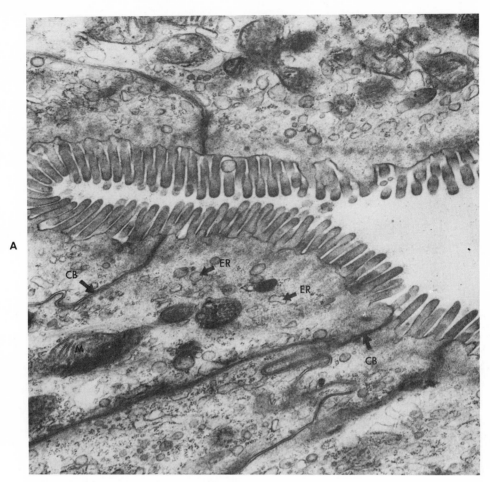

Fig. 12-53. A, Apical portions of epithelial cells of mouse's small intestine showing "forest" of microvilli at surface. *CB,* Cell boundary; *ER,* smooth endoplasmic reticulum; *M,* mitochondrion. **B,** Apical end of epithelial cell of villus in jejunum of mouse. Microvilli are seen with their characteristic "coating" material. One microvillus seems to have bifurcated, or given off a bud (arrow). (**A,** ×29,000; **B,** ×57,600.)

cells, with ovoid nuclei in their lower third and striated borders over their apical ends. The electron microscope reveals this border to consist of immense numbers of microvilli (Fig. 12-53).

The epithelial cells on the villi are of two main types: (1) the absorbing cells, which take in the products of chemical digestion from the lumen of the gut and pass them on to the blood and lymph, and (2) the goblet cells, mucus-secreting structures in which a "theca" (like the bowl of a goblet) is formed by the fusion of droplets of mucus. The goblet cells generally become more numerous in the posterior ileum.

At the bases of the villi, the epithelium dips down to form intestinal glands, also called the **crypts of Lieberkühn** (Fig. 12-54). These are simple tubules contained entirely within the mucosal layer. Here the cells become shorter, and the striated layer is less prominent; the microvilli are shorter than on the epithelial cells of the villi. Thus these crypt cells appear to be less adapted for absorption. They seem to have two main functions: secretion and replacement of cells on the villi.

The important secretion is the **succus entericus,** intestinal juice, which contains several digestive

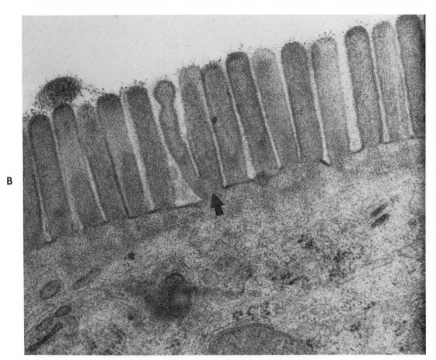

Fig. 12-53, cont'd. For legend see opposite page.

enzymes. Some goblet cells also secrete mucus into the lumen of the crypt. The process of replacement is evidenced by the presence of numerous mitotic figures among the cells of the crypt. Leblond and Stevens (1948) calculated that individual epithelial cells have a life of only 1.35 to 1.57 days. Thus rapid replacement by some means is obvious.

In the center of each villus is a "core" of the highly cellular lymphoreticular tissue of the lamina propria. The same kind of tissue fills the areas between the crypts of Lieberkühn. The framework is primarily reticular connective tissue with great numbers of wandering cells, including lymphocytes, eosinophils, and plasma cells.

Each villus shows small blood vessels in its core, and each also has a central lymphatic vessel. Because the lymphatic vessels frequently contain a milky-appearing fluid, they have been called the "lacteals," or milky vessels. Actually, the fluid is lymph, containing innumerable minute globules of fat, the chylomicrons. The lacteals pass into larger lymphatics, which lie more deeply in the wall of the intestine.

In life, the villi show a considerable amount of mobility. Strands of smooth muscle from the muscularis mucosae extend up into the richly cellular connective tissue of the core of the villus and, by their contraction, shorten and thicken the entire structure as well as cause it to sway in rhythmic fashion. Such motions are of importance for passage of lymph out of the villus through the central lacteal vessel.

Certain special cells with acidophilic granules are prominent in the crypts of many mammals. These cells are usually wedge-shaped, although when well-laden with secretory granules, they are spheroidal. These are the cells of Paneth; they have no microvilli at all. Peculiarly, they seem to be lacking in the intestines of carnivorous mammals. Their function remains obscure.

Another special type of cell, more widely distributed in the mammalian intestine than the cell of Paneth, is the argentaffin cell or "yellow cell." These are found not only in the crypts but also on the villi, along the whole length of the intestine including the vermiform appendix and in the submucosal glands of Brunner in the duodenum. The

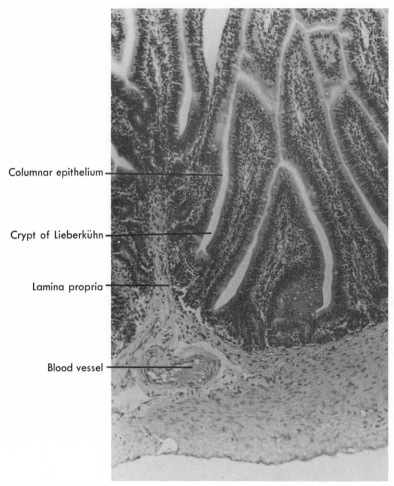

Columnar epithelium

Crypt of Lieberkühn

Lamina propria

Blood vessel

Fig. 12-54. Section through jejunum of rabbit's small intestine, showing crypts of Lieberkühn between villi. (×100.)

argentaffin cells seem to secrete serotonin, a vasoconstrictor substance.

Lymphocytes are found in the intestinal epithelium of all mammals that have been studied. In the crypts they appear to undergo degeneration. Lymph nodules may occur all along the course of the small intestine as solitary structures in the lamina propria. In the ileum especially, there are aggregations of nodules that cause a patchy appearance. These are the so-called **Peyer's patches.** Where nodules occur, the villi are usually short or absent. The nodules frequently bulge through into the submucosa and may even cause a bulging of the outer surface of the intestinal wall, visible to the naked eye.

Outside the lamina propria, some mammals (e.g., the dog and cat) show a stratum compactum. This region is essentially acellular and consists primarily of collagenous and/or elastic connective tissue fibers lying parallel to the circumference of the tube. Such a layer will be recalled as also conspicuous in the stomach and intestine of many lower vertebrates.

The muscularis mucosae has been described for many species of mammals. Generally, as in man, it consists of an inner circular and outer longitudinal layer. Many variations have been reported, but it is not always clear whether these are species or individual differences.

A distinctive feature of the small intestine of

class Mammalia is the presence of glands in the submucosa of the duodenum. These are the **glands of Brunner** and are found all the way from the lowest to the highest mammalian types. In the lower orders, however, they occupy a relatively small area and often closely resemble the pyloric glands of the stomach. The cells of these glands are cuboidal and show a pale-staining cytoplasm and spheroidal nuclei. Excretory ducts of these glands of Brunner empty into the bases of the crypts of Lieberkühn.

In higher mammals the glands of Brunner are relatively poorly developed in carnivores but are highly developed among the ungulates. In man their degree of development is intermediate. These glands often contain two different kinds of alveoli: one composed of serous cells and the other of mucous cells. They secrete mucus, at least, and probably digestive enzymes as well.

Large intestine. As already noted, the mucosa of the large intestine lacks villi (Fig. 12-55). Its crypts of Lieberkühn are generally longer than those of the small intestine. The epithelium between the crypts consists of columnar cells with well-defined striated borders and large numbers of goblet cells. The submucosa of the large intestine often shows masses of adipose tissue in the areolar connective tissue.

The two layers of muscle coat vary in relative thickness and pattern in different groups and species of mammals. In man the conspicuous feature is the presence of three bands or "ribbons," the **taeniae coli.** These are actually thickenings of the outer longitudinal muscle layer. Such bands are also present in apes, sloths, ungulates, and a few other mammals.

The serosa of the large intestine consists of connective tissue, generally with many elastic fibers. In its cellular population are lymphocytes, mast cells, fibroblasts, plasma cells, and eosinophilic leukocytes.

At the junction of the small and large intestines, a blind pouch, or cecum, is found in mammals generally. It is lacking in some, however, such as the Insectivora. In the ant-bear *(Myrmecophaga),* an edentate, there are two ceca; in *Hyrax,* the cony, there are three. The herbivorous mammals usually have well-developed intestinal ceca in which a large amount of digestion (e.g., of

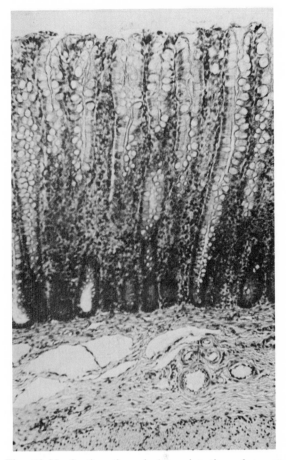

Fig. 12-55. Section through large intestine of a cat. Note numerous goblet cells. (×100.)

cellulose) occurs by symbiotic bacteria. In carnivores the cecum is much reduced or may be absent.

The vermiform appendix of man and some other mammals seems to represent a degenerated distal part of the cecum. However, it may have some remaining functional significance. A striking feature of the wall of the cecum is the presence of large amounts of lymphoid tissue. This includes numerous nodules with prominent reaction (germinal) centers. Bacteria are normally seen in these reaction centers in some species (Shimizu and Andrew, 1967). They appear to be in the process of being phagocytized there by macrophages. Whether this indicates a defense reaction or possibly a **nutritional process** for the animal remains to be determined!

References

Ackerman, G. A., and R. A. Knouff. 1959. Lymphocytopoiesis in the bursa of Fabricius. Am. J. Anat. **104**:163-205.

Ballmer, G. W. 1949. The comparative histology of the enteron of some American turtles. Pages 91-100 *in* Papers of the Michigan Academy of Science, Arts and Letters.

Bensley, R. R. 1902. The cardiac glands of mammals. Am. J. Anat. **2**:105-156.

Bleakney, J. S. 1967. Marine turtles as introductory material to basic biology. Turtox News **45**(3):82-86.

Bulger, R. E. 1963. Fine structure of the rectal (salt secreting) gland of the spiny dogfish, *Squalus acanthias*. Anat. Rec. **147**:95-107.

Bulger, R. E. 1965. Electron microscopy of the stratified epithelium lining the excretory canal of the dogfish rectal gland. Anat. Rec. **151**:589-608.

Burger, J. W., and W. N. Hess. 1960a. Function of the rectal gland in the spiny dogfish. Science **131**:670-671.

Burger, J. W., and W. N. Hess. 1960b. On the function of the rectal gland in the spiny dogfish, *Squalus acanthias*. Anat. Rec. **136**:173.

Dumont, J. 1965. Prolactin-induced cytologic changes in the mucosa of the pigeon crop during crop-"milk" formation. Z. Zellforsch. **68**:755-782.

Glick, B. 1970. The bursa of Fabricius: a central issue. BioScience **20**:602-604 (May 15).

Kleerekoper, H., and G. A. Van Erkel. 1960. The olfactory apparatus of *Petromyzon marinus* L. Can. J. Zool. **38**:209-223.

Krause, R. 1923. Mikroskopische Anatomie der Wirbeltiere in Einzeldarstellungen. IV. Walter de Gruyter & Co., New York.

Leblond, C. P., and C. E. Stevens. 1948. The constant renewal of the intestinal epithelium in the albino rat. Anat. Rec. **100**:357-371.

Norris, J. L. 1960. The normal histology of the esophageal and gastric mucosae of the frog *Rana pipiens*. J. Exp. Zool. **141**:155-171.

Patten, S. F., Jr. 1960. Renewal of the intestinal epitheleum of the urodele. Exp. Cell Res. **20**:638-641.

Radinsky, L. 1961. Tooth histology as a taxonomic criterion for cartilaginous fishes. J. Morphol. **109**(1):73-92.

Rogick, M. D. 1931. Studies on the comparative histology of the digestive tube of certain teleost fishes. II. A minnow (*Campostoma anomalum*). J. Morphol. Physiol. **52**:1-25.

Shimizu, Y., and W. Andrew. 1967. Studies on the rabbit appendix. I. Lymphocyte-epithelial relations and the transport of bacteria from lumen to lymphoid nodule. J. Morphol. **123**(3):231-250.

Snell, G. D., ed. 1941. The laboratory mouse. McGraw-Hill Book Co., New York.

Young, J. Z. 1962. Life of vertebrates. Oxford University Press, New York.

13 ACCESSORY DIGESTIVE GLANDS: LIVER AND PANCREAS

Liver

The liver is an organ of major importance in all vertebrates. In man it is the largest gland of the body. Its roles in the processes of digestion, food storage, intermediary metabolism, and detoxification of products of red cell breakdown are of great importance. Its secretion, bile, has an emulsifying action on fat, separating it into immense numbers of tiny droplets that can be readily attacked by the enzyme lipase from the pancreas. The liver transforms sugar and proteins into animal starch, or glycogen, which is stored in the parenchymal cells (Fig. 13-1) and released as needed in the form of blood sugar. In the adult and larval forms of many lower vertebrates, blood cells are formed in the liver. However, in the liver of mammals, including man, this hemopoietic function is confined to the embryo stages.

CYCLOSTOMATA

The liver of *Petromyzon* arises at about the level of the center of the heart. Indeed, the liver is very large near the apex of the heart and surrounds it in such a way that the cardiac apex appears sunk into the rostral end of the liver.

The general histologic structure of the liver in cyclostomes already shows the pattern seen in higher vertebrates. The organ is essentially a great mass of parenchyma cells divided into tubular lobules by rather scanty amounts of collagenous connective tissue. The portal vein, which drains the food-laden blood from the wall of the intestine, enters the liver and ramifies among the lobules, sending this blood into primitive sinusoids running between plates, or laminae, of the parenchymal cells, to converge at central veins in the interior of each lobule. These, in turn, join into larger veins, which eventually form the hepatic veins and empty into the systemic circulation.

The secretion of the liver cells passes into the lumina of tubules that are relatively wide compared with the analogous, but not necessarily homologous, intercellular bile canaliculi of higher vertebrates. These tubules join a small bile duct that leads to the intestine. The hepatic cells show a definite orientation or polarization, a condition almost completely lacking in higher vertebrates. The nuclei are basally located, and microvilli on the apical ends project into the lumen (Mugnaini and Harboe, 1967).

The tubules anastomose, thus beginning to form a complex epithelial structure within the mesenchymal stroma, as seen in the higher forms. Irregular blood channels, the primitive sinusoids, lie between and among the epithelial tubules.

CHONDRICHTHYES

In the elasmobranch fishes, the liver consists of two long right and left lobes with a small median, cystic lobe between. The liver of fishes and higher vertebrates is a modified tubular gland, a continuous mass of cells forming laminae, or plates (Elias and Bengelsdorf, 1951). Among the cellular laminae are the blood sinusoids lined by the reticuloendothelial cells (Fig. 13-2), some of which lie flattened upon the laminae, others of which bridge the sinusoids. Red blood cells are destroyed in these sinusoids, and the products ingested there by the reticuloendothelial cells.

The vertebrate liver, in all except the cyclo-

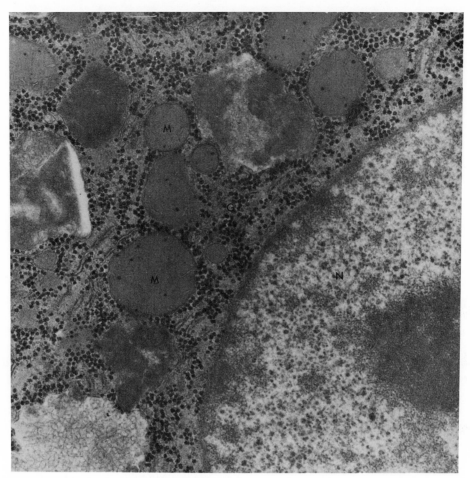

Fig. 13-1. Portion of hepatic parenchymal cell of salamander, *Ambystoma mexicanum.* Nucleus *(N)* shows a large nucleolus. Mitochondria *(M)* have rather inconspicuous cristae and dense matrix. Glycogen granules *(G)* are abundant as small mulberry-like bodies between mitochondria. (Lead citrate stain; ×28,400.)

stomes, is a kind of labyrinth. However, the lower vertebrates differ from the higher forms in several respects. The walls (laminae) separating neighboring sinusoids are generally two cells thick, whereas in birds and mammals they are but one cell thick. In many of the lower vertebrates, pigment is conspicuous in patches of cells throughout the liver, and the content of fat and oil in the liver of many species of fishes is higher than in other vertebrates.

The hepatic parenchymal cells (Fig. 13-3) range from cuboidal to polyhedral in shape and have distinct spherical nuclei. They generally contain the specific hydrocarbon, squalene, and large amounts of both fat and glycogen. Bile secreted by the parenchymal cells is collected in bile canaliculi, which converge into larger and larger bile ducts and finally into one main hepatic duct.

This hepatic duct joins with the cystic duct from the gallbladder to form the common bile duct, which empties into the duodenum. Connective tissue with small nuclei may be seen along the ducts and blood vessels of the liver. The gallbladder is a diverticulum that functions in storage and concentration of the bile.

OSTEICHTHYES

The liver of the trout (Fig. 13-4) appears similar in general histologic structure to that of the elasmobranchs. A separation into lobules is not

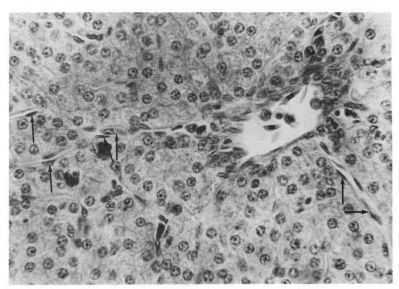

Fig. 13-2. Section of liver of *Squalus acanthias,* showing large sinusoid with corpuscles and debris. Arrows point to reticuloendothelial cells of von Kupffer. (×820.)

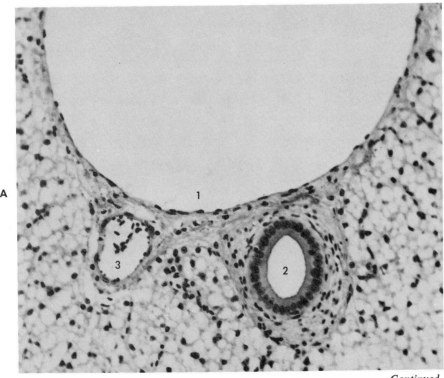

Continued.

Fig. 13-3. Liver of shark, *Scylliorhinus canicula.* **A,** Elements of a portal triad are seen: *1,* portion of a large branch of the portal vein; *2,* a bile duct lined by simp*l*e columnar epithelium; and *3,* a branch of the hepatic artery. Cells of hepatic parenchyma in this elasmobranch fish appear very pale because the large amounts of fat ín them were dissolved out in preparation of the tissue. **B,** Details of hepatic cells and hepatic artery, showing some smooth muscle (sausage-shaped nuclei). (**A,** ×100; **B,** ×400.)

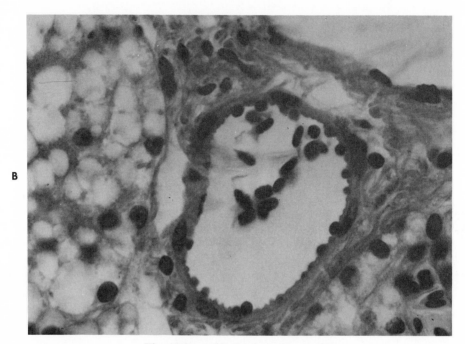

Fig. 13-3, cont'd. For legend see p. 299.

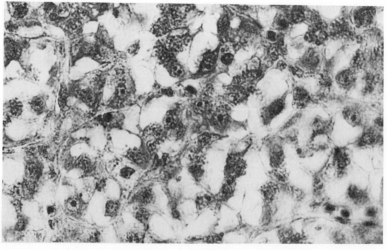

Fig. 13-4. Liver of *Salmo gairdneri*. Appearance of hepatic parenchyma is more heterogeneous than in higher vertebrates, with larger amounts of oily substances present. (Masson's stain; ×400.)

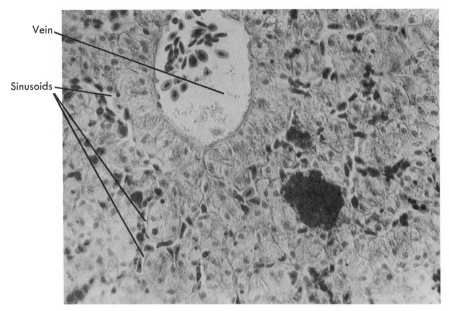

Fig. 13-5. Section of frog liver. Note sinusoids and prominent limiting membrane of hepatic cells around large vein. (×390.)

clearly seen. A rather homogeneous mass of cells forms the liver parenchyma, separated by narrow sinusoids, showing here and there a cross section or longitudinal section of a blood vessel. The plates of cells can be seen far more distinctly near the periphery of the organ than in its interior, and they can be distinguished most clearly in the region of the gallbladder.

The fish liver is a great depot for storage of fat. In the ordinary histologic sections the cytoplasm of the hepatic parenchymal cells shows many vacuoles, whereas in frozen sections such areas can be stained with lipid stains.

The parenchyma cells may be described as pyramidal or cone-shaped, with the narrow ends facing the bile canaliculus. Thus the canaliculus is seen in the fishes as the lumen of a glandular tubule, and its relation to the epithelial cells in higher vertebrates can be understood readily when one has comprehended this more primitive condition. Another characteristic reminiscent of the lower forms is the cell polarity indicated by the location of the nuclei closer to the sinusoids than to the intercellular bile canaliculi.

The sinusoids, like the liver cell tubules, form a great anastomosing network. Their lumina are always separated by hepatic cells from the lumina of the bile canals. They are lined by reticuloendothelial cells and their associated reticular fibers.

AMPHIBIA

The histologic structure of the amphibian liver follows the general vertebrate pattern, with "lobules" of parenchymal tissue separated by small amounts of connective tissue. Between the plates of liver cells are wide blood sinusoids lined by the phagocytic, reticuloendothelial Kupffer cells. Frog liver does not show the radiating sinusoids and true lobular structure of higher vertebrates. The rather large cells form a network separated by the sinusoids (Fig. 13-5). The prominent nuclei are located near the center of the more or less clear cytoplasm.

Unlike mammals, the majority of amphibians show scattered groups of liver cells containing large quantities of yellow-brown to black pigment (Fig. 13-6). Recently, in studies by light and electron microscopy, Andrew (1969) has identified these pigment-containing cells in *Necturus,* as **macrophages** of the reticuloendothelial system. In *Ambystoma mexicanum,* a form very similar

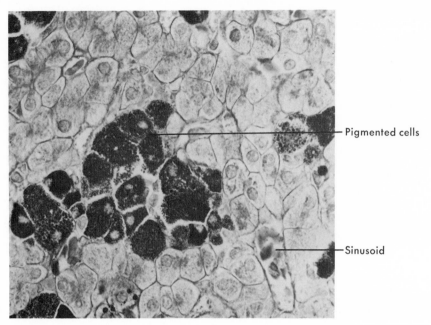

Fig. 13-6. Liver of young specimen of *Ambystoma mexicanum*. The hepatic parenchyma of many amphibians contains scattered groups of pigment-bearing cells. (×100.)

to *Necturus,* they show a marked increase in numbers with advancing age of the animal (Fig. 13-7).

Bile capillaries ramify between the cells in an irregular manner. Lateral ones may penetrate the cell bodies. The bile capillaries are simply spaces between adjacent liver cells and have no special lining epithelium. The bile ducts are composed of fibrous connective tissue walls and are lined with simple cuboidal and simple columnar epithelia. The common bile duct is lined with stratified columnar epithelium.

The gallbladder (Fig. 13-8) is a spherical greenish sac for temporary storage of bile. It is lined with ciliated simple or stratified columnar epithelium and consists mainly of fibrous connective tissue with a vascular supply. Bile secreted by the hepatic cells contains no digestive enzymes. It is a complex mixture serving many functions such as emulsifying fats, facilitating absorption, inhibiting growth of bacteria, and initiating peristaltic contractions (Haslewood, 1955). Its composition appears to be specific for each species.

The liver receives its blood from two sources. One of these is the hepatic artery, which carries arterial blood. The other is the hepatic portal vein

from the stomach, intestine, pancreas, and spleen. This blood is carried by way of the sinusoids, which form an extensive spongework between the plates of hepatic cells (Fig. 13-5). Blood enters the sinusoids from interlobular branches of the portal vein and hepatic artery. The materials absorbed by the blood from the digestive organs thus pass through the liver before they enter the hepatic veins and the general circulation. Many large lymph vessels form perivascular lymph spaces around the capillaries.

It is interesting to note that the polarity of hepatic cells exhibited by the fishes is partially retained in the amphibians. These cells are usually of relatively symmetrical structure with centrally located nuclei, etc. However, at certain seasons of the year, polar changes occur. For example, the toad *Bufo arenarum* presents an accumulation of fat toward the sinusoidal side, with the Golgi apparatus toward the canalicular side (DeRobertis and Magdalena, 1935).

REPTILIA

The liver of reptiles is similar to, but generally more elaborate and lobular than amphibian liver, and has a richer vascular supply. In it, the secret-

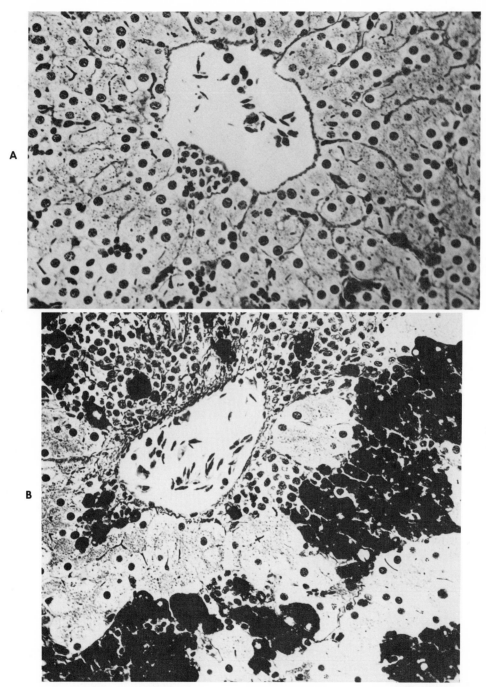

Fig. 13-7. Liver of a female axolotl, *Ambystoma mexicanum.* **A,** The generally polyhedral form of the hepatic parenchymal cells is seen. Along the borders of the sinusoids are the Kupffer cells. In a young animal such as this, little or no pigment is present in the liver. **B,** There are large quantities of pigment in this very old specimen (over 8 years of age). (×400.)

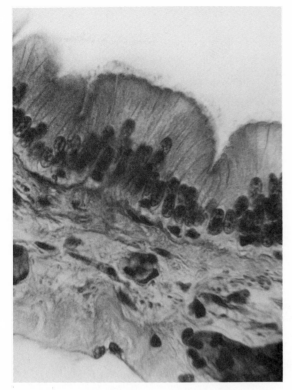

Fig. 13-8. Portion of wall of frog's gallbladder, showing ciliated epithelial lining. (×800.)

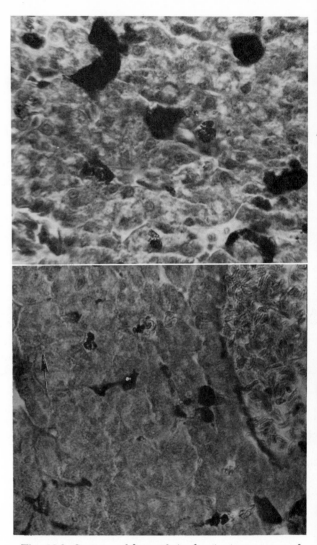

Fig. 13-9. Sections of liver of *Anolis*. **A,** Aggregation of pigmented cells. **B,** At medium power, shows hepatic lobules and great vascular supply. A bile canaliculus is shown by the arrow. (**A,** ×1070; **B,** ×200.)

ing end-pieces are composed of large, polyhedral cells (Fig. 13-9, *A*). The connective tissue between lobules bears the vascular sinusoids and bile capillaries. The general anastomoses of the liver laminae or plates in lizards are usually easy to see (Fig. 13-9, *B*).

The liver cells of reptiles generally show a polarity, as evidenced by the granular content of the portion of the cell closest to the intercellular bile capillaries. Pigmented cells are often scattered through the liver and frequently found within lymphatic vessels.

AVES

The avian liver is bilobed, and its histologic complexity approaches that of the mammals. The lobular microstructure described below for mammals is well developed in birds. The hepatic cell plates separating the sinusoids are one or two cells thick. A gallbladder is often lacking.

MAMMALIA

The liver in birds and mammals has a truly lobular architecture. The nature of the lobules and their limitations may be properly open to discussion (Rappaport and associates, 1954). However, the microscopic picture of the lobule as a repeating structural and functional unit is a useful one for orienting the viewer to the various structures of the liver.

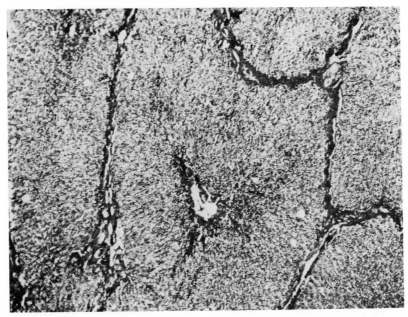

Fig. 13-10. Liver of camel. This liver, like that of the pig, has lobules much more sharply defined by interlobular connective tissue than the human liver. (Masson's connective tissue stain; ×35.)

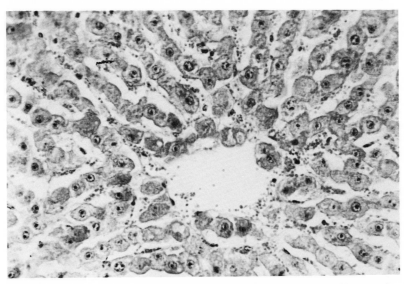

Fig. 13-11. Central vein of liver lobule of impala, *Aepyceros melampus*. Note cords of cells with intervening sinusoids. Cords seem to be made up of single rows of cells, but liver cells are arranged in plates of two or more cells in thickness. (Masson's stain; ×250.)

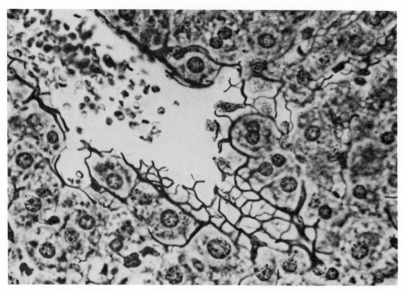

Fig. 13-12. Liver of mouse. A portion of venous channel, containing red corpuscles, is seen. Reticular fibers support the parenchymal cell masses. (Wilder's reticular stain; ×600.)

Each lobule is a polygonal prism or cylinder about 1 × 2 mm. The lobules are separated more or less distinctly by connective tissue, depending on the species (Figs. 13-10 and 13-11). This connective tissue covering is sometimes referred to as Glisson's capsule. A central vein, a tributary of the hepatic vein, runs through the center of each lobule. The matrix consists of masses of hepatic parenchyma cells, infiltrated by sinusoids radiating outward from the central vein. From a three-dimensional view, this arrangement has been described variously as anastomosing cords or as fenestrated plates of hepatic cells (Fig. 13-12). In any event, the system seems to be one of rather freely communicating blood channels in a spongework of glandular tissue.

The branches of the hepatic portal vein and hepatic artery form interlobular veins and arteries around the outside of the lobules. These interlobular vessels send branches directly to the liver sinusoids, thus completing the communication between the hepatic portal vein (and hepatic artery) and central veins that eventually empty into the inferior vena cava.

The portal vein carries food-laden blood from the walls of the small intestine. This blood, together with oxygen-laden blood in the branches of the hepatic arteries, is coursing through the portal areas between the lobules. From there the blood enters the sinusoids and converges toward the central vein. In the sinusoids the blood is in close contact with single layers of hepatic parenchyma cells and has much of the food stuff removed from it for storage as glycogen in those cells. The reticuloendothelial cells lining the sinusoids are macrophages, and they pick up the debris of red corpuscles as well as other materials.

The hepatic parenchyma cells appear rather similar to those seen in lower vertebrates. They are large polyhedral cells with spheroidal nuclei. In addition to Golgi apparatus and numerous mitochondria, their cytoplasm often shows large quantities of stored glycogen synthesized by these cells from the sugars taken from the blood that enters the liver through the portal vein. Each parenchyma cell has several distinct functions: to store glycogen and release it as sugar when needed; to secrete bile; to transform uric acid into urea; and to carry on various other metabolic and detoxification activities.

The cells lining the sinusoids are very important cells. They are a part of the reticuloendothelial system of the body, a far-flung defense line of phagocytic cells found in many places

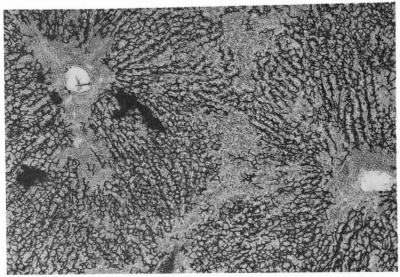

Fig. 13-13. Bile capillaries in liver injected with ink. Portions of two lobules are shown. Anastomosing system of bile capillaries drains toward periphery of lobule—the opposite to flow of blood in the sinusoids toward the central vein. (×200.)

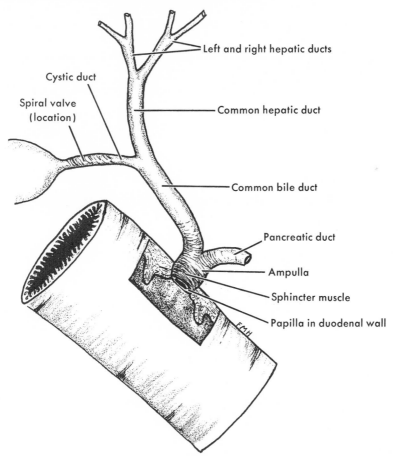

Fig. 13-14. Diagram of bile duct system in mammal. Various bile ducts pick up bile from the bile capillaries and pass it into large paired hepatic ducts that store it in gallbladder until discharged into duodenum. In rabbit, common bile duct opens separately from pancreatic duct, but most mammals have a common opening (papilla) for the two ducts.

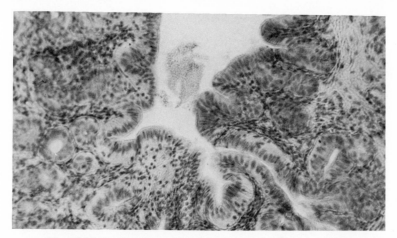

Fig. 13-15. Section through common bile duct of rabbit. (×100.)

such as the spleen, lymph nodes, bone marrow, and endocrine organs. The lining cells actually are of two types: (1) undifferentiated reticular cells with small, deeply staining, elongated nuclei; and (2) Kupffer cells, which are larger cells with clearer nuclei and which often contain pigment and other phagocytized material in their cytoplasm.

Electron-microscope study reveals a space, the so-called **space of Disse,** separating the endothelium of the sinusoids from the hepatic parenchyma cells. Processes of the liver cells extend into the space, and fine reticular fibers occur in portions of it. The endothelial lining is traversed by openings through which blood fluid apparently can filter from the lumen of the sinusoid. It is thought that this fluid then finds its way from the spaces of Disse to the blind lymphatic capillaries in the portal areas (see below), accounting for the rather copious lymph that constantly drains from the liver.

While the blood courses from the periphery of each lobule to its central vein, bile is secreted by the parenchyma cells into the canaliculi and drains in the opposite direction, toward the periphery. The bile canaliculi are not true tubules; they have no lining, as such. Electron microscopy reveals them to be merely spaces between apposed faces of hepatic cells (Fig. 13-13). The canaliculi are not easily seen by ordinary histologic methods but are revealed by the electron

microscope and by injection and experimental biliary obstruction techniques (Wachstein and Zak, 1949). These small canals lead into biliary ducts, which merge into larger bile ducts, collect into a common hepatic duct, and eventually empty into the duodenum (Figs. 13-14 and 13-15).

The gallbladder is a storage sac lined by simple columnar epithelium (Fig. 13-16). With a thin lamina propria, this epithelium forms a mucosa that shows numerous folds when the gallbladder is not distended. Irregularly oriented smooth muscle forms a thin layer, outside which is the serosa or adventitia. The cystic duct connects the gallbladder to the common bile duct.

In the interlobular connective tissue, at the angles of the lobules, three types of structures can be found: (1) a branch of the hepatic portal vein, (2) a branch of the hepatic artery, and (3) a biliary duct. The vein and the artery bringing blood from the intestines and oxygenated blood from the heart, respectively, empty into the sinusoids; the biliary duct collects bile from the numerous bile canaliculi of the lobules.

Outside, along the periphery of the lobules, are the portal canals or portal areas. Traversing these portal areas are branches of the portal vein and hepatic artery. The portal canals also carry the small bile ducts and lymphatic vessels (Fig. 13-17). This complex of ducts and vessels is supported by fibrous connective tissue, which is an

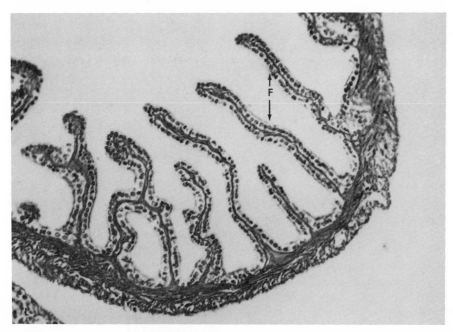

Fig. 13-16. Gallbladder of mouse. Numerous folds of the mucosa are covered by simple columnar epithelium. *F,* Mucosal folds of gallbladder; *L,* liver. (Wilder's reticular stain; ×125.)

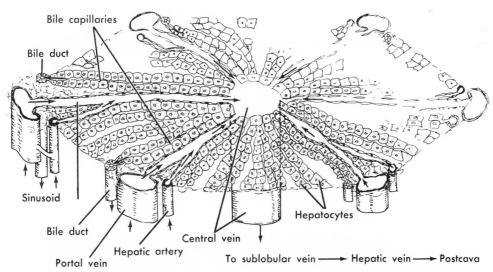

Fig. 13-17. Diagram of hepatic lobule, showing arrangement of blood vessels and bile duct. A classic lobule is indicated, but the region of the portal vein, hepatic artery, and bile duct represents a portal lobule or area that may be considered a functional unit. Such a triangular portal area has a central vein peripherally at each corner. Other criteria may be used to denote functional units.

outer extension of the liver. The blood vessels of the canal are thin-walled, and the epithelium of the bile duct is cuboidal to columnar, like that of the gallbladder.

Pancreas

The vertebrate pancreas is a double kind of organ. The pancreatic juice contains digestive enzymes. It is secreted by groups of large pyramidal cells clustered around central lumina and forming **alveoli** or **acini.** A duct telescopes into the lumen of each acinus and carries the digestive secretion out and into the lumen of the intestines. This process represents **exocrine** secretion.

The pancreas also consists of other types of cells clustered in islets. The groups of cells are called "islets" because they are usually scattered among the other tissue. The islet cells secrete hormones such as glucagon and insulin. Islets tend to be highly vascularized, and secretion takes place directly into the blood. This is **endocrine** secretion—without ducts.

CYCLOSTOMATA

Most cyclostomes have no separate pancreas as such. The exocrine pancreatic tissue (small alveoli, or groups of secreting cells) is widely disseminated and occurs in the submucosa of a considerable portion of the intestine and even in parts of the liver. The secretions are zymogen granules—rounded, eosinophilic bodies that contain proenzymes. In the lumen of the intestine these proenzymes become hydrolytic enzymes that act to digest the food there.

In the submucosa of the anterior part of the intestine of larval lampreys and hagfishes *(Myxine),* there are follicles of granular cells with no openings to the lumen of the intestine. These follicles were described by Langerhans, the discoverer of the mammalian islets, in 1873. Apparently they represent **endocrine** (islet) pancreatic tissue. In adult hagfishes the endocrine follicles are usually concentrated near the opening of the bile duct into the intestine and may show a fairly marked development of a connective tissue capsule. Capillaries are numerous.

By special staining methods (Schirner, 1959; Falkmer and Hellman, 1961) both A (alpha) and B (beta) types of cells have been identified

in *Myxine.* They seem to correspond to these same cell types in mammals, the B cells being the producers of insulin. The A cell to B cell ratio is approximately 2 to 3. Electron microscopy has also shown a third type of cell, without secretion granules but rich in mitochondria.

The pancreas of *Petromyzon* is peculiar in that its endocrine portion is a separate structure lying close to the bile duct. It is several millimeters in length. Each follicle of cells within the organ is separated from its neighbors by connective tissue, and a capsule of such tissue surrounds the entire organ as well. An interesting feature is the presence of spheroidal cavities in the centers of many of the follicles. The lumina contain granules of the B (beta) type.

This complete separation of endocrine and exocrine parts of the pancreas makes conditions ideal for experimental study of islet tissue. It also casts light on the question of whether the islets in the pancreas of higher forms are of fundamentally similar nature to the exocrine tissue, since in their development, the islet cells appear to bud off from the cells of the exocrine ducts. The fact that two distinct organs form in this primitive vertebrate would seem to indicate that the pancreatic endocrine tissue is basically **distinct** from the exocrine, morphologically as well as physiologically.

CHONDRICHTHYES

The elasmobranch and bony fishes both show variation in pancreas structure among the different species. Sometimes the pancreas is diffuse or disseminated, and the masses of gland cells are found along the course of blood vessels within the liver or mesentery or in two separate glands. Sometimes the tissues of both types are together, forming a definite gland. It may consist of discrete, grapelike masses, or it may be relatively compact.

The shark pancreas is a somewhat compact, double-lobed gland. It has a connective tissue sheath and lies close to the stomach. The gland is made up of masses of secreting alveoli (Fig. 13-18). Each exocrine alveolus consists of several pyramid-shaped cells with spherical nuclei. Prominent nucleoli are usually observed in the nuclei. The cells of the alveoli have a more regu-

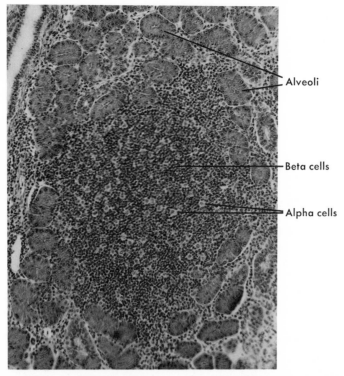

Fig. 13-18. Section through shark's pancreas, showing numerous secreting alveoli surrounding an islet of Langerhans, which contains two types of cells. Note that the islet tissue is invading the region of the alveoli in places, a characteristic of the shark's pancreas. Sinusoids and blood vessels are also visible. (×125.)

lar arrangement than do the endocrine islets that are scattered among them. Some lymphoid and hemopoietic tissue (small round cells) may also be seen. It will be noted that the darkly stained alveoli have large lumina and are widely separated from each other by hemopoietic tissue.

OSTEICHTHYES

In the trout the pancreas is a compound branched alveolar gland. The lumina of the acini are very narrow and set off from the secretory cells by the centroacinar cells, which are a feature of the pancreas through all the vertebrate classes. The centroacinar cells are squamous cells lining the ducts and lumina of the acini. The secretory cells in the trout are pyramidal. The nuclei are situated in the basal parts of the cells, and the cytoplasm of all the remaining part of the cell body is usually filled with large secretion granules.

The ducts seem to telescope into the secretory alveoli; they are first lined by thin, squamous centroacinar cells. The duct system of the pancreas may be said to begin with the centroacinar cells. The reason for this peculiar "telescoping" of a duct into the secretory alveolus remains obscure. The ducts leaving the alveoli consist of a low cuboidal epithelium and are inconspicuous. On entering the interlobular connective tissue, the epithelium becomes taller, and in the larger ducts it is a low columnar type. The main duct of the pancreas empties into the small intestine in close relation with the common bile duct.

The endocrine portions of the pancreas, the islets of Langerhans, stand out clearly from the exocrine portion by their lighter stain and non-alveolar structure. The islets are larger and more numerous in the anterior part of the pancreas. The islets of Langerhans are consistently present **for the first time in the bony fishes** and then continue to be seen in all the vertebrate series. The essential structure, as seen in the trout, is one of anastomosing cords of cells separated by sinusoids.

AMPHIBIA

The pancreas of amphibians is a well-developed, compound tubuloalveolar gland. This group shows considerable gross variation among the species. In the frog the pancreas is divided into dorsal and ventral portions, with the dorsal part lying in the dorsal mesentery and the ventral part between the intestine and liver; in *Necturus,* the gland is well developed and has five lobes.

The exocrine portion is made up of great masses of alveoli that secrete the pancreatic juice with its digestive enzymes. The acini are long and tubular, with large zymogen granules in the apical portion of the cells. The cells have spheroidal nuclei in which the nucleolus (usually only one per nucleus) is a conspicuous element. The cytoplasm has intensely basophilic portions, usually in the basal region. Often this material is concentrated in a "paranuclear body" *(Nebenkern)*— a single, large, striated mass near the nucleus. The basophilic material, sometimes called ergastoplasm, represents areas of rough endoplasmic reticulum in which ribosomes are abundant. Eosinophilic leukocytes are abundant among the duct cells.

The islets of Langerhans are seen as regularly occurring cords or masses of cells in close association with the intralobular ducts of the exocrine pancreas. With appropriate fixation and staining, three or four different kinds of cells may be found in the islets. Small blood vessels are abundant, into which the endocrine secretions are discharged. The islet tissues may be distinguished from the alveolar tissue by their lighter staining and by the absence of large granules. Alpha, or A, cells may have fine granules and large, lightly staining nuclei; beta, or B, cells are smaller, with nuclei that contain coarsely granular chromatin. Some forms (e.g., *Taricha*) apparently contain only one type of actively secreting cell, corresponding to the beta cell of the mammalian islets (Wurster and Miller, 1960).

REPTILIA

In the pancreas of the lizard, the connective tissue of the capsule gives rise to septa that divide the gland into lobes and lobules, into which numerous vessels and nerves ramify. The exocrine part, which makes up most of the gland, consists of secreting end-pieces of pyramidal cells. Delicate reticular connective tissue surrounds each secretory end-piece.

The characteristic acini resemble those of amphibians, but hemopoietic tissue is less common. The cells contain coarse zymogen granules, forerunners of the enzymic secretions. The nuclei lie near the basal parts of the cells. Many sections of the gland may show centroacinar cells in the lumen of the acini along the inner ends of the secreting cells. The pancreatic cells in the snake (Thomas, 1942) show particularly well the various stages of transformation of the duct cells to cells of exocrine, alveolar type, on the one hand, and to those of the endocrine islet type on the other.

The islets of Langerhans are generally abundant in the reptilian pancreas. Masses of the lightly staining cells are arranged in anastomosing cellular cords scattered among the acini of the exocrine glandular cells. The alpha cells have a finely granular cytoplasm and are found chiefly near the central portion of an islet. Smaller cells with coarser granules found near the periphery are the insulin-secreting beta cells. Other types have also been described in some reptiles.

AVES

The bird pancreas has about the same anatomy and functions as does that of other vertebrates. It is relatively compact, although in many species it is bilobed, even trilobed. As an endocrine gland, it produces two hormones; glucagon from the alpha cells and insulin from the beta cells. Islet tissue is not as sharply delineated by connective tissue as it is in the mammals. The exocrine portion is a compound racemose gland, furnishing the digestive enzymes and buffering compounds for the neutralization of chyme in the intestines.

An interesting departure seen in birds is the occurrence of two kinds of islets of Langerhans. One is pale, consisting of beta cells, the other a dark type made up of alpha and delta cells (Nagelschmidt, 1939). Delta cells have special granules, small and staining blue with Mallory-azan, whereas the granules of alpha cells are red and those of beta cells are brownish orange.

MAMMALIA

The exocrine pancreas of the mammal, a typical serous gland, secretes enzymes for chemical digestion of fats (lipase), carbohydrates (amylase), and proteins (trypsin). The endocrine portion of this double gland produces the two major hormones glucagon and insulin, both involved in the regulation of carbohydrate metabolism.

The structure of the gland involves a framework of areolar connective tissue that forms thin septa, subdividing the pancreas into many lobules. The septa carry nerves, blood vessels and capillaries, lymphatics, and ducts. The lobules are made up of acini and intralobular ducts.

The acini are irregularly pyramidal serous cells around a small central lumen. A basement membrane surrounds the secretory end-piece of each cell. In the apical ends of the acinar cells, there are acidophilic zymogen granules, whereas the basal parts are strongly basophilic because of an abundance of RNA granules and ribosomes, as revealed by the electron microscope. The nuclei are spherical and relatively clear with one or two conspicuous nucleoli.

The highly branched, intercalated intralobular ducts go between the secreting acini. They are lined with low cuboidal or flattened epithelial cells (Fig. 13-19). The acini are not always at the

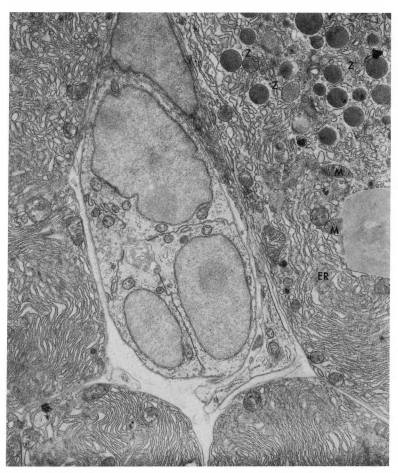

Fig. 13-19. Intralobular duct of mouse pancreas. Three epithelial cells in center comprise a "wall" in this section. The lumen is partially filled by microvilli of these cells. Portions of several glandular cells of the exocrine pancreas are shown. In their cytoplasm are seen abundant rough endoplasmic reticulum *(ER)*, scattered mitochondria *(M)* with their characteristic cristae, and at upper right, large spheroidal zymogen granules *(Z)*. (Lead citrate stain; ×10,000.) (Courtesy Dr. Ralph Jersild, Indianapolis, Ind.)

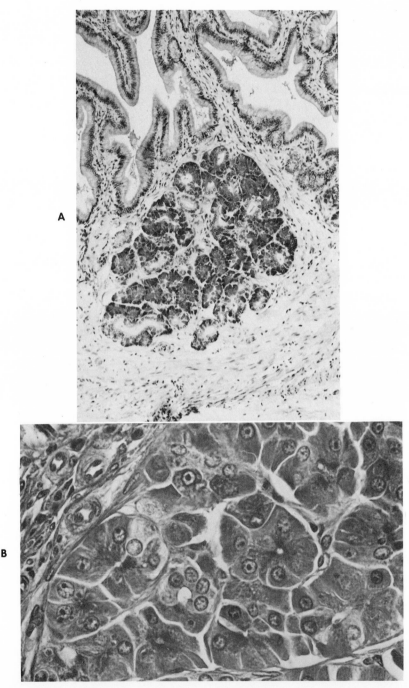

Fig. 13-20. Wall of mink gallbladder. **A,** Cross section. Note large pocket of ectopic pancreas in lamina propria. **B,** Cross section of a pocket of ectopic pancreas tissue from animal shown in **A.** Note prominent nucleoli in the acinar cells and their absence in centroacinar cells. (**A,** Hematoxylin-eosin, ×200; **B,** Mallory's trichrome stain; ×800.) (Courtesy Dr. R. F. Bostrom, East Lansing, Mich.)

end of the ducts but may bulge from the side of the duct or overlay the duct along one side. This makes it appear, in certain sections, as though the duct cells were in the lumen of the acinus. Such duct cells are called centroacinar cells, and are nonsecretory. The intercalated ducts enter the interlobular ducts, which are lined with simple columnar cells interspersed with a few goblet cells. These major ducts are surrounded by a dense layer of collagenous connective tissue. The interlobular ducts eventually empty into the main pancreatic duct, which drains into the duodenum (Fig. 13-14). The pancreatic duct has taller epithelium lining its lumen and is surrounded by a prominent connective tissue.

The endocrine portion of the mammalian pancreas consists of many islets of Langerhans. They are irregular cords of cells in close association with numerous capillaries. The cells are polyhedral, rounded, or sometimes cuboidal, and the nuclei stain rather lightly. Special stains show that there are several distinct kinds of secreting cells in any one islet. The beta cells are the most numerous type.

Histologic structure of the pancreas of mammals is generally uniform through the class, but its arrangement and macroscopic structure vary. The lower orders, the monotremes and marsupials, generally show a less compact arrangement than the higher forms. Parts of the gland may be found in the mesentery or in the greater or lesser omentum. Bostrom (1968) has described ectopic pancreatic tissue in the mink (Fig. 13-20). It resembles in most respects the organ as found in other mammals.

References

Andrew, W. 1969. Comparative study of aging of the liver. Acta Anat. Nippon. **44**(2):1.

Bostrom, R. E., and M. L. Calhoun. 1968. Ectopic pancreas—a common finding in the mink gall-bladder. Anat. Rec. **160**:320. (Abstr.)

DeRobertis, E. D. P., and A. Magdalena. 1935. Nota sobre la citología hepatica del *Bufo arenarum* (Hensel). Rev. Soc. Argent. Biol. **11**(3):179-185.

Elias, H., and H. Bengelsdorf. 1951. Die Struktur der Leber der Wirbeltiere (vorläüfige Mitteilung). Anat. Nachr. **1**:273-281.

Falkmer, S., and B. Hellman. 1961. Identifications of the cells in the endocrine pancreatic tissue of the marine teleost *Cottus scorpius* by some silver impregnation procedures. Acta Morphol. Neerl. Scand. **4**: 145-152.

Haslewood, G. A. D. 1955. Recent developments in our knowledge of bile salts. Physiol. Rev. **35**:178-196.

Mugnaini, E., and S. B. Harboe. 1967. The liver of *Myxine glutinosa:* a true tubular gland. Z. Zellforsch. Mikrosk. Anat. **78**:341-369.

Nagelschmidt, L. 1939. Untersuchungen über die Langerhans'schen Inseln der Bauchspeicheldrüse bei den Vögeln. Z. Mikrosk. Anat. Forsch. **45**:200-232.

Rappaport, A. M., Z. J. Borowy, W. M. Lougheed, and W. N. Lotto. 1954. Subdivision of hexagonal liver lobules into a structural and functional unit: role in hepatic physiology and pathology. Anat. Rec. **119**: 11-33.

Schirner, H. 1959. Über die Verwendung von Pseudoisocyanin zur Pankreasuntersuchung der Cyclostomen und Darstellung von Insulin. Z. Naturforsch **14B**: 690-691.

Thomas, T. B. 1942. The pancreas of snakes. Anat. Rec. **82**:327-342.

Wachstein, M., and F. G. Zak. 1949. Intracellular bile canaliculi in the rabbit liver. Proc. Soc. Exp. Biol. Med. **72**:234-236.

Wurster, D. H., and M. R. Miller. 1960. Studies on the blood glucose and pancreatic islets of the salamander *Taricha torosa*. Comp. Biochem. Physiol. **1**:101-109.

14 ENDOCRINE ORGANS

The endocrine glands are organs of great importance and interest. With the nervous system, they share in the functions of regulation and control of the other systems and, indeed, of the development and behavior of the organism itself. The endocrine glands do not form a uniform system, either morphologically or histologically, but, mostly for convenience, they are treated as a unit. Many of the glands have various definite interrelationships with each other. Although the endocrine glands are often considered more primitive than the nervous system, it is doubtful that they were evolved before the nervous system in some forms. They probably originated from tissue that had the power of external secretion and were modified to influence target tissue at a distance. Their origin in all vertebrates is very much the same. They have three types of embryologic origin: (1) from the wall of the alimentary tract, (2) from the nervous system and its neural crest, and (3) from the gonads and adjacent coelomic wall. Hormones naturally vary in their chemical constitution, but those in glands derived from the alimentary canal and nervous tissue are generally protein in nature, whereas hormones from the gonads are steroids.

Endocrine glands, in contrast to exocrine glands, have no ducts to carry away their manufactured material to their various targets. They must eliminate their secretions (hormones) by diffusion into the blood, which carries the chemical mediators throughout the body. On this account they are often called ductless glands or glands of internal secretion.

The phylogeny of the endocrine glands as we know them has been one of flux. They have changed their positions frequently. Some have acquired endocrine functions secondarily. Some are duplex and have arisen from tissues of wholly different origins. They have a marked correlation among themselves, so that the hormones of one gland may profoundly influence the functioning of other endocrine glands. Their exact relationships and functioning have never been completely determined for any vertebrate, although much is known about individual glands.

In the lower chordates—tunicates and *Amphioxus*—two organs are found that may be homologized with considerable certainty to the endocrine glands of vertebrates. The **subneural gland** appears to correspond at least in part to the **hypophysis;** the **endostyle,** or **subpharyngeal gland,**

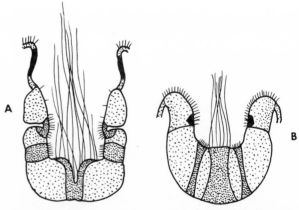

Fig. 14-1. Endostyles of *Ciona,* a simple ascidian **(A)** and of *Amphioxus* **(B),** showing the relationship of iodine-binding cells (black) to the structure as a whole. In both cases these cells lie close to the lip of the endostyle and just above the tracts of glandular cells. Note long flagella in the center of the endostyle. (Redrawn from Barrington, 1959.)

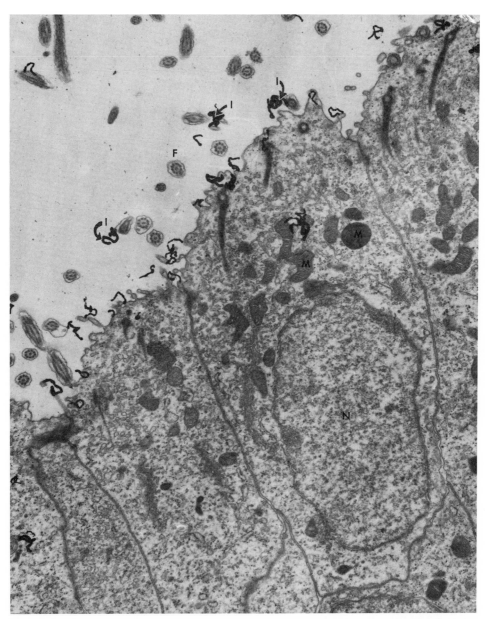

Fig. 14-2. Electron microscope radioautograph of cells of endostyle of *Petromyzon* larva 1 hour after injection with 200 millicuries of [125]I. Localization of the radioactive iodine is seen at apices of the cells. *F*, Flagellum (cross section); *M*, mitochondria; *N*, nucleus; arrows, location of radioactive iodine. (×17,800.) (Courtesy Dr. H. Fujita, Hiroshima, Japan.)

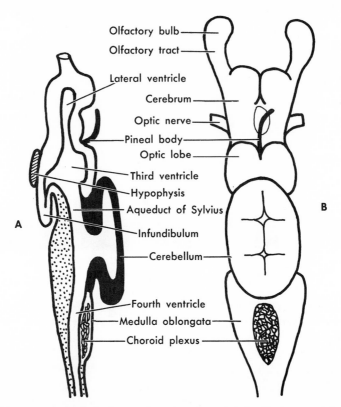

Fig. 14-3. Diagram of shark brain to show location of infundibulum and hypophysis. **A,** Longitudinal section. **B,** Dorsal view.

a ciliated groove in the floor of the pharynx, corresponds to the **thyroid gland** (Figs. 14-1 and 14-2).

Pituitary gland

The pituitary gland, or **hypophysis,** is found underneath the diencephalon division of the brain at the tip of the infundibular stalk, in the sphenoid area of the skull (Fig. 14-3). The gland in higher vertebrates is made up of two major subdivisions of different embryonic origins: a **neurohypophysis** as a downgrowth from the diencephalon floor of the brain and an **adenohypophysis** from the dorsal wall of the embryonic oral cavity. Each of these major divisions is further subdivided. The subdivisions of the posterior lobe (neurohypophysis) are the **median eminence,** a highly vascular region of the infundibulum, and the **pars nervosa,** a **neuro-hemal** organ for hormones produced in the hypothalamus of the brain. The adenohypophysis (an-terior lobe) is divided into the **pars tuberalis,** a dorsal, elongated extension of the gland, and the **pars distalis,** the largest and distinctly glandular portion. The **pars intermedia,** a distinct layer of cells and (sometimes) follicles, lies between the pars nervosa and the pars distalis. Variations of these divisions are found throughout the vertebrate groups.

Functionally, the pituitary gland is very complex, and many aspects remain obscure. The neurohypophysis is a neurohemal organ. Such an organ is a part of the nervous system, containing terminations of neurosecretory cells whose cell bodies are elsewhere in the nervous system. These neurosecretory cells produce neurosecretions, or hormones, which are eventually discharged into blood sinusoids for distribution throughout the body. The neurohypophysis **produces no hormone** but does serve as a neurohemal organ for **oxytocin** (a smooth-muscle stimulator) and **vasopressin** (a blood-vessel constrictor and antidiuretic agent).

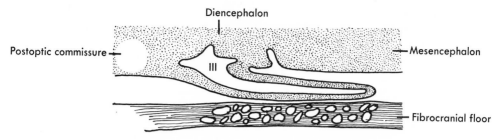

Fig. 14-4. Diagram of sagittal section of pituitary gland and its surroundings, in hagfish, *Myxine.* III = Third ventricle. (Redrawn from Gorbman and Bern, 1962.)

Both these hormones are made in the hypothalamus and travel down nerve fibers to the pars nervosa where they are finally released into the blood. The pars nervosa contains thick networks of unmyelinated nerve fibers and branching neuroglia cells (pituicytes).

The adenohypophysis (actually the pars distalis, the glandular portion of the pituitary) secretes a whole battery of potent regulatory chemicals into the blood. These include the growth hormone (STH); the hormone for stimulating the adrenal, or suprarenal, cortex (ACTH); thyroid-stimulating hormone (TSH); follicle-stimulating hormone (FSH); the hormone that works on the corpus luteum of the female, where it is called the luteinizing hormone (LH), or stimulates the interstitial cells of the male testis (ICSH); and a third gonadotropin hormone (prolactin or LTH), which stimulates milk production in female mammals and drives terrestrial salamanders to water for breeding. Evolution has produced species-specific differences in each of these hormones.

The pars intermedia lies in close intimacy with the neurohypophysis, and in birds and some mammals there is no cytologically recognizable intermedia. The chief function of this intermediate lobe is the production of the melanocyte-stimulating hormone (MSH), which causes the pigment in the chromophores to disperse and darken the skin.

CYCLOSTOMATA

The neurohypophysis (Fig. 14-4) is a hollow, elongated sac made up chiefly of ependymal-type cells. The cell bodies lie close to the ventricle, and their processes extend out to unite at the peripheral glial membrane. In the outer portions of the neurohypophysis is the termination of the **neurosecretory tract,** a group of nerve fibers that convey secretions into the pituitary gland. These fibers arise from nerve cell bodies in the hypothalamic part of the brain and travel through the pituitary stalk, or infundibulum, to reach the posterior lobe. The secretion probably enters into blood vessels soon after leaving the axons (Olsson, 1959).

The anterior, glandular portion of the pituitary (adenohypophysis) of cyclostomes has been a subject of some dispute, in particular as to whether it is homologous with the anterior or the intermediate lobe of other vertebrates. The hagfish adenohypophysis differs greatly from that found in other vertebrates. It consists only of islets of cells embedded in dense connective tissue on the floor of the cranial cavity near the infundibulum. There are two main kinds of cells, the clear cells, or chromophobes, which have a very lightly staining cytoplasm and seem to be nonsecreting cells, and the basophils, which show evidence of secretion. According to Olsson (1959), the basophil cells probably correspond to the beta cells of higher vertebrates. They resemble certain pharyngeal mucous cells in *Myxine* so closely that they may be the prototypes from which beta cells of vertebrates were derived.

The adenohypophysis of the lamprey more closely resembles the general vertebrate form. It is a glandular mass extending forward as a flattened, somewhat fingerlike process beneath the forebrain. Embryologically, its origin, as in other vertebrates, is by an ectodermal invagination. In the lamprey, however, it retains a close connection with the olfactory organ for a long period,

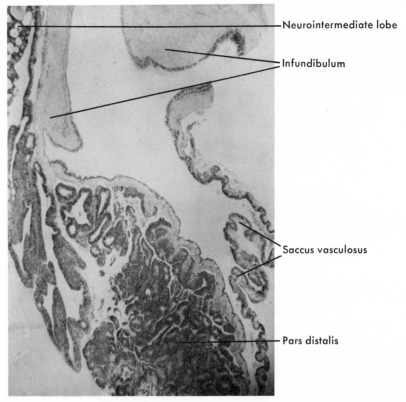

Fig. 14-5. Longitudinal section through pituitary gland (hypophysis) of shark *Squalus acanthias,* showing most of the complex. The pars ventralis is not shown. (×60.)

remaining attached to the nasal sac by a solid stalk for as long as five or six years.

The pituitary gland of lampreys has been studied by cytologic and histochemical methods by a number of investigators in recent years (Bargmann, 1953; Scharrer, 1953; Roth, 1956 and 1957; Sterba, 1961). Adam (1963) has reviewed these studies, as well as several on the hagfishes.

CHONDRICHTHYES

The elasmobranch pituitary is basically similar to the gland of the higher vertebrates (Figs. 14-5 and 14-6), but there are some differences. The adenohypophysis consists of the pars distalis (two zones), the **pars ventralis,** and the pars intermedia. There is no pars tuberalis. The pars ventralis, or ventral lobe, is found in no other vertebrate group. It is commonly bilobed and attached to the pars distalis by a stalk. Through the pars ventralis the internal carotids penetrate into the endocranium. Part of the pars distalis is in contact with the large infundibulum for its full length. No distinct pars nervosa is present, but the terminals of the nerve fibers from the preoptic nucleus of the hypothalamus spread over the pars intermedia and are distributed among its cells; this complex is often called the neurointermediate lobe. These terminals, like those in other vertebrates, store neurosecretions from the hypothalamus (Scharrer, 1953).

The pars distalis, which produces most of the pituitary hormones, is made up of cords and tubules between which are numerous blood vessels. Acidophilic cells occur in the peripheral part of the cords and tubules near the blood vessels (Fig. 14-7). Small, weakly staining **chromophobes** appear to be the innermost cells. The acidophils are most common in growing sharks, but when the gonadotropic hormones are secreted in excess, the chromophobes increase and the acido-

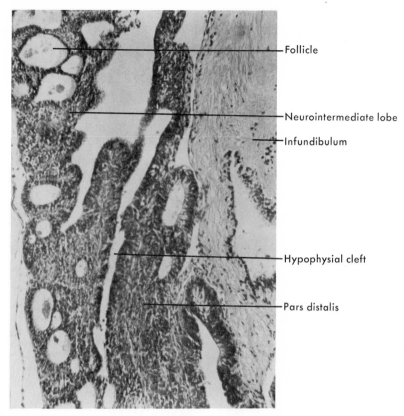

Follicle

Neurointermediate lobe

Infundibulum

Hypophysial cleft

Pars distalis

Fig. 14-6. Section through neurointermediate and anterior lobes (pars distalis) of the pituitary of *Squalus acanthias.* Note numerous follicles, or vesicles, in neurointermediate lobe. (×190.)

phils are forced to the extreme periphery of each tubule. According to Comes (1936), the growth hormone is secreted by the acidophils and the gonadotropic hormones by the chromophobic cells.

OSTEICHTHYES

In contrast to the cartilaginous fishes, teleosts always show a distinct neurohypophysis and adenohypophysis (Fig. 14-8). The neurohypophysis (or its pars nervosa) is closely associated with the pars intermedia, which is the posterior part of the adenohypophysis; in fact, strands of tissue from the neurohypophysis often extend far into the adenohypophysis. The gland cells of the adenohypophysis include types similar to those of higher vertebrates, including basophils, acidophils, and chromophobes. This is true of the main mass of the adenohypophysis, but the anterior portion is difficult to compare with anything in higher vertebrates.

AMPHIBIA

The amphibian anterior lobe, the adenohypophysis, is an epithelioid mass with the cells arranged in cords. At least three types of cells are recognized: basophils (beta), acidophils (alpha), and chromophobes (chief or reserve cells). The proportions of these cells vary in different animals. Alpha cells have many granules that stain strongly with acid dyes. Chromophobes show no granules and have a clear cytoplasm. The pars intermedia is made up of basophilic cells, which are nongranular. They produce a colloid material, which is found in gland cavities. Among the amphibians this lobe is best developed in the frogs. The pars tuberalis contains granular basophilic cells. The posterior lobe (neurohypophysis) resembles nervous tissue. It has both nonmyelinated nerve fibers and neuroglia cells. A particular cell type, the pituicyte, with long processes is found. From the preoptic nuclei (hypothalamus) a pair

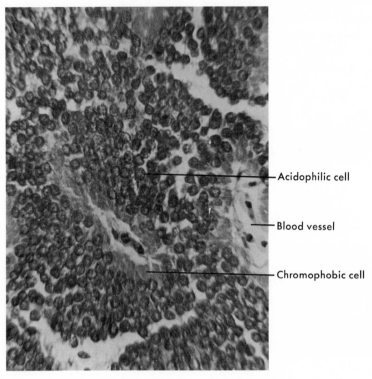

Fig. 14-7. Section of anterior lobe (pars distalis) of pituitary gland of *Squalus acanthias,* showing cell types. Most cells are acidophilic, as is common in young sharks. The lighter-staining cells are chromophobes. (×720.)

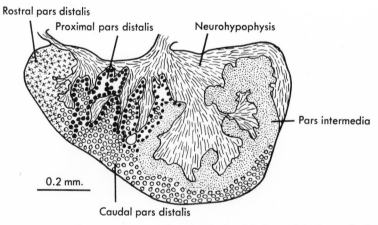

Fig. 14-8. Sagittal sections through pituitary gland of adult bony fish *Perca fluviatilis.* (Redrawn from Kerr, 1942.)

of fiber tracts extend into the neurohypophysis for carrying neurosecretions (Scharrer and Scharrer, 1954).

REPTILIA

A pars tuberalis appears to be lacking in the reptilian pituitary. The anterior lobe has the same three types of cells: acidophils, basophils, and chromophobes. The cells are commonly arranged in flattened cords. The chromophobes by differentiation give rise to the other two types of cells. The most common type is the acidophil, or alpha cell, which can be recognized by its smaller size and oval nucleus. Alpha cells stain with acid stains. The ovoid or angular basophils increase in the female lizard during the breeding season. Their granules are smaller than those of the alpha cells and stain with basic dyes. Chromophobes are small, with scanty cytoplasm that stains weakly. They are usually arranged in clusters near the center of a cord.

The pars intermedia of reptiles is very large in comparison with that of other classes of vertebrates, the greater bulk lying in the lateral portions. There usually is a cleft separating the pars intermedia from the pars distalis. The pars intermedia is almost surrounded by the pars nervosa. It contains large columnar acidophilic cells in contact with the capillaries of sinusoids. Vesicles with colloid droplets have been derived from acidophilic granules. Cords of cells as well as vesicles are commonly found. The pars intermedia is unusual in being the largest part of the pituitary (Poris and Charipper, 1938). The pars nervosa is a fibrouslike tissue with some scattered cells. The spherical, colloidlike masses cannot be traced to any cellular component of the pars nervosa, since there are no secretory cells in this part of the pituitary (Altland, 1939).

AVES

The pituitary gland of adult birds is distinctive by its lack of a pars intermedia. There are anterior and posterior lobes and a pars tuberalis. A connective tissue separates the small anterior lobe from the globular neurohypophysis. The adenohypophysis contains the typical three cell types: chromophobes, acidophils, and basophils. Payne (1942) describes two types of acidophils: A_1

cells in the caudal region and A_2 cells in the cephalic region. There are two cytologically distinct areas in the anterior lobe (Rahn and Painter, 1941). Only A_1 type cells appear to be found in mammals. They have red or orange granules when stained with acid fuchsin and methyl green. The smaller A_2 cells have scarlet granules when so stained. The cellular picture changes with age as chromophobes are gradually transformed into basophils and acidophils (Payne, 1946). Histologically, the neurohypophysis consists of blood sinuses and pituicytes with branches and granules.

MAMMALIA

The pituitary of mammals does not differ remarkably from that of reptiles or birds. In some mammals, as in the birds, the pars intermedia is absent. In others, for example, in certain aquatic mammals, it is greatly reduced. The human hypophysis has a pars intermedia well developed in early life but decreasing in prominence with age. The cleft of the hypophysis generally is present in mammals.

The anterior portion of the adenohypophysis, the pars distalis, is made up of acidophilic, basophilic, and chromophobic cells (Fig. 14-9). These cells are usually arranged in cords between which are sinusoids lined with macrophages. Many of the cells may also be arranged in irregular clusters. The proportions of cells vary, but the chromophobes, which are the smallest and have little affinity for dyes, make up about 50% of the total. Special staining techniques may show additional cell types among the **chromophils** (acidophils and basophils). The chromophobe cells are rounded or polygonal with little cytoplasm and often appear in groups in the center of the cords. Acidophils (alpha cells) are large, with distinct cell boundaries. Their cytoplasm has many granules that are stained with eosin, acid fuchsin, azocarmine, etc. The largest type, and usually the fewest in number, are the basophils (beta cells). Their granules are smaller than those in the acidophils, and they are best identified by the periodic acid–Schiff technique because of the glycoproteins in their granules.

The alpha and beta cells are largely responsible for the numerous hormones produced by the anterior lobe. However, attempts to associate the

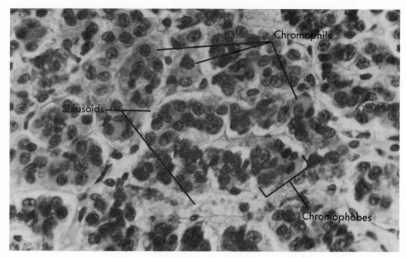

Fig. 14-9. Section through anterior lobe of cat pituitary gland. Special staining methods (acid fuchsin, orange G, or Azocarmine) are often employed to distinguish the three major types of cells—chromophobes and chromophils (acidophils and basophils). The problem is complicated by different subtypes within the major types. (Azocarmine; ×480.)

secretion of the hormones with one of the specific cell types have been only partially successful. For instance, alpha cells appear to produce the growth hormone, and the beta cells the follicle-stimulating and luteinizing hormones. As noted previously, these various hormones are concerned chiefly with the promotion of growth, the regulation of metabolism, and the control of the functions of the gonads and adrenals.

The pars tuberalis has its cells arranged in cords around the infundibular stalk in close association with blood vessels. Its cuboidal cells contain some granules but are weakly basophilic. The function of the pars tuberalis is unknown. The intermedia may contain follicles with colloid material. It contains large numbers of nongranular basophilic cells and some small, weakly staining cells. In some mammals it is not sharply separated from the anterior lobe (Dawson, 1937).

The pars nervosa of the neurohypophysis produces no known hormones but serves as a neurohemal organ for hormones made in the hypothalamus. As in the lower vertebrates, it is modified nervous tissue lacking nerve cells. Its cells (pituicytes) are related to neuroglia. A tract of unmyelinated nerve fibers from the hypothalamus passes down the stalk of the pituitary into the neurohypophysis. These nerve fibers originate

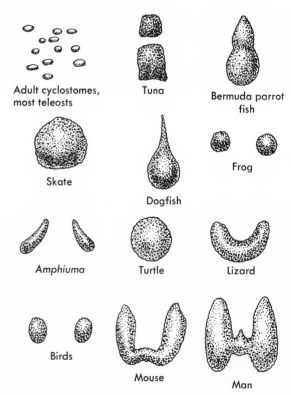

Fig. 14-10. Varied shapes of thyroid glands in different kinds of vertebrates. (Redrawn from Gorbman and Bern, 1962.)

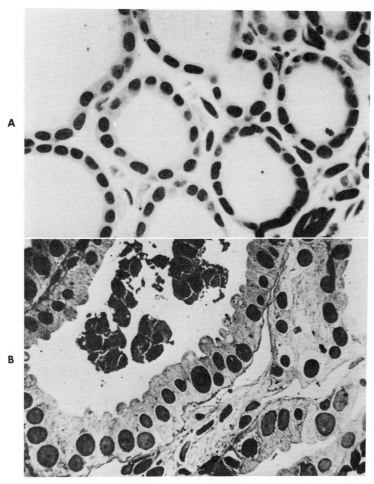

Fig. 14-11. Control and experimentally treated thyroid of *Ambystoma tigrinum.* **A,** Untreated gland. Follicles are oval or round with low cuboidal cells. Cytoplasm is scant. **B,** Gland treated for two days with TSH (thyroid-stimulating hormone from the anterior pituitary gland). The epithelium has become high columnar. (×320.) (Courtesy Dr. Lawrence Herman, Brooklyn, N. Y.)

from neurosecretory cells whose secretions are carried by the fibers through the stalk to the neurohypophysis. The hypothalamus secretes two distinctive hormones in this manner: oxytocin, which stimulates smooth muscle of the uterus, and vasopressin, which constricts blood vessels and has an antidiuretic effect.

The hypothalamus of the brain also controls the secretions of the anterior lobe, but in a different way. Although the actual hormones of the adenohypophysis are produced in the gland itself by specific cells (not in the hypothalamus), the adenohypophysis is unable to release its hormones into the bloodstream without the stimulation of neurosecretions from cells in the hypothalamus, delivered to the adenohypophysis by way of the hypophysial portal blood system.

Thyroid gland

The specific functions of the thyroid gland (Fig. 14-10) are the accumulation of iodine, the combining of this halogen with tyrosine to form thyroxine, and the controlled liberation of this metabolism-stimulating hormone into the bloodstream. Any discussion of thyroid gland structure

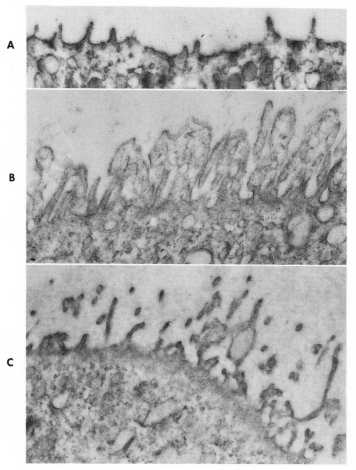

Fig. 14-12. Effects of administration of TSH (thryoid-stimulating hormone of pituitary) on the free border of thyroid cells in urodele amphibians. **A,** Untreated control, *Ambystoma tigrinum.* Microvilli are few in number and are short and stubby. **B,** Experimental, *Triturus viridescens.* Received 0.1 mg. of TSH daily for twelve days. Microvilli are long and numerous, some showing bulbous enlargements. **C,** Experimental, *Triturus viridescens.* Received 0.1 mg. of TSH daily for twelve days. Microvilli here show a somewhat different kind of alteration, becoming long and fingerlike. There appear to be droplets of secretion in some of them. (×27,000.) (Courtesy Dr. Lawrence Herman, Brooklyn, N. Y.)

in different kinds of vertebrates must take into account the remarkable morphologic changes in this gland that are correlated with increased functional activity (Figs. 14-11 and 14-12). Chiefly from studies of pathologic change and experimental stimulation or suppression in higher vertebrates, morphologic criteria of increased function have been identified. These include (1) increase in height of epithelial cells, (2) appearance of vacuoles in the colloid within the follicles, and (3) increased basophilia of the colloid. These, as well as more subtle changes within the cells, can be observed in the thyroid at certain stages in the life histories of some of the fishes, amphibians, and reptiles.

CYCLOSTOMATA

The primordium of the thyroid gland in the lamprey is a long segment of the groove located in the midventral part of the pharynx. A tubular structure is formed from the part of the groove between the second and fifth branchial arches.

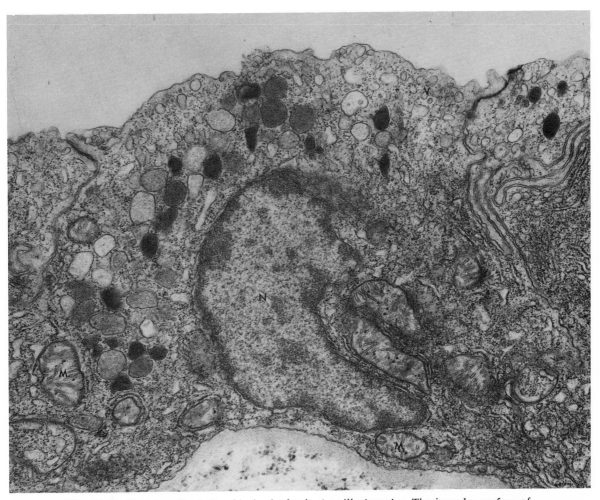

Fig. 14-13. Follicle cell of thyroid gland of eel, *Anguilla japonica*. The irregular surface of the cell lacks microvilli. There are numerous small vesicles, some droplets of moderate density (gray), and dense granules (black) in the apical cytoplasm. *M,* Mitochondria; *N,* nucleus; *V,* vesicles. (×15,000.) (From Fujita, H., H. Suemaso, and Y. Honma. 1966. Arch. Histol. Jap. **27:**153-163.)

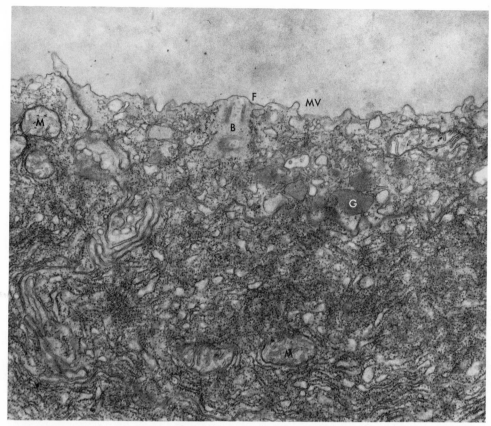

Fig. 14-14. Apical surface of a follicle cell of the eel, *Anguilla japonica*. Basal body of the central flagellum is seen. *B,* Basal body; *F,* flagellum; *G,* granule; *M,* mitochondria; *Mv,* microvillus. (×18,000.) (From Fujita, H., H. Suemaso, and Y. Honma. 1966. Arch. Histol. Jap. **27:**153-163.)

This tube is the subpharyngeal gland of the larval lamprey. It has a duct that opens onto the pharyngeal floor at the level of the fifth visceral arch.

Although this elongated site of origin differs from the embryologic formation of the thyroid gland in all other vertebrates, the evidence is convincing that the subpharyngeal gland represents the primitive thyroid. In addition, this seems to be a case in which an exocrine gland (with a duct) develops into an endocrine (ductless) gland. A large part of this evidence lies in the further development of the gland in *Petromyzon* itself. Differential growth of the epithelial lining of the tube causes the originally simple structure to become a complicated one. Some of the cells lose their cilia. Physiologic and biochemical studies have shown that at that early stage the

gland is able to metabolize iodine. It also is highly responsive to the thyroid-stimulating hormone from the anterior pituitary gland. The subpharyngeal gland, or thyroid, then remains unchanged throughout the remainder of the larval stage, which in some species may be five or six years. As the organism undergoes metamorphosis, the duct is closed and the gland becomes "endocrine" in nature.

Studies on the thyroid follicles of young adult lampreys show remains of ciliated cells of exocrine glandular type with orange or yellow granules such as are found in the exocrine gland. Fujita and associates (1966) have made careful studies of the ultrastructure of the thyroid gland of the silver eel *Anguilla japonica* (Figs. 14-13 and 14-14). They find the cisternae of the endoplasmic reticulum to be more flattened than in higher vertebrates, giving the cytoplasm a more compact appearance. A **central flagellum** was observed in some follicle cells, reminiscent of the endostyle stage.

The thyroid gland of the adult cyclostome consists of numerous individual saclike follicles, which do not form a definite organ. They are found scattered singly or in small groups in the fatty tissue in which the gill pouches occur and along the entire length of the pharynx. Each follicle consists of a layer of epithelial cells surrounding a lumen filled with colloid. Some follicles are almost 1 mm in diameter in *Myxine,* perhaps the largest seen in any of the vertebrates.

A functional study of the thyroid follicles in the cyclostome *Eptatretus stoutii* was made by Tong and associates, 1961. They injected ^{131}I and found that this isotope becomes bound to protein in the lumen of the follicle. Studies by hydrolytic and radiochromatographic procedures identified the iodinated organic compounds seen in other vertebrates: thyroxine, monoiodotyrosine, and diiodotyrosine.

CHONDRICHTHYES

The single-lobed thyroid of sharks is made up of a series of lobular follicles (Fig. 14-15) lying side by side in a compact encapsulated organ. It is suspended in a large venous sinus formed by the fusion of venous and lymphatic channels (Norris, 1918). Each follicle is hollow and lined by an epithelium that varies from cuboidal to columnar cells. The cells have round nuclei located near the basement membrane. The cytoplasm of the cells usually contains several granules in different stages of development of the thyroid secretion. From these cells the secretion passes into the lumen of the follicle where it is retained as a colloid substance.

In *Mustelus canis* and probably in other species of shark many of the follicular cells bear cilia. These cilia have no apparent adaptive value but may indicate the ancestry of the thyroid from a flagellated epithelium like that in the endostyle of lower chordates. Since the follicular lumen is the ancestral duct that originally communicated with the pharynx, these cilia may have had a

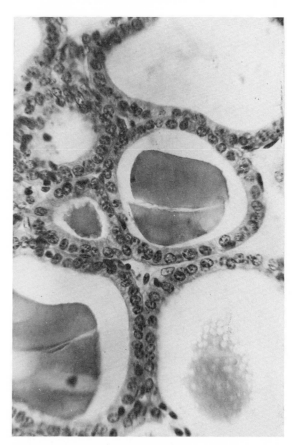

Fig. 14-15. Section through thyroid gland of *Squalus acanthias,* showing several follicles with colloid secretion. The middle follicle on right shows homogeneous, heavily staining colloid; follicle just below has many vacuoles, which some consider as artifacts of fixation methods. The separation of the colloid from epithelial lining of follicles is due to shrinkage in preparation. (×700.)

definite function in the movement of the secretion (Cowdry, 1921).

The general function of the thyroid hormones in warm-blooded vertebrates is to regulate metabolism and heat production, but there is little evidence for such functions in cold-blooded (poikilothermic) animals such as the shark. Some investigators do not believe that thyroxine has such functions in these forms. The available evidence indicates, at least, that thyroid cells take up inorganic iodine from the blood, store it in the form of thyroxine, and enzymatically release it later into the blood. What and how target tissues

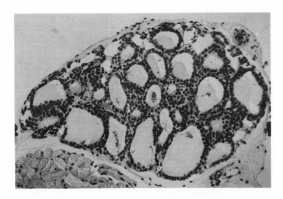

Taricha

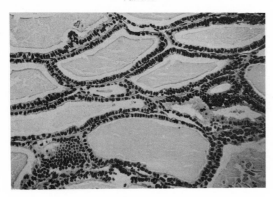

Amphiuma

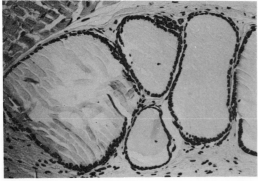

Necturus

Fig. 14-16. Thyroid glands of three species of urodele amphibians. The thyroid of *Taricha* presents picture of most active secretion; that of *Amphiuma* is intermediate; and that of *Necturus,* a "permanent larval" form, appears least active and resembles the human thyroid in colloid goiter. (×100.) (From Kerkof, P. R., W. Tong, and I. L. Chaikoff. 1963. Endocrinology 73[2]:185-192.)

react to it remains a contradictory theoretical matter.

OSTEICHTHYES

In fishes the thyroid usually occurs as several small masses of follicular tissue along the course of the ventral aorta. Sklower and Murr (1928) found a great enlargement of the thyroid in the larva of the eel at the beginning of metamorphosis. The enlargement begins as an increase in the quantity of colloid, but later, formation of new follicles also occurs. In the salmon, hypertrophy of the gland takes place at the time when the **parr** stage transforms into the **smolt,** the migratory form in which the salmon travels downstream to the sea.

The thyroids of poikilothermic animals generally show histologic evidence of inactivity or slight activity during the cold months and of maximum activity in the warmer ones; the breeding cycle rather than temperature as such seems to be an important factor.

AMPHIBIA

In urodeles the thyroid gland of the **larval stage** is of the low-activity or storage type with distended follicles, generally homogeneous, acidophilic colloid, and low cuboidal or squamous epithelium. In the larval stages of some urodeles the gland consists of one enormous follicle with a compact colloid lacking vacuoles. The beginning of metamorphosis in urodeles, as indicated by the first molting of the skin, is marked by an increase in overall weight of the gland, but a **resorption** of as much as 60% of the colloid. The walls of the follicle become folded, giving rise to star-shaped or irregular lumina (Uhlenhuth, 1927).

There are considerable differences in the histologic aspect of the thyroid gland in different genera of amphibians (Fig. 14-16) as shown by the work of Kerkof and associates (1963). The "follicular" pattern, however, is remarkably constant for this gland in the vertebrate series. It is one of three fundamental architectural patterns of the endocrine glands (Fig. 14-17).

The thyroid gland in frogs shifts position a great deal during the metamorphosis of the tadpole (D'Angelo and Charipper, 1939). In the young tadpole it is located anteriorly on the ventral surface of the hyoid cartilage; later it moves

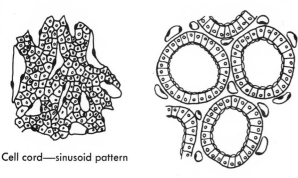

Cell cord—sinusoid pattern

Thyroid follicular pattern

Neurosecretory neurons

Fig. 14-17. Three basic types of architecture of endocrine glands of vertebrates. Each type shows a different relation between secreting cells and vascular channels that receive their products. (Redrawn from Gorbman and Bern, 1962.)

posteriorly to the adult position on either side of the hyoid apparatus between the posterior lateral and thyrohyoid processes. The gland arises as a median outgrowth from the ventral wall of the pharynx. It was originally an unpaired organ; but it became bilobed, and the two parts separated. Some thyroid tissue has also been described on the hypoglossal muscle and elsewhere (Stone and Steinitz, 1953).

The glands reach their greatest size towards the end of metamorphosis. Thyroid tissue consists of a mass of rounded follicles held together by connective tissue with a rich blood supply (Fig. 14-18). Each follicle is a closed sac lined by a single layer of cuboidal epithelium, which may show some seasonal variation. The nuclei are large and round, and cell boundaries are indistinct. The inner border of the cells may be cuticular. The lumen of the sacs holds the yellowish colloid secretion produced by the follicle cells.

Removal of the gland prevents metamorphosis; Allen (1918) also found that extirpation of the thyroid prevented limb development and inhibited the hind limbs from emerging. On the other hand, the feeding of thyroid extract to tadpoles inhibited growth and promoted early metamorphosis, so

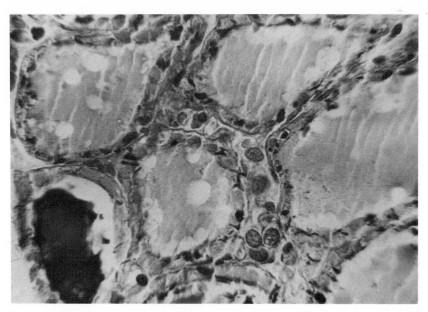

Fig. 14-18. Section through thyroid gland of *Plethodon glutinosus*. Note follicles filled with colloid containing the thyroid hormone. (×700.)

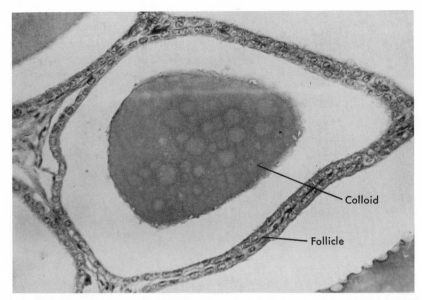

Fig. 14-19. Section through turtle thyroid. Note that follicles lie close together, with little interfollicular stroma. As a rule, the more active the gland is, the taller the cells appear and the more basophilic is the colloid. When the colloid does not completely fill the follicular lumina, it may have shrunk during the fixation process. Vacuoles and other irregularities in colloid may be due to the activity of the gland or may be artifacts caused by preparation. (Bouin's fixative; ×700.)

that tadpoles became frogs of diminutive size (Gudernatsch, 1912). The thyroid hormones contain a high percentage of iodine, which is considered to be the essential substance in promoting metamorphosis. Many other tissues are also known to accumulate iodine, including the notochord, gastric glands, and chloride-secreting cells of teleosts. Apparently only the thyroid cells have the power to synthesize thyroxine, the thyroid hormone.

REPTILIA

The thyroid gland of the lizard is made up of follicles lined with cuboid epithelium (Fig. 14-19). It is a single gland and lies ventral to the trachea, just anterior to the divergence of the carotid arches (Adams, 1939). However, the gland undergoes many seasonal changes and has a wide variety of forms even in the members of the same family (Lynn, 1960). In its storage state during the winter, the follicle contains a viscid colloid and is lined with low cuboid cells. At the height of activity, the follicles may have columnar epithelium. This is usually early summer, although

the cycle varies with different species (Miller, 1955).

The gland has an external connective tissue capsule with deep penetration of the fascia into the interior. Septa of fibroelastic tissue divide the lobes into lobules. The follicles nearly always contain colloid. The gland exerts a control over the rates of metabolism and heat production in birds and mammals, but according to present evidence, it seems to have no control over the respiratory rate of lizards. However, general body activity and behavior appear to be affected to some extent. Injection of ovarian hormones stimulates the thyroid in both sexes (Evans and Hegre, 1938).

AVES

The thyroid of the bird consists of two oval organs, one located on each side of the trachea near its bifurcation into the two bronchi. There is no isthmus connecting the two. They receive blood from branches of the carotid artery and are drained by veins into the jugular. Each thyroid gland consists of a capsule enclosing rounded

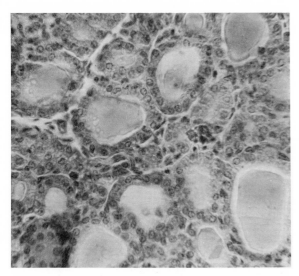

Fig. 14-20. Section of thyroid gland of *Columba livia*. Lumen of each follicle has colloid that normally occupies the whole lumen, but there has been shrinkage in preparation. Follicles vary in size, depending on the degree of distention by secretion. Cells of epithelium (usually cuboid) are high when the gland is hyperactive. Colloid contains active hormones. (×470.)

follicles lined with simple endodermal epithelium (Fig. 14-20). The size and shape of these epithelial cells depend on the activity of the gland. When active, the follicles contain colloid. Between the follicles is connective tissue containing interfollicular cells. Thyroxine is formed in the same manner as in mammals and contains iodine for which the gland shows a strong affinity. The effect of the hormone is also similar to that in mammals, but in addition, along with other factors it exerts an influence on molting (Fleischmann, 1947) and on pigmentation and growth of feathers (Schneider, 1939).

MAMMALIA

The thyroid gland of most mammals is located in front of the trachea at about the level of the second and third tracheal cartilage rings. It consists of two lobes connected by an isthmus located at the posterior ends of the lobes. It is surrounded by a firm capsule of fibroelastic connective tissue from which extensions or septa extend inward to divide and ensheath the lobules, which are composed of groups of follicles. These lobules interconnect, so that no one lobule is completely enclosed with connective tissue.

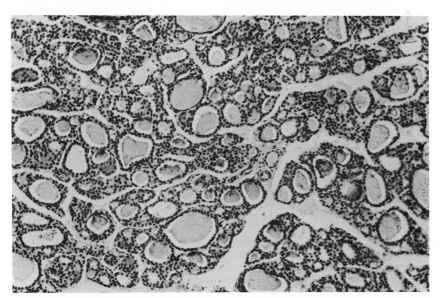

Fig. 14-21. Section through thyroid gland of rabbit in a fairly active condition. Note different sizes of follicles and the connective tissue septa that divide the gland into indefinite lobules. (Low power.)

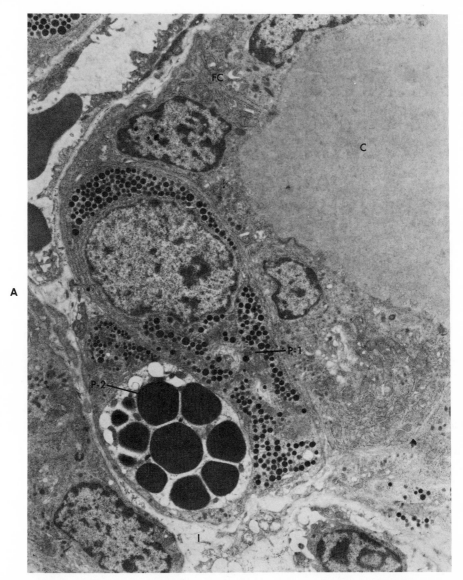

Fig. 14-22. Electron micrographs of the thyroid gland of bat. **A,** From a bat caught in late August. Parafollicular cells with small *(P-1)* and large *(P-2)* solid dense granules are found in the most basal regions of the thyroid follicle. *C,* Colloid; *FC,* follicular cells; *I,* intercellular space. **B,** Partly degranulated parafollicular cells of thyroid gland from a hibernating bat collected in December. Many small granules now contain myelin and rodlike subunits. **C,** Degranulated parafollicular cell from a hibernating bat collected during January. These cells are now characterized by agranular whorls of cytoplasmic membranes, which enclose a mass of cytoplasmic material (arrows). **D,** Agranular parafollicular cells found in thyroid glands of January bats. These cells *(C-1)* are characterized by cytoplasmic matrix packed with medium-sized rounded profiles of the rough-surfaced endoplasmic reticulum *(ER). C,* Colloid; *FC,* follicular cells; *DB,* dense lysosome-like bodies; *PC,* granular parafollicular cell. (**A,** ×13,000; **B,** ×20,000; *inset,* ×55,000; **C,** ×30,000; **D,** ×20,000.) (A from E. A. Nunez, R. P. Gould, D. W. Hamilton, J. S. Hayward, and J. S. Holt. 1967. J. Cell Sci. **2:**401-410; **B** courtesy Dr. E. A. Nunez and Dr. R. P. Gould, New York, N. Y.; **C** and **D** from E. A. Nunez, R. P. Gould, and J. S. Holt. 1969. J. Cell Sci. **5:**531-559.)

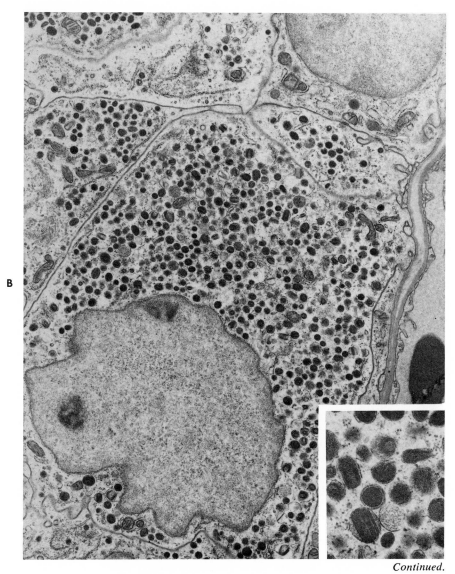

B

Continued.

Fig. 14-22, cont'd. For legend see opposite page.

The gland as a whole consists of a large number of closed epithelial sacs, or follicles, of varying size (Fig. 14-21). Each follicle lies separately in a common, connective tissue stroma. The epithelium of the follicle is usually simple cuboidal, although these cells change with age and with the secretion cycle. They are low when the follicles are stored with colloid, high when the gland is active and the follicles have little colloid. Near the follicular epithelium, there may be large clear **parafollicular** cells with argyrophilic granules.

They are larger than the epithelial cells and correlated with the follicular activity, but their real significance is unknown. The follicular cells have a cuticular border, or terminal bars, near their free surface. A basement membrane is a doubtful entity. The nuclei are large and vesicular; the finely granular cytoplasm is basophilic. Around the follicles are plexi of blood vessels and lymphatics. In fact, the thyroid is considered the most highly vascularized organ in the body.

Among mammals that hibernate (e.g., the

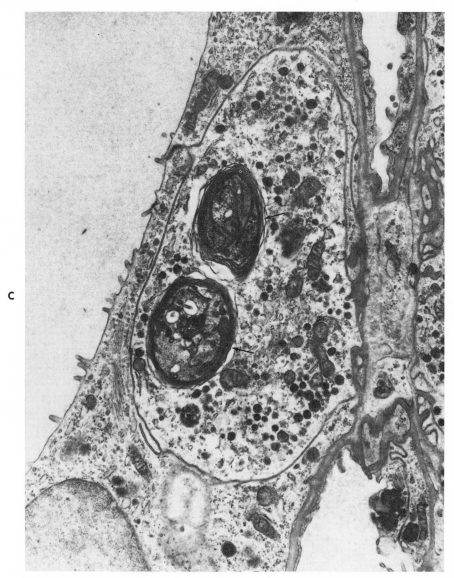

Fig. 14-22, cont'd. For legend see p. 334.

ground squirrel, hedgehog, and marmot) the thyroid is in the inactive storage state during hibernation. It has long been known (Adler, 1920) that the thyroid gland of the bat undergoes a marked atrophy, even to complete destruction, during hibernation. The body of the bat reaches lower temperatures and undergoes more variations in temperature than do those of other hibernating mammals.

The bat's parafollicular cells undergo drastic changes during hibernation (Nuñez and Gould, 1968). In active (nonhibernating) bats these cells generally contain large numbers of large and small dense granules (Fig. 14-22, *A*). In early hibernation many of the granules lose their dense cores and reveal rodlike, membranous subunits. Numerous cells are completely degranulated (Fig. 14-22, *B*). Later, before the gland becomes active again there is autophagic digestion of the residues remaining from the old granules (Fig. 14-22, *C*), followed by an increase in the amount of rough endoplasmic reticulum (Fig. 14-22, *D*), which is generally associated with protein synthesis. Eventually the endoplasmic

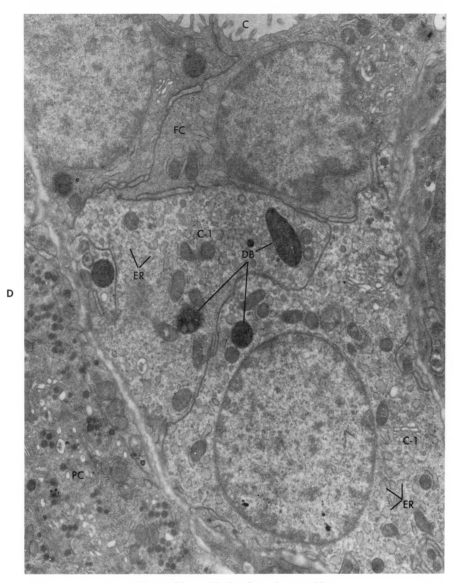

Fig. 14-22, cont'd. For legend see p. 334.

reticulum profiles are reduced in number and size. During this time, the Golgi apparatus, a secretory organelle, is progressively expanded, and small dense granules reappear throughout the matrix of the cytoplasm of these parafollicular cells.

Parathyroid glands

These glands are not recognized as definite entities until the Amphibia, although there have been reports of tissues with parathyroid-like functions in the larvae of certain teleost fishes. Indeed,

parathyroids are the only example of an endocrine gland not found in **all** the classes of vertebrates.

AMPHIBIA

In frogs and toads parathyroid glands occur as small round bodies, generally four in number with a pair lateral to the thyroid gland on each side. They are closely associated with branches of the jugular veins. The parathyroids are surrounded by connective tissue from which a meshwork extends inward separating the epithelioid cells into cords. Between the cords of cells are sinusoids (Waggener, 1929). Two types of cells have been described in the cord: large, pale chief cells with vesicular nuclei and even larger acidophilic (chromophilic) cells with granular cytoplasm. Some of the cells appear to be arranged in small follicles with colloid material.

During the colder part of the year, important changes in microscopic structure occur. The cells show degenerative changes with pyknosis of nuclei and eventual extensive cytolysis. In the bullfrog *(Rana catesbeiana)* these changes are marked, and masses of "colloid" appear as a result of the cytolysis. In the spring there is a regeneration of gland cells, presumably from some cells that have persisted beneath the fibrous capsule.

AVES

The parathyroids of the bird appear to contain only one kind of cell—the chief cells, which are usually arranged in anastomosing cords (Benoit, 1950). The cords are separated by capillary-containing connective tissue. The appearance of the chief cells depends on the physiologic condition of the gland, distinct granules appearing during the active state.

MAMMALIA

Parathyroid glands in mammals are usually found in two pairs. A pair are attached to the capsule of the thyroid at the back of each lateral lobe. Each parathyroid has its own thin capsule from which septa extend inward. It contains cords of epithelioid cells separated by connective tissue fibers and sinusoids. There are usually two main cell types in the gland: chief cells, which are large and pale and usually have relatively large vesicular nuclei; and oxyphil cells, which are less abundant but larger than chief cells. Also, the oxyphils show numerous large mitochondria (Trier, 1958).

In higher vertebrates the vital function of parathyroid glands was recognized when it became clear that the fatal effects of thyroid removal usually were actually due to the removal of the parathyroid glands along with the larger gland. The secretion of the parathyroid, called parathormone, is necessary for the maintenance of nitrogen equilibrium and metabolism of carbohydrates. Animals that have lost their parathyroids soon develop tetany, a constant tremor of the muscles, and eventually die. Parathormone also has important effects on the physiology of bone and kidney.

Ultimobranchial body

These structures are found in all vertebrates except cyclostomes. The ultimobranchial bodies are usually paired and associated with the thyroid gland. Histologically, the gland is made up of tubules and vesicles surrounded by a rich plexus of tiny veins that drain into the inferior jugular. It also has lymphatics that communicate with the veins. Its vesicles are lined by a single layer of columnar cells that produce mucuslike secretions. The elongated nuclei are found in the basal part of the cell (Camp, 1917).

The function of the ultimobranchial glands is obscure. Some investigators consider that they have a parathyroid-like function in regulating calcium metabolism. (There is no readily identifiable parathyroid tissue below the Amphibia.) In mammals, the glands become incorporated into each lateral lobe of the thyroid gland, and some think that the ultimobranchial body may take some part in the functioning of the thyroid gland and may actually give rise to accessory thyroid tissue (Van Dyke, 1959).

The ultimobranchial body differentiates into thyroidlike vesicles. These vesicles enlarge and produce a colloid substance. One study (Robertson and Swartz, 1964) considers them active secretory glands that may respond to environmental fluctuations in some way. However, this study showed that their secretion has nothing to do with the thyroid colloid; rather, it is a coagulum of carbohydrate-protein complex. Some authorities regard these glands as the terminal members of a series of pharyngeal epithelioid derivatives that have no known function or else have lost whatever function they may have had.

Adrenal gland

Two types of tissue are combined in higher vertebrates to form the adrenal gland, although in lower vertebrates they may be entirely separate from each other. These tissues are embryologically, histologically, and physiologically distinct. In the higher vertebrates such as the mammals the compact adrenal consists of a cortex, homologous to the interrenals of lower vertebrates, and near its center a medulla, corresponding to the chromaffin cells of lower forms. The chromaffin cells are so called because they give an intense color with chromic acid.

The cells of chromaffin tissues are derived from the neural crest like the cells of the sympathetic ganglia. They produce adrenaline and noradrenaline, which, in higher forms at least, increase cardiac output and vasoconstriction, respectively. The cortex of the adrenal gland, or interrenal tissue in lower forms, is derived from mesoderm of the coelomic wall near the differentiating gonads. This tissue produces important cortical steroids, such as androgenic sex hormones, deoxycorticosteroids that regulate carbohydrate metabolism. These substances and functions have been shown in mammals and some higher forms, but little is known about the function of interrenal tissue in the lower vertebrates.

CYCLOSTOMATA

It has not been possible to identify adrenocortical cells in the cyclostomes, although small islets of a few cells each, in the vicinity of the cardinal veins, have been suggested as representing this cell type (Sterba, 1955). In addition, corticol and corticosterone, products of adrenal cortex in higher vertebrates, have been found in the blood of *Myxine* (Jones, 1963). Both lampreys and hagfishes possess a chromaffin cell system. These cells are found not only along the cardinal veins and dorsal aorta but also within the heart. They represent an early stage of the medullary component of the adrenal glands of higher vertebrates.

CHONDRICHTHYES

In sharks, most other elasmobranchs, and to some extent in other fishes, the cortex and medullary parts are completely separated (Fig. 14-23). The chromaffin tissue, scattered in small segmented bodies along the dorsal aorta and extending below the caudal half of the opisthonephroi, is made up of irregularly arranged masses of epi-

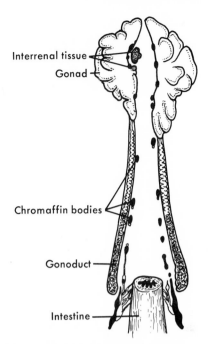

Fig. 14-23. Adrenal tissues of *Raja diaphanes,* the big skate. Chromaffin bodies (black) form an interrupted chain on either side. Anterior masses are associated with sympathetic ganglia. Interrenal (cortical) tissue forms two masses, both on left side. (Redrawn from Hartman and Brownell, 1949.)

thelial cells separated by sinusoids (Fig. 14-24). Most of the cells contain fine cytoplasmic granules that stain with chromic acid. The chromaffin tissue forms part of each ganglion (paraganglia) of the autonomic nerve chain.

The interrenal tissue is located in one or more compact masses between the opisthonephroi. This tissue appears to have only one type of cell (Dittus, 1940). In higher vertebrates the cords of epithelial cells are divided into three zones according to the way the cells are arranged, but this zonal arrangement is not apparent in the shark.

OSTEICHTHYES

The bony fishes generally have a mingling of the chromaffin and interrenal tissue. Lymphoid or hemopoietic tissue often is associated with the adrenal aggregates. In some of the more primitive teleosts (e.g., the salmon and trout) the interrenal tissue is diffuse and separate from the chromaffin cells.

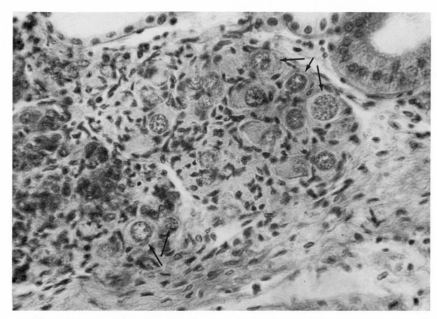

Fig. 14-24. Section showing composite of chromaffin and kidney tissue (shark). Sometimes a body of chromaffin tissue (arrows) is found near to, or embedded in, connective tissue of kidney. A portion of kidney tubule is seen in upper right corner. (×720.)

AMPHIBIA

The chromaffin and interrenal tissues are associated together into more or less definite organs. In the urodele amphibians, the masses of tissue do not form discrete structures. They are small, irregular bodies, often scattered between or on the ventral surfaces of the kidneys. In the anurans, the adrenal glands are distinct elongated organs lying on, or partly embedded in, the kidneys.

In the frog, the adrenal bodies are thin, golden yellow bands along the middle of the ventral surface of the kidneys. They lie close to branches of the renal arteries. The cells are arranged in branching and anastomosing cords around the sinusoids. There is no morphologic demarcation between cortex and medulla as occurs in the adrenal of higher forms, although both cortical and chromaffin cells are present in the frog adrenal. Cytologic studies (Burgos, 1959) indicate that there are at least three distinctive cell types: (1) lipid cells, large and polygonal with round nuclei and lipid droplets, representing the cortical cells of higher forms; (2) chromaffin cells of a brownish color, scattered among the cortical cells or arranged along the cortical cords, corresponding to the medullary, or central, cells of the higher vertebrates; and (3) eosinophilic cells (Stilling cells), staining intensely red with eosin, with no counterpart in the adrenal glands of mammals. The eosinophils seem to be present only during the summer months. Other types of cells—phagocytic leukocytes, lymphoid cells, etc. —may occur, especially in the winter and spring. The cortical cells are derived from the peritoneum, and the chromaffin cells from modifications of the sympathetic ganglia.

REPTILIA

The adrenal gland in the Reptilia shows many variations of gross and microscopic structure in the different groups and species of the class. In a number of the lizards the chromaffin cells form a sort of covering capsule over the dorsum of the interrenal mass of each gland—a relationship almost the reverse of that with which we are familiar in mammals. The crocodiles show a scattering of groups of chromaffin cells within the interrenal tissue, much as occurs among the birds. Although the two adrenals generally are symmetrically located in the reptiles, they lie at tandem in many of

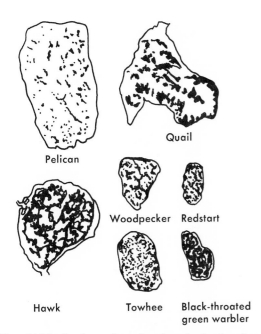

Fig. 14-25. Sections through adrenal glands of seven species of birds to show amount and distribution of chromaffin tissue (black) and interrenal (cortical) tissue (white). In the black-throated green warbler, relative positions of two kinds of tissue are such that interrenal tissue has an aspect of being distributed within the chromaffin tissue. (Redrawn from Hartman and Brownell, 1949.)

the snakes. The common garter snake, *Thamnophis,* has the right adrenal well anterior to the left one. In some turtles the adrenal is a single structure as in the majority of members of the genus *Thalassochelys.* In others, as in *Pseudemys,* it is **diffusely** distributed over the ventral surface of the kidney.

AVES

In the birds the adrenal glands have a position similar to that in mammals and a structural arrangement very different from the mammalian one. The location is at the anterior end of the kidneys. The structure is not at all like the corticomedullary one in mammals but shows a complete **dispersal** of the chromaffin amidst the interrenal tissue (Figs. 14-25 and 14-26). Histologically, the adrenals in pigeons (Miller and Riddle, 1942) consist of thin capsules containing blood vessels and nerves, interrenal cells arranged in cords, and the dark-staining chromaffin cells in islets among the interrenal tissue.

MAMMALIA

The paired adrenals in mammals lie near the kidneys but have no functional connection with them. One adrenal is found at the cranial pole of

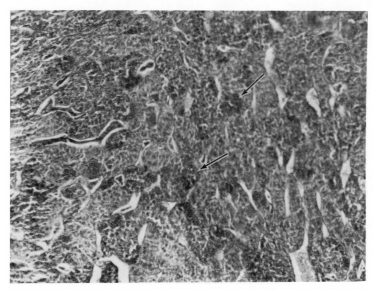

Fig. 14-26. Section of adrenal gland of chicken. Note intermingling of the chromaffin, or medullary, tissue (in dark areas indicated by arrows) with interrenal, or cortical, tissue that makes up most of the gland. Compare with mammal. (×275.)

Fig. 14-27. Photomicrograph of adrenal gland, including capsule, of cat. Supporting septa are seen as extensions from the capsule. (Wilder's reticular stain; ×125.)

each kidney. In man they are called suprarenals because of their location. Each gland has a capsule of thin, tough fibrous connective tissue with an inner areolar vascular zone. Extensions (trabeculae) form sheaths of reticular fibers, which radiate throughout the cortex region (Fig. 14-27). Blood vessels accompany the delicate trabeculae.

In almost all mammals the cortex (interrenal tissue) surrounds the medulla (chromaffin tissue) (Fig. 14-28), but the relative amounts vary greatly. The cortex makes up the bulk of the gland. The cortex consists of epithelial columns, or cords, usually two cells thick and arranged parallel to each other. These cords are knotted and looped at the periphery and branch and anastomose at their inner ends next to the medulla. The cords are separated from each other by thin trabeculae and elongated sinusoids. Blood enters the cortex at its periphery and leaves at its deep border.

The cortex is divided into three zones on the basis of different patterns of cell arrangement. Just beneath the capsule is the **zona glomerulosa** with tall columnar cells and oval nuclei. The cells are arranged in balls and loops. The next layer is

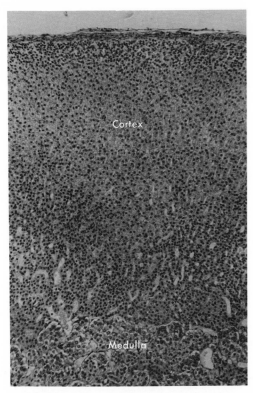

Fig. 14-28. Section through adrenal gland of rabbit, showing both cortex and medulla with its chromaffin cells (lower part of illustration). Line of demarcation between two zones is somewhat irregular. Cortex is composed mostly of cuboidal cells arranged in long, radial cords. Cytoplasm is largely basophilic. (×100.)

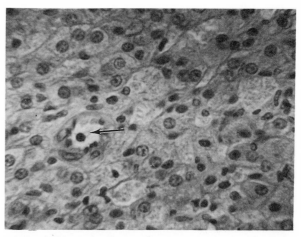

Fig. 14-29. Adrenal medulla with chromaffin cells in rabbit. Cells occur in groups surrounded by venules, capillaries, and connective tissue. Note ganglion cell (arrow) with dark nucleolus. Parenchymal cells of medulla originate from same group of cells that form sympathetic ganglion cells. (×860.)

the **zona fasciculata** consisting of straight parallel cords of polyhedral epithelioid cells. This is the thickest zone. Its cells are larger than those of the first zone. The cell cords of the zona fasciculata are continuous with the branching and anastomosing cords in the narrow **zona reticularis,** which lies next to the medulla. The cords of this region are usually one cell thick. Many of the cells are degenerating; as the cells die they are replaced by cells from the zona fasciculata. Cells are constantly being formed in the zona glomerulosa, and daughter cells are being shifted down toward the medulla. The adrenal cortical cells are filled with fatty droplets containing neutral fats, steroids, and phosphatides. The fasciculate zone, in particular, has many of these droplets.

The boundary between the cortex and medulla is irregular. The medulla lies near the center and is made up of interlacing cords or rounded groups of cells (Fig. 14-29). Within its spaces are capillaries and venules. The cells are polyhedral in shape and may appear stellate in places. Their nuclei are large and vesicular, and their cytoplasm is filled with granules that react with chromium salts to give the yellowish brown color characteristic of chromaffin cells. The medulla also contains sympathetic ganglion cells.

The cortex and medulla have separate functions. The cortex is essential to life. In mammals more than forty steroid compounds have been isolated from the cortex. It appears to play an important role in the functioning of other organs and tissues; it influences the permeability of cells, carbohydrate metabolism, salt exchange, nature of connective tissue, process of inflammation, secondary sex characters, and many others. Although the medulla is not essential to life, it produces the important hormones epinephrine (adrenaline) and norepinephrine, which play important roles in blood pressure. The direct innervation of the medullary cells by preganglionic sympathetic nerve fibers indicates an emergency mechanism.

It should be noted that several of the above substances have been shown to be produced by the adrenals of lower forms, but their functions have not been clearly established in those animals.

Pineal body

The pineal body (epiphysis) of the fishes is a slender stalk extending upward from the dorsal part of the diencephalon, usually with an enlarged vesicle on its end (Fig. 14-3, *A*). It is very small in the shark but appears to be a hollow organ. Little is known about its histologic structure in the lower forms; almost nothing is known about its function.

Over the years there has been considerable interest and mystery over the pineal body (frontal organ, Stieda's organ, pineal eye) of the frog and toad. Its location is indicated externally by the **brow spot** just in front of the eyes on the dorsal side of the head. Some have considered it to have a secretory function; some, a sensory one; and still others thought that it might be functionless. Whatever, its function, the electron microscope has made it possible to say definitely that this is, or was, a structure of photoreception. Eakin (1961) has studied the frontal organ in the tadpole of the Pacific treefrog *Hyla regilla*. He found highly specialized cells, clearly similar to the rods and cones of the retina of ordinary "lateral" eyes, protruding into the tiny lumen of the organ. Among other features, these cells show inner and outer segments. The inner segment has a connecting piece, closely packed mitochondria, and vesicles; the outer segment has a highly ordered array of disks as in the lateral eye of *Necturus*. The frog pineal body also contains epithelial cells and some neuroglia elements. Some cells resemble multipolar nerve cells without processes. The body appears to be innervated by fibers from the autonomic nervous system. Among the Reptilia the parietal or pineal eye of the the lizard has also been shown to contain specialized cells similar to rods and cones (Steyn, 1959).

In mammals this gland is a small, cone-shaped body attached by a stalk to the top of the third ventricle. The pia mater above it sends septa into the gland, dividing it into lobules. Within the lobules are epithelioid cells and neuroglia. The epithelioid cells are irregularly shaped cells with long branching processes, and their cytoplasm may contain granules and lipoid droplets. In some animals are found lamellated concretions (**acervuli** or **brain sand**) (Fig. 14-30). The nerve fibers

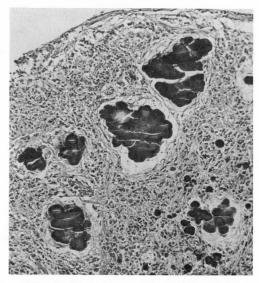

Fig. 14-30. Section of mammalian pineal gland showing brain sand (dark masses). Brain sand tends to accumulate with age and may be indicative of degenerate changes. Little is known about brain sand. (×165.)

in the pineal are from the sympathetic system, with the nerve terminals ending directly on the pineal cells.

The gland in rat and some other mammals is known to contain melatonin and serotonin, which may undergo circadian rhythms. Injection of melatonin into rats slows their estrus cycles and produces smaller ovaries (Lerner and Case, 1960). Increased activity of the gland (as a result of tumors, etc.) delays puberty in children.

Gonads

The general structure of vertebrate gonads is considered in Chapter 11. In addition to giving rise to gametes, the testes and ovaries of most vertebrates also produce steroid hormones. These sex hormones perform several functions such as regulating the cycles of sex behavior and developing and maintaining the secondary sex characteristics. The endocrine parts of the gonads are the interstitial cells of Leydig in the testes, and the graafian follicle, corpus luteum, and interstitial tissue in the ovary.

The major male hormones are androgens; the female hormones are estrogens. The chief androgen, testosterone, appears to be secreted by clusters of connective tissue cells between the seminiferous tubules; these are the Leydig or interstitial cells. They are cells with a heterogeneous cytoplasm that often contains crystals. Estradiol is the chief estrogen; it is produced by the theca interna of developing ovarian follicles. Another steroid, progesterone, comes from the corpora lutea that form after an egg has been discharged from the graafian follicle. Small amounts of androgens are also produced by females, and some estrogens by males. It is thought that the Sertoli cells in the seminiferous tubules of the testes secrete estrogenic hormones in some animals.

Islets of Langerhans

The pancreas is both an exocrine gland (with duct) and an endocrine gland (ductless). As described in Chapter 13, the exocrine portion secretes pancreatic juice containing many digestive enzymes. Its secretions are discharged through a duct into the intestine. The endocrine portion of the pancreas is made up of many distinct entities, the islets of Langerhans, scattered among the exocrine glandular tissue. The pancreatic hormones are insulin and glucagon, which regulate the storage of glycogen in the liver, the sugar level of the blood, and the oxidation of sugar by body cells.

The islets are highly vascular epithelial masses or cords. Each one is separated from pancreatic acini by a reticular membrane. The islet tissue is usually organized into irregular, anastomosing plates between which are blood capillaries. The polyhedral cells are paler than alveoli, and in many preparations they show very indistinct cell boundaries. Two well-defined cell types are distinguishable in the islet tissue (Fig. 13-18): alpha (A) cells with large vesicular nuclei and granules soluble in water but not in alcohol and beta (B) cells with small dense nuclei and granules soluble in alcohol. The beta cells are far more numerous than the other type and are the first to appear in development (Bencosme, 1955). A third type of cell (delta) has been described in mammals. These have a light cytoplasm and few granules, but they are poorly defined entities and have no known function. The important hormone, insulin, is produced by the beta cells, and glucagon by the alpha cells. Special

staining techniques (Gomori aldehyde fuchsin and others) are required to bring out distinctly the different cell types.

References

Adam, H. 1963. The pituitary gland. Pages 459-480 *in* A. Brodal and R. Fänge, eds. The biology of *Myxine*. Universitetsforlaget, Oslo, Norway.

Adams, W. E. 1939. The cervical region of the Lacertilia. J. Anat. **74**:57-71.

Adler, L. 1920. Schilddrüse and Wärmerregulation (Untersuchungen an Winterschläfter). Arch. Exp. Pathol. **86**:159-224.

Allen, B. M. 1918. The results of thyroid removal on the larvae of *Rana pipiens*. J. Exp. Zool. **24**:499-520.

Altland, P. D. 1939. Cytology of the hypophysis of the fence lizard. Anat. Rec. **74**:109-125.

Bargmann, W. 1953. Über das Zwischenhirn-Hypophysensystem von Fischen. Z. Zellforsch. Mikrosk. Anat. **38**:275-298.

Barrington, E. J. W. 1959. Some endocrinological aspects of the Protochordata. *In* A. Gorbman, ed. Comparative endocrinology. John Wiley & Sons, Inc., New York.

Bencosme, S. A. 1955. The histogenesis and cytology of the pancreatic islets in the rabbit. Am. J. Anat. **96**:103-151.

Benoit, J. 1950. Les glandes endocrines. Vol. 15, pp. 290-310 and 314-329 *in* P. P. Grassé, ed. Traité de zoologie. Masson & Cie., Paris.

Burgos, M. H. 1959. Histochemistry and electron microscopy of the three cell types in the adrenal gland of the frog. Anat. Rec. **133**:163-185.

Camp, W. E. 1917. The development of the suprapericardial (post branchial, ultimobranchial) body in *Squalus acanthias*. J. Morphol. **28**:369-415.

Comes, O. C. 1936. Sulla funzione delle celluie eosinofile e cromofobe nei lobo anteriore dell' ipofisi. Boll. Soc. Ital. Biol. Sper. **11**:216-218.

Cowdry, E. V. 1921. Flagellated cell in the dogfish thyroid (*Mustelus canis*). Anat. Rec. **22**:289-299.

D'Angelo, S. A., and H. A. Charipper. 1939. The morphology of the thyroid gland in the metamorphosing *Rana pipiens*. J. Morphol. **64**:355-373.

Dawson, A. B. 1937. The relationship of the epithelial components of the pituitary gland of the rabbit and cat. Anat. Rec. **69**:471-485.

Dittus, P. 1940. Histologie und Cytologie des Interrenalorgans der Selachier unter normalen und experimentellen Bedingungen. Z. Wiss. Zool., Part A **154**:40-124.

Eakin, R. M. 1961. Photoreceptors in the amphibian frontal organ. Proc. Natl. Acad. Sci. USA **47**:1084-1088.

Evans, L. T., and E. Hegre. 1938. The effects of ovarian hormones and seasons on *Anolis carolinensis*. I. The thyroid. Anat. Rec. **72**:1-9.

Fleischmann, W. 1947. Comparative physiology of

the thyroid hormone Q. Rev. Biol. **22:**119-140.

Fujita, H., and Y. Honma. 1966. Electron microscopical studies on the thyroid of a cyclostome, *Lampetra japonica,* during its upstream migration. Z. Zellforsch. Mikrosk. Anat. **73:**559-575.

Fujita, H., H. Suemaso, and Y. Honma. 1966. An electron microscopic study of the thyroid glands of the silver eel, *Anguilla japonica.* (A part of phylogenic studies of the fine structure of the thyroid.) Arch. Histol. Jap. **27:**153-163.

Gorbman, A., and H. A. Bern. 1962. A textbook of comparative endocrinology. John Wiley & Sons, Inc., New York.

Gudernatsch, J. F. 1912. Feeding experiments on tadpoles. 1. The influence of specific organs given as food on growth and differentiation; a contribution to the knowledge of organs with internal secretion. Arch. Entwklgsmech. Organ. **35:**457.

Hartman, F. A., and K. A. Brownell. 1949. The adrenal gland. Lea & Febiger, Philadelphia.

Jones, I. C. 1963. Adrenocorticosteroids. Pages 488-502 *in* A. Brodal and R. Fänge, eds. The biology of *Myxine.* Universitetsforlaget, Oslo, Norway.

Kerkof, P. R., W. Tong, and I. L. Chaikoff. 1963. I[131] utilization by salamanders: *Taricha, Amphiuma* and *Necturus.* Endocrinology **73**(2):185-192.

Kerr, T. 1942. A comparative study of some teleost pituitaries. Proc. Zool. Soc., Series A **112:**37-46.

Lerner, A. B., and J. D. Case. 1960. Melatonin. Fed. Proc. **19:**590-592.

Lynn, W. G. 1960. Structure and function of the thyroid gland in reptiles. Amer. Midl. Natu. **64:**309-326.

Miller, M. R. 1955. Cyclic changes in the thyroid and interrenal glands of the viviparous lizard, *Xantusia vigilis.* Anat. Rec. **123:**19-32.

Miller, R. A., and O. Riddle. 1942. The cytology of the adrenal gland in normal pigeons and in experimentally induced atrophy and hypertrophy. Am. J. Anat. **71:**311-341.

Norris, E. H. 1918. The morphogenesis of the thyroid gland in *Squalus acanthias.* J. Morphol. **31:**187-223.

Nunez, E. A., and D. V. Becker. 1970. Secretory processes in follicular cells of the bat thyroid. I. Ultrastructural changes during the pre-, early and mid-hibernation periods with some comments on the origin of autophagic vacuoles. Am. J. Anat. **129:**369-397.

Nunez, E. A., and R. P. Gould. 1968. Fine structural changes in the follicular cells of the bat during early hibernation. Anat. Rec. **160:**320. (Abstr.)

Olsson, R. 1959. The neurosecretory hypothalamus system and the adenohypophysis of *Myxine.* Z. Zellforsch. Mikrosk. Anat. **51:**97-107.

Palay, S. L., and G. E. Palade. 1955. The fine structure of neurons. J. Biophys. Biochem. Cytol. **1:**69-88.

Payne, F. 1942. The cytology of the anterior pituitary of the fowl. Biol. Bull. **82:**79-111.

Payne, F. 1946. The cellular picture in the anterior pituitary of normal fowls from embryo to old age. Anat. Rec. **96:**77-91.

Poris, E. G., and H. A. Charipper. 1938. Studies on the endocrine system of reptiles. I. The morphology of the pituitary gland of the lizard *(Anolis carolinensis)* with special reference to certain cell types. Anat. Rec. **72:**473-490.

Rahn, H., and B. T. Painter. 1941. A comparative histology of the bird pituitary. Anat. Rec. **79:**297-311.

Robertson, D. R., and G. E. Swartz. 1964. Observations on the ultimobranchial body in *Rana pipiens.* Anat. Rec. **148:**219-230.

Roth, W. D. 1956. Some evolutionary aspects of neurosecretion in the sea lamprey *(Petromyzon marinus).* Anat. Rec. **124:**437. (Abstr.)

Roth, W. D. 1957. The pars distalis of the adenohypophysis of the sea lamprey *Petromyzon marinus.* Anat. Rec. **127:**445. (Abstr.)

Scharrer, E. 1953. Das Hypophysen-Zwischenhirnsystem der Wirbeltiere. Verhandl. Anat. Gesellsch. 51. Versamml., Mainz. Erg.-Bd. Anat. Anz., Vol. 100, pp. 5-27.

Scharrer, E., and B. Scharrer. 1954. Neurosekretion. Vol. III, part 5, pages 953-1066 *in* W. von Mollendorff, ed. Handbuch der mikroskopischen Anatomie des Menschen. Springer Verlag, Berlin.

Schneider, B. A. 1939. Effects of feeding thyroid substances. Q. Rev. Biol. **14:**289-310.

Sklower, A., and B. Murr. 1928. Untersuchungen über die inkretorischen Organe der Fisch. I. Das Verhalten der Schilddrüse in der Metamorphose des Aales. Z. Vergl. Physiol. **7:**279-288.

Sterba, G. 1955. Das adrenal- und interrenal-System in Lebensablauf von *Petromyzon planeri* Bloch. Zool. Anz. **155:**151-168.

Sterba, G. 1961. Fluoreszenzmikroskopische Untersuchungen über die Neurosekretion beim Bachneunauge *(Lampetra planeri* Bloch.). Z. Zellforsch. Mikrosk. Anat. **55:**763-789.

Steyn, W. 1959. Ultrastructure of pineal eye sensory cells. Nature **183:**764-765.

Stone, L. S., and H. Steinitz. 1953. Effects of hypophysectomy and thyroidectomy on lens regeneration in the adult newt, *Triturus viridescens.* J. Exp. Zool. **124:**469-504.

Tong, W., P. R. Kerkof, and I. L. Chaikoff. 1961. I[131] utilization by thyroid tissue of the hagfish. Biochim. Biophys. Acta **52:**299-304.

Trier, J. S. 1958. The fine structure of the parathyroid gland. J. Biochem. Biophys. Cytol. **4:**13-22.

Uhlenhuth, B. 1927. Die Morphologie und Physiologie der Salamander Schilddrüse. I. Histologish-embryologische Untersuchung des Sekretionsprozeses in den verschiedenen Lebensperioden der Schilddrüse des Marmersalamanders, *Ambystoma opacum.* Roux Arch. **59:**611-749.

Van Dyke, J. H. 1959. The ultimobranchial body. *In* A. Gorbman ed. Comparative endocrinology. John Wiley & Sons, Inc., New York.

Waggener, R. A. 1929. A histological study of the parathyroids in the Anura. J. Morphol. Physiol. **48:**1-44.

15 NERVOUS SYSTEMS

In both invertebrates and vertebrates the nervous system is considered to be ruler and controller of the other systems and, to a large extent, of the whole organism. Although part of this role certainly is an automatic type of control, it is recognized that animals' "awareness" of their internal and external environments, and even of their own existence, is also a function of this system. Surprisingly, however, nervous tissue, as found in histologic preparations, is not very complex. The real complexity of the system lies in the way in which the cellular units, the **neurons,** are organized and interrelated.

In nervous tissue, two basic properties of all protoplasm are developed to a great degree: irritability and conductivity. The first of these is concerned with the capacity to respond to physical and chemical agents or with the initiation of an impulse; the second is the ability to transmit such an impulse from one region to another. The development of these two protoplasmic properties requires a special type of cell of great diversity and shape, the neuron. The cell bodies and their processes distinguish the neuron from all other cell types.

Most cell bodies are located within the central nervous system itself (spinal cord and brain) or in ganglionic centers scattered strategically throughout the body. However, the neuron processes, which carry impulses, may lie entirely within the central nervous system, may extend from the central nervous system to other parts of the body, or may lie entirely outside the spinal cord and brain. All neurons make contact with other neurons by specialized connections called **synapses.** This intimate association between neurons makes possible the exciting of other neurons, that is, the conduction of an impulse from one nerve cell to another.

Nervous tissue is composed of millions of nerve cells and their processes with coverings, or sheaths, and connective tissue investments with associated blood vessels. The sheaths are specialized supportive and nutritive cells, called **neuroglia** in the central nervous system and Schwann cells (**neurolemma**) in the peripheral system. Nerve tissue is composed of compactly arranged cells with little intercellular matrix; a superficial inspection gives the impression of few cells. The basic microscopic structure of nervous tissue is similar throughout the vertebrate world and will be discussed before moving on to phylogenetic differences in its organization.

Basic structure of nervous tissue

The neuron. The neuron consists of the cell body and all its processes (Fig. 15-1). It is the structural and functional unit of the nervous system. The cell body, also called the **perikaryon,** is usually large in comparison with other cells, although there is a great range in size (4 to 135 μm). The nucleus is large, spherical, and centrally located in the cell. The nuclear chromatin is scanty, but there is usually a single, prominent nucleolus. A distinct nuclear membrane can be seen, whereas around the entire cell body is the indistinct plasma membrane. The shape of the perikaryon may be round, ovoid, or elliptical, with a smooth or irregular cell surface. Its shape is dependent chiefly on the nature and arrangement of the cell processes.

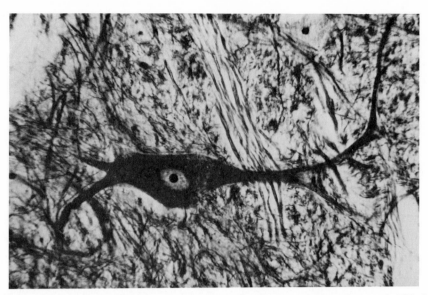

Fig. 15-1. Motor neuron from the anterior horn of gray matter of the spinal cord. This is an example of a multipolar neuron, some of which in this region are the largest of all neurons. (×900.)

The unipolar cell has only one process, and the perikaryon is globular. This type is found in the ganglia alongside the brain and spinal cord. A bipolar cell is a spindle-shaped cell with a process at each end. Olfactory neurons, some retinal neurons, and cells of the acoustic ganglia are examples of this type. The multipolar cell has many processes arising from it. Its shape may be quite variable (stellate, pear-shaped, etc.), and it is represented by cells in the central nervous system and autonomic ganglia.

FINE STRUCTURE OF THE PERIKARYON. The electron microscope has revealed the nerve cell body as similar in many respects to the cells of other tissues, yet differing from them in some important details of structure and distribution of the organelles. Mitochondria are present in all nerve cells. They appear to vary from small spheroids to long filaments. Their cristae are similar to those found in other cells (Fig. 15-2). The Golgi apparatus, an irregular network of wavy threads under the light microscope, appears in electron micrographs as isolated collections of stacked lamellae, vacuoles, and vesicles. Three-dimensional reconstructions of these collections are helping to correlate the light- and electron-microscope pictures.

The rough endoplasmic reticulum is a distinc-

tive feature in the great majority of nerve cells. This "organelle" is well developed and occurs in discrete, sharply bounded regions. The parallel folds, or cisternae, show numerous ribosomes attached to their membranes as well as free ribosomes between them (Fig. 15-3). The large motor neurons such as those in the ventral horns of the spinal cord, the motor nuclei of cranial nerves, and the cells of Purkinje show an especially well-developed arrangement of rough endoplasmic reticulum.

These "regions" of rough endoplasmic reticulum are actually the **Nissl bodies,** or **ergastoplasm,** of the light microscopist (Fig. 15-4) and are related to the high degree of protein synthesis carried out by the nerve cell. It has been recognized for many years that changes in amount of Nissl substance occur during fatigue, starvation, etc. and that this **chromatolytic change** is an early alteration in the cell body when the axon is severed or crushed. This phenomenon makes it possible for investigators of the nervous system to locate the "nuclei," or groups of nerve cells, from which the fibers of individual nerves arise. Nissl bodies are present in dendrites, but not in axons nor in their clear, conical bases in the perikaryons (axon hillocks).

The nature of Nissl bodies as well-defined

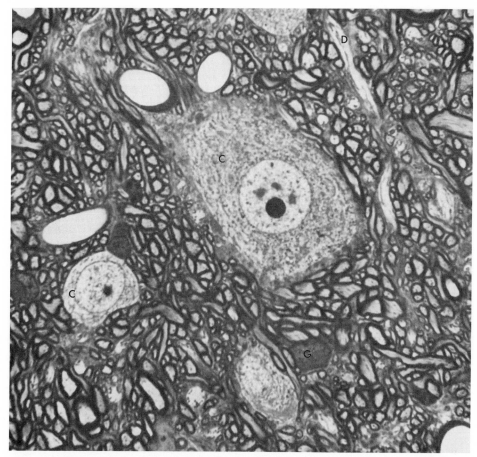

Fig. 15-2. Section from interposed dentate nuclear mass. Cell bodies *(C)* of large and small neurons are seen, lying in a neuropil composed of dendrites *(D)*, glial cells *(G)*, and myelinated fibers. Some myelinated fibers lying close to the surface of cell bodies and dendrites are of minute size. (Toluidine blue stain; ×1400.) (Courtesy Dr. R. P. Eager, New Haven, Conn.)

regions of rough endoplasmic reticulum is of comparative significance when we recall the presence of discrete **basophilic bodies** in the cytoplasm of various other types of cells. The gland cells, especially those of the pancreas, salivary glands, and liver of cold-blooded animals, often have one large body of this type, the paranuclear body (Nebenkern). It is, in one sense, simply a large Nissl body in a different location, that is, outside the nervous system. It seems probable that a functional parallelism also exists.

The question of neurotubules has occupied an important place in the research literature on the fine structure of the nerve cell (Palay, 1955). These appear to be real structures—hollow tubules, each with a "lumen" approximately 100 Å thick in higher vertebrates. The wall itself shows

even more ultrastructural detail; it contains filaments that have a dense periphery and a light center. **Neurofibrils** are apparently composed largely of great masses of these neurotubules. The neurofibrils form bundles. From these they branch and anastomose as a complex network throughout the cell body and its processes, the dendrites and axon (Fig. 15-5). Neurofibrils were demonstrated originally by methods of silver impregnation and have been shown to be present in living nerve cells in tissue culture (Weiss and Wang, 1935).

Cytoplasmic **(neuroplasmic)** inclusions frequently also include fat droplets and pigment granules. A **centriole** is usually absent from mature nerve cells.

NERVE CELL PROCESSES. The neuroplasm of

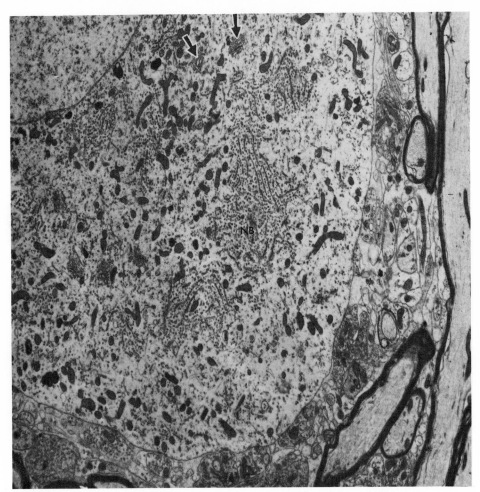

Fig. 15-3. Portion of cell body of large neuron of interposed dentate nuclear mass. Portions of several Nissl bodies *(NB)* are seen. Also shown are mitochondria and several areas of Golgi apparatus (arrows). Portion of nucleus is seen in the upper left; several synaptic terminations are located on the surface of the cell body. (×9500.) (Courtesy Dr. R. P. Eager, New Haven, Conn.)

the nerve cell body is drawn out into processes of various sizes, numbers, and arrangements. Typically, there are two types: **dendrites** (dendrons), one or more in number, which conduct impulses toward the cell; and a single **axon,** or axon cylinder, which conducts impulses away from the cell. Multipolar cells (the most common nerve cells) have several dendrites, whereas unipolar and bipolar cells each have only one. A typical dendrite may branch freely from its stem, producing contorted and varicose arborizations. The single axon is usually longer and more slender than the dendrites and has a uni-

form size throughout its length. Its contour is smooth, and when it branches, it does not decrease in size. It may give rise to lateral branches called **collaterals.** Usually the axon terminates in a twiglike arborization **(telodendria).**

Unlike the dendrites, the axons are frequently provided with an accessory sheath that may be of two kinds: an inner fatty **myelin** sheath and an outer cellular sheath of Schwann cells, or the **neurolemma.** The neurolemma is found only outside the central nervous system. The cytoplasm of an axon is called the **axoplasm** and is continuous with the neuroplasm of the cell. The

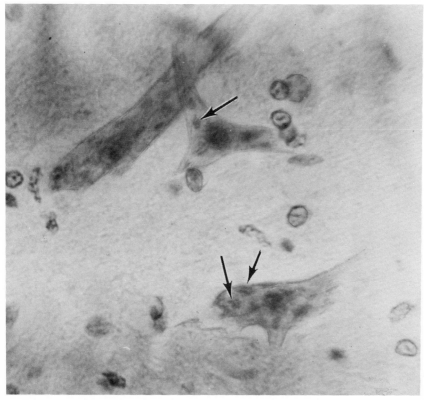

Fig. 15-4. Cell bodies of neurons from the anterior horn of the spinal cord of rabbit, prepared to show Nissl bodies (arrows) in cytoplasm. These bodies may be involved in protein synthesis.

axon is covered by a membrane **(axolemma).** An axon and its sheath are called a **nerve fiber.** Bundles of nerve fibers are called **tracts** in the central nervous system, and **nerves** in the peripheral system. Both dendrites and axon are intrinsic parts of the neuron and must be considered as thread-like extensions of the cell body. New protoplasm is synthesized by the cell body and flows down the nerve cell processes to replace protoplasm used in metabolism (Weiss and Hiscoe, 1948).

Nerve fibers are classified as myelinated (with a myelin sheath) and unmyelinated (without a sheath). In myelinated nerves the axon is surrounded by a tubular sheath of myelin, which is the **white matter** of the brain and spinal cord. The term **gray matter** refers to those parts of the nervous system that contain the **cell bodies** of the neurons. Electron-microscope observations have shown that myelin consists of a series of

many concentric lamellae. These lamellae actually are formed by an extension of the plasma membranes of neuroglia cells (within the central nervous system) or by an extension of the neurolemma, or Schwann cells (in peripheral nerves) (Fig. 15-6). Myelin is definitely not a structureless, secreted material. In cross section, the concentric divisions of myelin consist of dense, dark portions approximately 30 Å in thickness that alternate with clear, light layers about 100 Å thick. A fine dark line often is seen in the center of the light band.

The studies on the fine structure of myelin offer a good example of the correlation of data from different disciplines in modern science. X-ray diffraction studies indicate that myelin has the chemical nature of a lipoprotein. This is the combination (lipid and protein) found in cell membranes. Thus the x-ray diffraction data and the electron-microscope demonstration of the

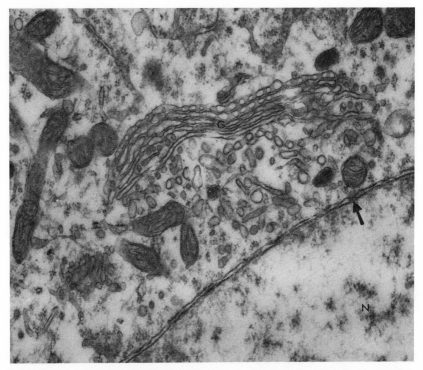

Fig. 15-5. High-magnification electron micrograph of a portion of cell body of a large neuron, with details of the Golgi apparatus *(G)* lying close to the nucleus *(N)*. A fenestration in the nuclear envelope is indicated by the arrow. (×36,225.) (Courtesy Dr. R. P. Eager, New Haven, Conn.)

sheath as essentially an extension of a cell membrane are mutually confirmatory.

If a neurolemma is present, the tubular myelin sheath is interrupted at intervals by regions called **nodes of Ranvier.** At these nodes the continuous neurolemma dips inward to only partially cover the axon. There is one Schwann cell between each two nodes, that is, in each internodal segment. Most of the peripheral cranial and spinal nerves are of this type. In myelinated nerve fibers without a neurolemma, the myelin sheath seems to be an uninterrupted cylinder, but indistinct nodes and incisures are actually present. This latter kind of fiber occurs in the central nervous system, where the sheath is formed by neuroglia cells (oligodendroglia).

In unmyelinated peripheral nerve fibers (Remak's fibers), the neurolemma is a syncytial nucleated sheath showing no nodes of Ranvier. The axons of autonomic ganglia are of this type. Some unmyelinated nerves have no neurolemma. These

are the axons of the gray matter of the cord and brain. Many tracts of the brain and spinal cord lack sheaths, and the terminations of **all** nerves have no sheaths.

Organization of nerve trunks. Nerve trunks are bundles of many nerve fibers, held together by fibroconnective tissue (Fig. 15-7). Surrounding the nerve trunk is a connective sheath, the **epineurium,** with blood vessels and lymphatics. The nerve trunk consists of a number of smaller bundles **(fascicles)** each of which is surrounded by a **perineurium** of collagenous fibers and elongated fibroblasts. From the perineurium, strands of fine connective tissue **(endoneurium)** extend between individual fibers. Most peripheral nerves (cranial and spinal nerves) are a mixture of both myelinated and unmyelinated fibers. Small peripheral nerves are made up wholly of unmyelinated fibers.

The synapse. The junction of neurons where the axon of one neuron is in contact with the nerve cell body or dendrites of another neuron

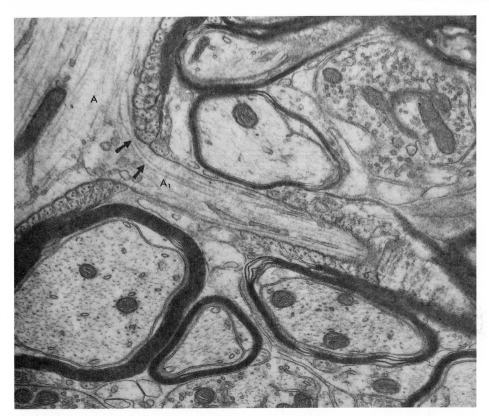

Fig. 15-6. Branching of an axon in cerebellum. Branch *(A₁)* becomes myelinated shortly after leaving main trunk *(A)*, leaving only a small portion unmyelinated. Long segments of microtubules (arrows) are seen passing from the trunk into the branch. (×39,600.) (Courtesy Dr. R. P. Eager, New Haven, Conn.)

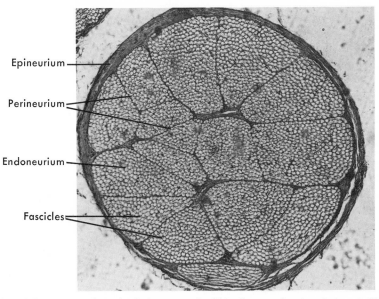

Epineurium

Perineurium

Endoneurium

Fascicles

Fig. 15-7. Cross section of sciatic nerve of rabbit. See text for description. (×50.)

is the synapse. In rare cases the contact may be from one axon to another axon. A given neuron may be involved in a few to thousands of synapses. Physiologically, a synapse is the place at which transmission of an impulse between neurons takes place. Although the structure of a synapse varies, the axon terminals usually end in small expansions known as **end bulbs,** or **boutons,** which are closely applied to the cell body or dendrites of the receiving neuron. In certain cases, the axonal terminals form a loose, basketlike arrangement around the cell body or dendrites. In every case, the association between two neurons is one of contact only, for there is a barrier between them consisting of two thin portions of their plasma membranes. This physical arrangement is observable with the electron microscope.

A measurable delay takes place in impulse-transmission at the synapse. Neurofibrils of the axons do not extend to the end bulbs and never across the gap at the synapse. As noted above, there is no cytoplasmic continuity between neurons. **Within a nerve fiber,** conduction in either direction is possible. However, a synapse can conduct in only one direction—from an axon to dendrites or perikaryon of the receiving neuron.

Fine structure studies show the end bulbs to contain mitochondria and hosts of much smaller **synaptic vesicles,** which measure about 500 Å in diameter. The synaptic vesicles are believed to contain acetylcholine, which is released at the synapse and is considered to be the transmitter substance. In postganglionic endings a different and larger type of vesicle has been observed. These large vesicles appear to be associated with catecholamines similar to those of the adrenal gland (e.g., epinephrine). These substances have been described as present also in the **nonneural** hearts of cyclostomes.

The interrelationships of neurons with each other are prodigious. A few sensory neurons may activate many motor neurons, or several thousand axonal terminals may bring impulses to only one neuron.

Ganglia. Ganglia are collections of cell bodies located outside the central nervous system. A similar collection in the central nervous system is called a **nucleus.** Most of the cranial and all the spinal nerves have ganglia near their connections with the central nervous system. There are two main types of ganglia: craniospinal and autonomic. They vary greatly in size. Besides nerve cells and fibers, the ganglia also contain neurolemma cells, fibrous connective tissue, and blood and lymphatic vessels. Each ganglion is surrounded by fibrous connective tissue, which forms its **epineurium** and **perineurium.**

The cells of the ganglia may be in clusters separated by trabeculae of connective tissue. Some clusters of cell bodies are located among the fascicles of nerve fibers. Each ganglion cell body is surrounded by a capsule formed by a prolongation of the neurolemma sheath of its process and made up of satellite cells. The nerve cells of most cranial and spinal ganglia are spherical and unipolar (Fig. 15-8). Their nuclei are vesicular, with prominent nucleoli. The nerve fibers of the larger cells are myelinated, but the smaller ones have unmyelinated fibers. The craniospinal ganglia occur on the posterior roots of the spinal nerves and on the sensory roots of certain cranial nerves (trigeminal, facial, vagus, etc.).

Autonomic ganglia are located as swellings along the sympathetic and parasympathetic trunks and within the walls of visceral organs. They are small, and the ganglion cells are usually multipolar and stellate in shape. They have connective tissue, blood vessels, lymphatic vessels, and nerve fibers, just as do spinal ganglia. The numerous dendrites of their cells branch, coil, and become entangled with those of adjacent ganglion cells to form **glomeruli.**

The axons of the cells are called postganglionic fibers and are unmyelinated. All the cells are **motor** in function. Some axons leave the ganglia in small nerves to innervate smooth muscle, blood vessels, and glands of the viscera; others join spinal nerves to be distributed to blood vessels, sweat glands, and smooth muscles of hair follicles.

Terminations of nerve fibers. The motor control of skeletal muscle is performed over neurons that have cell bodies in the spinal cord and brain. Their axons run out in spinal and cranial nerves. One nerve fiber may supply more than a hundred muscle fibers. Such a nerve fiber is usually myelinated and branches repeatedly as it enters a muscle fascicle. Each terminal branch forms a motor

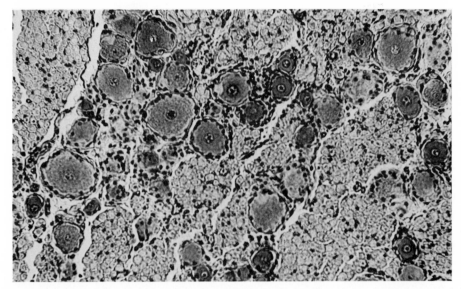

Fig. 15-8. Cross section through spinal ganglion. Note that bodies of nerve cells are rounded and vary in size. Their nuclei contain prominent nucleoli, and capsule or satellite cells surround them. Nerve fibers (axons and dendrites) are found running between them. (×100.)

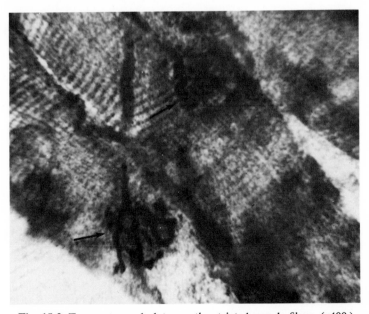

Fig. 15-9. Two motor end plates on the striated muscle fibers. (×400.)

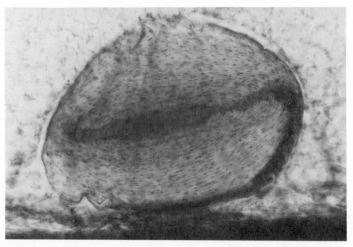

Fig. 15-10. Pacinian corpuscle in mesentery. These are sense organs of deep pressure and possibly of vibration. They consist of concentric layers of connective tissue and squamous epithelial cells surrounding a granular mass. A nerve fiber enters at one end and terminates in a small swelling. (×300.)

end plate on one muscle fiber (Fig. 15-9). As the nerve fiber approaches the muscle fiber, its myelin sheath is lost. The end of the axis cylinder breaks up into telodendria, and the neurolemma at the end of the fiber appears to merge with the sarcolemma of the muscle fiber.

At the junction of the nerve and muscle fibers, the sarcoplasm beneath the sarcolemma is elevated into a mass, the **motor plate.** Here the naked axon breaks up into a number of terminal ramifications. The deep layer of the motor plate adjacent to the contractile substance is called the **sole.** Muscle nuclei tend to accumulate here. The ramifications of the axon bear beads, or varicosities. Nerve terminations do not penetrate the sarcolemma, and the endings are called **epilemma.** Nerve fibers to smooth and cardiac muscle are of the unmyelinated type and terminate as nodular enlargements on the muscle fibers.

Sensory nerve endings are found in epithelia, connective tissue, muscles, and serous membranes. Some are free nerve ends and merely terminate in a treelike arborization. Others terminate in capsules or corpuscles of flattened connective tissue cells and fibers, such as pacinian corpuscles (Fig. 15-10), end bulbs of Krause (Fig. 15-11), or Meissner corpuscles (Fig. 15-12). Muscle spindles consist of one or more small muscle fibers (with sensory fibers wrapped as spirals

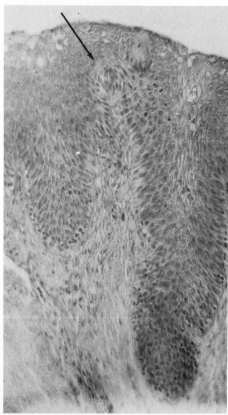

Fig. 15-11. Krause's end bulb (arrow) in the lip of a fish. Responds to the sensation of cold. This corpuscle varies in complexity but is usually bulblike with a granular mass in which the nerve terminates. Connective tissue capsule encloses it. (×300.)

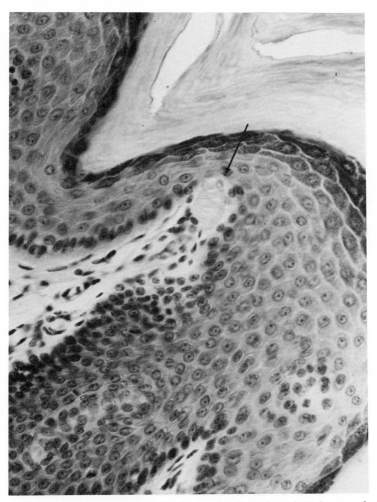

Fig. 15-12. Meissner's corpuscle (arrow). Produces sensation of touch. It is somewhat ellipsoidal and consists of a central mass of irregular cells transversely oriented, with curved nerve endings, all embedded in fibrous tissue. One or more nerve endings enter at one pole and spiral around the tactile cells. They are most common in connective tissue of skin and palmar surface of fingers. (×600.)

about them) plus blood vessels and a connective tissue capsule.

Neuroglia (glia). These interstitial cells and their processes form the supporting framework of the delicate nervous tissue (Fig. 15-13). The neuroglia tissue consists of scattered cells that far outnumber neurons. Special preparations are necessary to demonstrate neuroglia, which are considered to have great functional and dynamic significance in the role played by neurons in nervous integration. Glial cells represent an actively metabolizing tissue. Some of them are much-branched **(astrocytes),** some are rectangular with beaded, branched processes **(oligodendroglia),** and others are small cells with wavy, thorny processes **(microglia)** (Fig. 15-13). Not all the functions of neuroglia cells are known, but oligodendroglia help form myelin, and microglia are known to have phagocytic properties and form part of the reticuloendothelial system. The sheath of Schwann cells of the peripheral nerves and the capsular cells of peripheral ganglia may be considered peripheral neuroglia.

The electron microscope presents a picture of

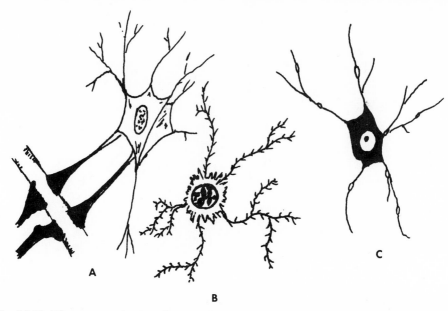

Fig. 15-13. Three types of neuroglia cells in brain of mammal. **A,** Fibrous astrocyte. **B,** Microglia. **C,** Oligodendrocyte. Note that processes in **A** spread out to surface of a blood vessel where they form "sucker feet."

the nervous system "interstitial material," which differs considerably from the impression obtained with the light microscope alone. The general concept before electron microscopy was that the intercellular space, between nerve cells and neuroglial cells, is occupied by neuroglial fibers and a matrix or relatively homogeneous ground substance. However, electron-microscope study reveals that almost every nook and cranny is filled by cytoplasm of the cell bodies and branching processes of the neuroglial cells, chiefly the astrocytes and the oligodendroglia. The astrocytes have **sucker feet** that spread out along the surfaces of capillaries. Oligodendroglia also contact capillaries but with less definitive areas of contact. More frequently they are satellites of the nerve cell bodies and also occur along the course of nerve fibers, having much the same relation to myelinated fibers within the central nervous system as do the Schwann cells to peripheral nerves.

The literature on fine structure of neuroglial cells, and indeed on the identification of the astrocytes and oligodendroglia and their distinction from each other, shows considerable difference of opinion (Schultz and associates, 1957; Dempsey and Luse, 1958).

Subdivisions of the nervous system. The comparative anatomy of the nervous system is a field in which many special investigations have been made and on which comprehensive treatises have been written (Ariëns Kappers, 1929; Ariëns Kappers and associates, 1936). Our discussion will concentrate on the more general and comparative aspects of the microscopic structure of certain of the subdivisions of the system.

The primitive vertebrate nervous system is a simple tube, seen even in the early human embryo. This plan is retained to a large extent in the spinal cord, except for a rather uniform thickening of the walls. In the brain, on the other hand, unequal thickenings and foldings of the wall of the tube are seen as development proceeds. These processes seem to occur in association with development of the special masses of cells (nuclei) and groups of fibers (tracts) in the various parts of the brain.

In general, the spinal cord and the medulla oblongata are the great reflex centers of the nervous system, whereas the upper and anterior portions of the brain are concerned with sensation, association, and highly coordinated and volitional motor activity. For each of the vertebrate groups, the microscopic anatomy of the spinal cord and the cerebellum will be considered, and in a more general way, the medulla oblongata, midbrain, thalamus, and cerebrum will be described.

Meninges, the coverings of the nervous system

In accordance with its great importance and the soft and vulnerable nature of its tissue, the central nervous system of vertebrates is protected by skeletal structures, the cranium and portions of the vertebrae, and by special connective tissue membranes, the **meninges.** The single membrane surrounding brain and spinal cord in cyclostomes and elasmobranchs is known as the **meninx primitiva.** Usually it is surrounded by masses of peculiar mucoid cells that appear to afford additional protection. In teleost fishes there is often only one meninx, but in some it is partly or wholly divided into two layers.

In urodele amphibians there is seldom a clear division of the membrane. In the Anura, however, there are definitely two membranes, and sometimes three, as in the higher forms. The meninges of the frog have been described as consisting of an inner pigmented **meninx secundaria** and a firm, tough **meninx primaria.** Palay (1944) has studied the meninges in the toad *(Bufo)* and finds that there are three membranes, similar to the organization found in the higher forms, reptiles and mammals.

The meninges consist of fibrous connective tissue and mesothelium, a form of simple squamous epithelium of mesenchymal origin. The outermost layer is the **dura mater,** which is fused to the internal periosteum of the cranium. It is made up of dense fibrous tissue with some blood vessels and nerves and an inner surface of simple squamous epithelium. Next to the dura mater is the narrow subdural space between the dura mater and the second layer, the **arachnoid.** This is a thin, transparent layer of nonvascular connective tissue, covered on both sides with simple squamous epithelium. Cellular trabeculae extend inward through the subarachnoid space to the **pia mater.** The pia mater is a delicate, highly vascular layer of connective tissue with its external sur-

face covered with simple squamous epithelium. The pia is in close contact with the outer surface of the brain and spinal cord, held down by astrocytes. This layer carries the blood vessels supplying the central nervous system.

Once the double arrangement (at least) has developed, the spaces between the membranes are filled with a clear cerebrospinal fluid. This fluid fills the subarachnoid space and is continuous, by way of specific apertures, with the fluid in the ventricles of the brain and the central canal of the spinal cord. The cerebrospinal fluid contains some proteins, glucose, inorganic salts, and a few lymphocytes and is similar to the aqueous humor of the eye. It serves as a shock absorber for the brain and spinal cord.

Within the major divisions of the brain in all vertebrates lies a series of fluid-filled cavities, the **ventricular system.** Like the central canal of the spinal cord, with which it is continuous, this system is lined by the **ependyma,** a kind of neuroglia. In lower vertebrates, and even in human children, much of the ependyma is ciliated. In the adult human the ciliation generally is lost, and in fact, in the spinal cord the central canal usually is obliterated. The ependymal cells in lower vertebrates often show conspicuous basal processes that pass for great distances into the substance of the nervous tissue. They may even reach the outer pia mater, under which they then spread out somewhat as do the sucker feet of glial astrocytes in this same location and upon the walls of capillaries.

Evidence has been presented recently by Pesetsky (1969) that in lower vertebrates the ependymal cells may have functions that in higher vertebrates are carried out only by the satellite glia found about neurons and along blood vessels. The action of a specific enzyme, carbonic anhydrase, was demonstrated by histochemical means in cells of the ependyma of fishes and amphibians. In lizards and rodents ependyma cells give a negative reaction, whereas perineural and perivascular neuroglia give a positive one. In all classes studied, the neurons had no carbonic anhydrase activity.

There have been suggestions that astrocytes, with their sucker feet spread out on capillaries, may be involved in regulation of electrolyte trans-

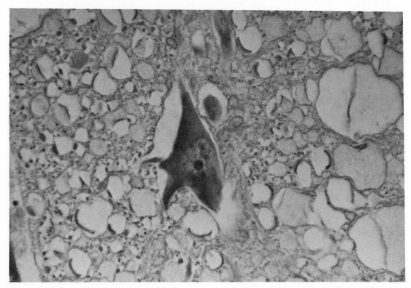

Fig. 15-14. Portion of the spinal cord of *Petromyzon*. Large cell corresponds to a motor horn cell and shows the relatively clear nucleus and conspicuous nucleolus characteristic of such cells in vertebrates, generally. Clear areas represent sections of nerve fibers of varying ages. (×400.)

fer between blood and brain. The high levels of carbonic anhydrase associated with cells of the ependyma in fishes and amphibians strongly suggest an involvement of these cells in a similar function—regulating the ionic milieu of the brain. It seems possible that in lower vertebrates the ependyma is engaged in selective ionic transport, whereas in the thicker-walled brain of higher vertebrates, glial cells, closer to the surface and thus farther removed from the ventricles, have taken over this function.

Spinal cord

CYCLOSTOMATA

The shape of the spinal cord in the cyclostomes is very different from the cylindrical one seen in the species more familiar to us. Also, the pattern of gray and white matter does not present the familiar H-shape of the central cord gray matter. Indeed, the color distinction is not a sharp one in the lowest vertebrates because the fibers are not myelinated. Nevertheless, the area containing the nerve cell bodies (Fig. 15-14) is well marked off from the outer, fiber-containing layer, and the **neuropil** (of naked nerve fibers and

neuroglial processes) around the nerve cells is distinctive in appearance. The peripheral fiber mass does not show a clear separation into tracts. There is a dorsal longitudinal septum but no ventral longitudinal septum.

In cross section, the cord appears in the form of a broad band. It shows **giant fibers (of Müller),** axons that conduct for long distances. The gray matter lacks the dorsal and ventral horns. In *Petromyzon* the ventral and dorsal "roots" of the spinal nerves do not unite but continue as individual nerves. In myxinoids, however, they do join to form a common trunk for each spinal nerve. The roots of the spinal nerves arise alternately on the two sides, as in the prevertebrate *Amphioxus*.

Both in the ammocoete larva and in the adult the dendrites ramify to form a conspicuous **peripheral dendritic plexus.** Medially directed dendrites of the largest cells of the spinal cord branch about the giant fibers of Müller (Fig. 15-15). Tretjakoff (1909) believed that the peripheral dendritic plexus is related to the need for nutrition of the nerve cells, since in a number of cyclostomes the spinal cord lacks blood vessels. The presence of similar plexi in various

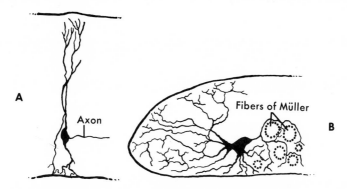

Fig. 15-15. Large motor neurons in the spinal cord of the cyclostome ammocoete larva. **A,** In longitudinal section of cord. **B,** In cross section of cord. The dendrites ramify to form a conspicuous marginal dendritic plexus. Medially directed dendrites ramify about the giant fibers of Müller. (Redrawn from Kappers, Huber, and Crosby, 1936; after Tretjakoff, 1909.)

animals where blood vessels are present would seem, however, to be against such a view.

The afferent fibers (dendrites), carrying messages of sensation from the body wall, are primarily processes from bipolar sensory cells in the dorsal ganglia; some belong to cells within the spinal cord **(intramedullary cells).** Central processes from the sensory cells branch in the dorsolateral part of the cord and give off short descending and ascending branches that travel in the white matter and send collaterals into the gray.

The spinal cord of cyclostomes functions almost entirely for local reflexes. Müller's fibers arise from cells of the medulla oblongata and midbrain and descend to even the most posterior segments of the cord. A striking feature of some of these large axons in the spinal cord of *Myxine* is the presence of conelike enlargements at their terminations, which lie at various places along the cord. These enlargements seem entirely comparable to the growth-cones seen in developing or regenerating fibers of higher vertebrates; it seems likely that growth is continuing in these fibers as a functional process.

Cyclostomes have no autonomic nerve trunks, but spinal nerves do give off visceral branches that go to blood vessels and various viscera. The lack of autonomic trunks in these forms may seem peculiar. Allen (1917) suggested that the nerve cells strung out along the spinal nerves may correspond to the migrating cells that in

higher vertebrates form the ganglia of the autonomic trunks.

CHONDRICHTHYES

The spinal cord of the shark extends the full length of the vertebral canal, which is enclosed by cartilaginous vertebrae. The cord has within it a central canal and consists of a central region of gray matter (of nerve cell bodies and nonmyelinated fibers) and an outer region of white matter (nerve tracts) (Fig. 15-16). The microscopic anatomy shows a number of features that remain as characteristics of the cord in all higher vertebrates. These include (1) myelinization of the fiber bundles (tracts), (2) H-shaped gray matter, and (3) union of dorsal and ventral roots to form mixed nerves. No chains of autonomic ganglia are seen, but the spinal nerve, just after its formation, gives off a small visceral efferent branch to the intestine and the walls of the blood vessels.

There are no cervical or lumbar enlargements of the cord; thus the gray matter is generally similar at all levels. Although **dorsal** and **ventral horns** are present, the shape is not a well-formed H, since the fibers of the dorsal **funiculi** (longitudinal columns of white tissue) do not separate sharply from the dorsal gray horns. Dendrites of the motor cells in elasmobranchs form a marginal plexus, as in the cyclostomes; an especially dense one is near the lateral surface of the cord. The afferent fibers come only from cells of the

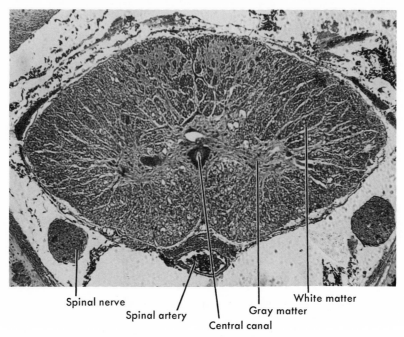

Spinal nerve

Spinal artery

Central canal

Gray matter

White matter

Fig. 15-16. Cross section of shark's spinal cord from tail region. Note gray matter in center of cord (surrounding the central canal) and gray strands extending out into the white matter. (×70.)

spinal ganglia—none from intramedullary cells (within the cord)—although in the embryo some arise from transient intramedullary cells corresponding to those found in cyclostomes.

Completion of the reflex arc seems to be by an arrangement opposite that found in higher vertebrates. In the shark the motor cells of the ventral horn send robust dendrites into the dorsal horn to synapse with the sensory fibers and receive stimuli from them. In higher vertebrates the sensory fibers terminate near or on the cell bodies of the large motor neurons in the ventral horn to complete the local reflex arc.

There are more tracts descending into the cord than in the cyclostomes. Such tracts originate from **reticular nuclei** in the medulla oblongata and from vestibular (inner ear) and lateral line centers. Since the optic lobes, the base of the midbrain, and other higher centers send fibers into the reticular formation of the medulla, these reticulospinal tracts form a "final common path" for impulses from these centers to the spinal cord.

The spinal cord of elasmobranchs (and tele-

osts) is unique in having at its posterior end an enlargement that contains masses of secretion produced by neurosecretory cells of the cord. This structure is the **urohypophysis** (urophysis), a **neurohemal** organ similar to the **neurohypophysis** of the pituitary body. In the urohypophysis, the neurosecretory cell bodies (called Dahlgren cells) are very large and are located along the posterior region of the cord. In the shark the Dahlgren cells discharge their secretions into a capillary plexus along the ventromedial surface of the spinal cord. C. C. Speidel (1919) was the first investigator to describe these cells as neurosecretory in function. Their function may be related to osmoregulation and changes in buoyancy.

OSTEICHTHYES

In the bony fishes the appearance of the spinal cord is variable. In the trout, the gray matter approaches the H-shape (Fig. 15-17), but in others it has the form of an inverted Y. In the latter cases, the dorsal gray matter forms a single median mass because of the lack of development of the posterior longitudinal columns of white mat-

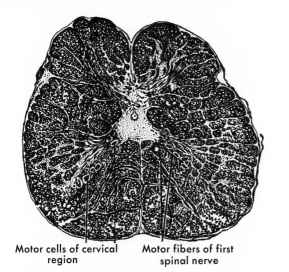

Motor cells of cervical region   Motor fibers of first spinal nerve

Fig. 15-17. Cross section of spinal cord of the brown trout *Salmo fario,* at the level of the first spinal nerve. Gray matter already approaches an H shape. (Redrawn from Kappers, Huber, and Crosby, 1936; after van der Horst, 1918.)

ter (funiculi). The tracts within the cord are well demarcated by size differences of the fibers. The ventral tracts, especially, have large fibers and include the giant structures known as **Mauthner's fibers.**

Among the bony fishes, some species show a peculiar tissue as a sort of covering over the dorsum of the cord in its anterior portions. This is a loose, glial mass in which giant **supramedullary ganglion cells** occur. Such a condition is seen in *Lophius piscatorius* (Fig. 15-18). In general, in contrast to the terrestrial vertebrates, the bony fishes show the greatest degree of development of the spinal cord in its anterior portion. In some bony fishes the head and brain remain well developed while the "body" and spinal cord undergo an enormous reduction, as in the moonfish, *Orthagoriscus mola* (Fig. 15-19).

An important innovation in teleosts is the appearance of a true, ganglionated autonomic chain,

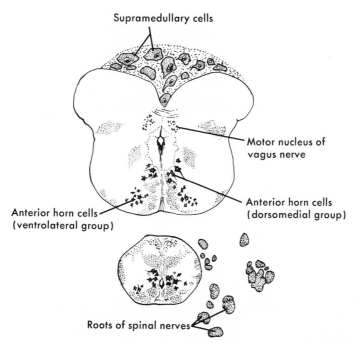

Supramedullary cells

Motor nucleus of vagus nerve

Anterior horn cells (ventrolateral group)

Anterior horn cells (dorsomedial group)

Roots of spinal nerves

Fig. 15-18. Spinal cord of the anglerfish, *Lophius piscatorius.* Upper cross section is near the junction of the medulla oblongata with the cord; the lower one is well caudad in the body. Conspicuous in the upper one are the large supramedullary cells, which far exceed in size the cells of the ventral horn. (Redrawn from Bolk, Göppert, Kollius, and Lubarsch, 1938.)

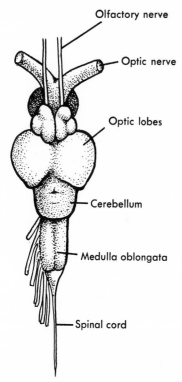

Olfactory nerve

Optic nerve

Optic lobes

Cerebellum

Medulla oblongata

Spinal cord

Fig. 15-19. Central nervous system of *Orthagoriscus mola,* the moonfish, dorsal aspect. Spinal cord is greatly reduced in size relative to the brain. "Plasticity" of organs of the nervous system in reduction or in hypertrophy is a remarkable fact of evolution. (Redrawn from Bolk, Göppert, Kollius, and Lubarsch, 1938.)

connected to the spinal nerves only by the **rami communicantes.** This is a phylogenetic "first," and such ganglionated trunks are found in all higher vertebrate classes. Ramifications of dendrites of the motor cells of the ventral horns form a marginal plexus in these fishes, as in cyclostomes and elasmobranchs.

In teleosts (in contrast to elasmobranchs) the sensory fibers often arise from intramedullary cells as well as from the cells of the spinal ganglia. The descending tracts of the cord are similar to those in elasmobranchs, and the reticulospinal tract (from the medulla oblongata) is again a final common path for impulses. Stimuli from olfactory, optic, trigeminal, and vestibular (inner ear) centers discharge to the reticular formation. The giant cells of Mauthner send their great fibers throughout the length of the spinal cord. Tail reflexes in response to vestibular stimuli are medi-

ated by these fibers. They also afford a path for responses to optic and trigeminal stimuli.

The urohypophysis was described for the cartilaginous fishes. In the Osteichthyes, the neurosecretory Dahlgren cells have long axons that terminate in the enlarged capsular body (urohypophysis) where the secretions are discharged.

AMPHIBIANS

The spinal cord is elongated in Gymnophiona and Caudata, shortened in Anura. In all amphibians with limbs, cervical and lumbar enlargements are present but vary considerably in their degree of development. Phylogenetically, this is the first appearance of such enlargements. In some anurans the spinal cord is remarkably short. In *Pipa pipa,* it reaches only to the third vertebra and is one sixth the length of the **filum terminale,** which anchors the cord to the coccyx. The filum terminale is primarily glial tissue but contains some nerve fibers.

The frog or toad has about ten pairs of nerves arising from the spinal cord. In comparison, the adult shark, for example, may have as many as a hundred pairs of spinal roots. Larval anurans have many more roots than the adults; such roots disappear when the caudal part of the body is lost and the posterior end of the cord is transformed into a long filum terminale.

Each pair of spinal nerves arises from the spinal cord by a dorsal root of sensory fibers leading into the cord and a ventral root of motor fibers leading away from the cord. The sensory fibers have their cell bodies in the spinal ganglia located on the dorsal roots; the cell bodies of the motor nerves are in the ventral horns of the spinal cord. In the amphibians the ventral roots arise a short distance cephalad to the corresponding dorsal ones. There is thus a tendency toward alternation, but not to the extent seen in the cyclostomes and *Amphioxus.*

The motor neurons of the ventral horn are in two main groups: a ventromedial group, which runs the length of the cord, and a ventrolateral group, which is more or less confined to the regions of the enlargements where the motor fibers of the extremities arise. The marginal plexus, which is still well developed in amphibians (Fig. 15-20) is formed primarily from dendrites of the

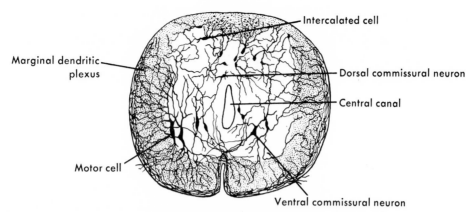

Fig. 15-20. Spinal cord in older larva of the toad *Bufo*. Large motor cells of ventral horn are shown only on the left side; the right side is used to show the ventral arcuate, or commissural, neurons that send fibers to the opposite side of the cord. Marginal dendritic plexus is still conspicuous. (Redrawn from Kappers, Huber, and Crosby, 1936.)

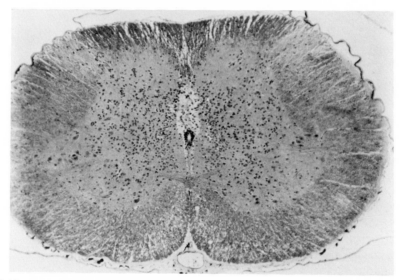

Fig. 15-21. Cross section of spinal cord of *Rana pipiens*. Note that gray matter resembles the letter H and contains nerve cell bodies. White matter is made up of nerve fibers and surrounds the gray matter. Note central canal in horizontal limb of the H. (×150.)

ventrolateral groups of cells. Sensory fibers, except in early development, are related only to cells of the spinal ganglia.

The gray matter surrounding the central canal of the amphibian cord is in a much more definite H, or butterfly, shape than is that of lower forms (Fig. 15-21). Increased numbers of fibers make the white dorsal funiculi larger than in fishes. At the sides the gray matter extends dorsally and ventrally as the dorsal and ventral horns. The gray matter on the two sides of the cord is connected above and below by the dorsal and ventral commissures. The spinal cord is partially divided by a shallow dorsal fissure and a deeper and wider ventral fissure that divide the right and left columns of white matter. The most dorsal portions of the horns are formed by the afferent somatic and visceral fibers from the spinal ganglia. As noted previously, the lower horns enclose the cell bodies of the visceral and somatic efferent, or

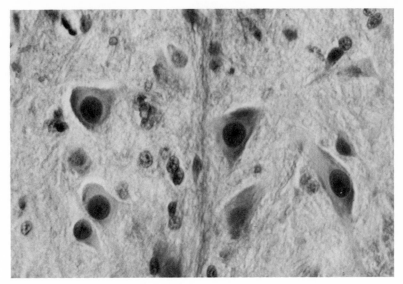

Fig. 15-22. Section through spinal cord of frog showing several motor nerve cell bodies. (×480.)

motor, neurons (Fig. 15-22). The axons from these latter centers form the surrounding white matter of the cord.

REPTILIA

The reptiles show some unusual distinctions of the spinal cord in the various groups (Figs. 15-23 and 15-24). In all except the snakes, cervical and lumbar enlargements are well developed. In the turtles the thoracic part of the cord is very slender, apparently in relation to the protection and isolation afforded by the carapace. The central canal of the crocodile lies far ventrally in the spinal cord, and the ventral fissure is shallow.

The marginal dendritic plexus is much reduced in reptiles, when compared with that in fishes and amphibians. Both reptiles and birds show border nuclei in some portions of the spinal cord. These are masses of nerve cells that bulge at the cord surface. They are well developed in crocodiles. The function of these groups of cells is not known, but it is thought that they may be commissural (**arcuate**) neurons that have made a secondary migration from the central gray matter.

The number of ascending fibers in the dorsal funiculi is increased in reptiles, as compared with those in amphibians. In the brain of the reptile the **nucleus gracilis** and **nucleus cuneatus** appear for the first time. These nuclei serve as termina-

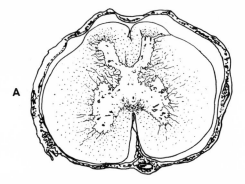

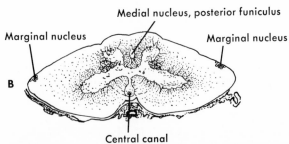

Medial nucleus, posterior funiculus

Marginal nucleus

Marginal nucleus

Central canal

Fig. 15-23. Spinal cord of reptiles, showing variations in shape. **A,** Cross section through cervical cord of *Python reticulatus,* with rounded form. **B,** Cross section through cervical cord of *Crocodilus porosus,* with flattened form, not unlike that seen in cyclostomes. (Redrawn from Kappers, Huber, and Crosby, 1936.)

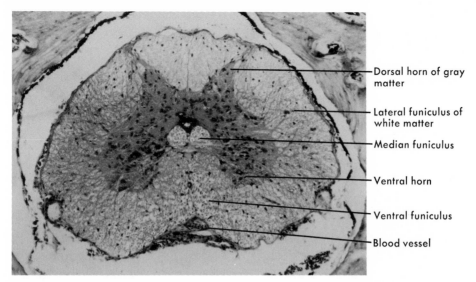

Fig. 15-24. Cross section of the spinal cord of the lizard *Anolis*. In general, it is similar to that of most vertebrates. (×820.)

tions for the **proprioceptive fibers** of the dorsal funiculi. Such receptors are involved in detection of muscle and joint positions and apparently are necessary in meeting the greater demands of terrestrial life for finer adjustments in movements of the body and the limbs.

AVES

In birds the tissue of the spinal cord shows another distinctive feature. For a length of five or six segments in the lumbosacral region, a great mass of loose neuroglial tissue occupies the mid-dorsal area, rising up above the neighboring parts of the cord. It may be thought of as a "glial septum" that has undergone a peculiar modification. The function of this tissue is not clear. It is reminiscent to some extent of the middorsal mass of the cord of some fishes, but that mass appears to be supramedullary, or above the cord, whereas the avian one seems to be an integral part of the spinal cord.

The ventromedial and ventrolateral groups of motor cells are very distinct in the bird. Some instances of special development are noteworthy. One example is the tremendous development of the ventrolateral group at the lumbar enlargement in the ostrich, in relation to the extremely powerful legs of this flightless bird. The ventrolateral group in the cervical region of many birds has

been called the "flying center" because of its relation to the muscles of flight. The marginal dendritic plexus is seen in birds only in the embryo. The reduction of this plexus in reptiles and its disappearance in birds may be related to development of the border nuclei in these two classes.

The dorsal funiculi of birds are generally smaller than those of reptiles, as are the nucleus gracilis and nucleus cuneatus. The smaller size of the funiculi probably is related to a lesser skin sensibility in these animals with their cloak of feathers. The smaller proprioceptive nuclei may have to do with the bird's being "less terrestrial" than other higher vertebrates. On the other hand, there is an increase of descending fibers from the cerebellum, from vestibular nuclei, and from optic lobes, as compared with fishes, amphibians, and reptiles. This increase is associated with the great development of flight.

In some birds the dorsal white columns (funiculi) are lacking, and then the dorsal horns form one mass of gray matter. Also, the lateral border of the cord often shows a furrow, the site of the embryonic **sulcus limitans.**

MAMMALIA

In the mammals the posterior part of the spinal cord shows a strong tendency toward regression, and the tail becomes a relatively minor appendage rather than a massive continuation of the trunk posterior to the origin of the hind limbs as it is in many lower vertebrates. For example, the horse has 17 to 19 tail vertebrae but only 5 to 6 pairs of spinal nerves allotted to the tail. The segments posterior to these have regressed and are represented only by the **conus terminalis,** which ends in the thin filum terminale. However, in some aquatic mammals whose tail musculature is well developed, such as the porpoise *Phocaena,* there is actually a caudal enlargement of the cord in addition to the cervical (brachial) and lumbosacral enlargements.

All mammals show **cervical enlargement** of the spinal cord. This includes the posterior segments of the cervical region and a small portion of the thoracic region. It is the site of origin of the vast numbers of nerve fibers going to and coming from the anterior extremities. Actually, the size is due more to the large amount of white matter in the long descending and ascending tracts at these levels. Lumbar enlargements for the lower extremities are present in the vast majority of mammals but are lacking in those that have no posterior extremities, such as the dugong and whale. Usually the lumbosacral enlargement is smaller than the cervical, but there are outstanding exceptions. In the kangaroo, the powerful hind limbs are reflected in a lumbosacral enlargement of much greater size than the cervical.

The dorsal white funiculi are well developed and constitute the medial **column of Goll** and the lateral **column of Burdach.** The nerve fiber tracts that are the internal representations of these columns are the **fasciculus gracilis** and **fasciculus cuneatus.** These carry proprioceptive impulses of position sense and vibration up to the brain, where they terminate in well-developed **nucleus gracilis** and **nucleus cuneatus.** In one group of mammals, the marsupials, Goll's columns are united to form a single median mass. In many others (e.g., the horse) the more laterally placed Burdach's columns are "overshadowed" by those of Goll and forced down away from the surface.

The dorsal funiculi of white matter show a progressive development in ascending from lower to higher mammals, already begun in reptiles and birds. This appears to be related to the evolution of a higher stereognostic sense.

Tracts descending from higher centers into the spinal cord also show an increased development as compared with submammalian animals. Besides a greater degree of development of phylogenetically older descending paths, the new and important **pyramidal tract** makes its appearance in mammals. It arises from giant pyramidal cells of the cerebrum, passes through various regions of the brain and becomes a discrete column on the ventral surface of the medulla, where it is called the "pyramid." Fibers from the pyramid that cross the midline form the **lateral corticospinal tract;** the other fibers make up the **ventral corticospinal tract,** supposed to be concerned largely with the trunk musculature.

A striking correlation exists between the relative size of the corticospinal tracts and the degree of control that the higher centers of the brain exert over the centers in the spinal cord. Approximate percentages of the total white substance these tracts compose include dog, 10%; monkey, 20%; and man, 30%. Conversely, the power of autonomous functioning of centers in the spinal cord is decreased in higher mammals and man. Thus lesions of the cerebral cortex in the human subject will lead to serious motor impairment, whereas in the lowest mammals they may cause little disturbance.

The **substantia gelatinosa** forms a cap over the tops of the dorsal horns of all vertebrates. It consists of unmyelinated tissue with a gelatinous aspect and myriads of small, spindle-shaped nerve cells with many dendrites. This tissue is especially well developed in the ungulates, where it often forms marked convolutions. In some species of monkeys and carnivores, the gelatinosa of the two sides is continuous centrally.

Reticular formation

In all vertebrates there is a mass of mingled gray and white matter, varying in its prominence, in the cervical region of the spinal cord lateral to the posterior column. This "reticular formation" continues into the medulla oblongata where it includes the so-called respiratory and vasomotor control centers. Its further relations in the higher regions of the brain become difficult to trace.

In general, the reticular formation consists of

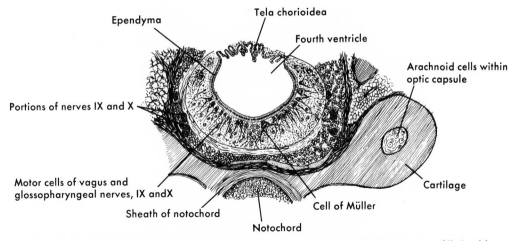

Ependyma

Tela chorioidea

Fourth ventricle

Arachnoid cells within optic capsule

Portions of nerves IX and X

Motor cells of vagus and glossopharyngeal nerves, IX and X

Sheath of notochord

Notochord

Cell of Müller

Cartilage

Fig. 15-25. Medulla oblongata of *Petromyzon.* Fourth ventricle is large space filled with cerebrospinal fluid and roofed by the tela chorioidea, a combination of pia mater and choroid plexus. In central portion of medulla is a huge cell of Müller.

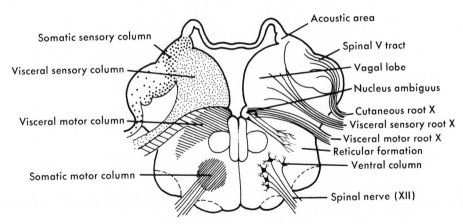

Somatic sensory column

Visceral sensory column

Visceral motor column

Somatic motor column

Acoustic area

Spinal V tract

Vagal lobe

Nucleus ambiguus

Cutaneous root X

Visceral sensory root X

Visceral motor root X

Reticular formation

Ventral column

Spinal nerve (XII)

Fig. 15-26. Cross section through medulla oblongata of sea-robin, *Pronotus carolinus,* a teleost fish. Arrangement of four chief functional columns, labelled on the left side of the figure, are shown. (Redrawn from Herrick, 1930.)

neuropil mingled with myelinated fibers. The neuropil is a feltwork of great numbers of short, delicate, unmyelinated fibers, with some neuroglial tissue. It is prominent in various parts of the nervous system of invertebrate types (Andrew, 1959). Neuropil is not well adapted to conduct nervous impulses for long distances but seems to have the capacity to **diffuse** such impulses widely.

Medulla oblongata

The medulla oblongata is that part of the brain directly continuous with the spinal cord (Fig. 15-

25). It contains a number of primary motor and sensory centers related to the cranial nerves and their receptors and effectors. The medulla is involved in the correlation of these centers in working reflex systems. It also contains a series of conduction pathways that connect it with the spinal cord on the one hand and with the higher centers of the brain—cerebellum, thalamus, basal ganglia, and cerebral cortex—on the other. The conduction pathways are both descending and ascending.

In the medulla oblongata of fishes (Fig. 15-26) the viscerosensory zone tends to be marked off

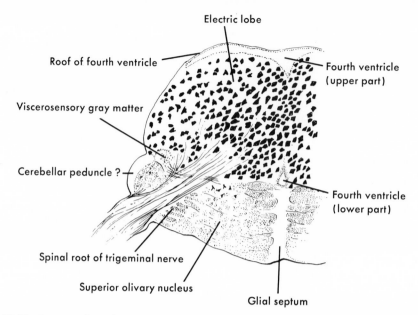

Fig. 15-27. Cross section of medulla oblongata of *Torpedo*. Large "electric lobes" with their giant visceromotor cells make up large portion of organ. Lobes from either side have come together and fused, separating fourth ventricle into upper and lower portions.

from the visceromotor zone by a furrow, the **sulcus limitans.** In higher vertebrates this furrow is seen only in the embryo. In selachians the viscerosensory area often forms a linear series of beadlike swellings. These, however, apparently have no truly segmental significance.

Torpedo shows a remarkable development of the nerve cell masses that control the electric organs. The growth, or hypertrophy, of these masses on each side is so great that, bulging upward, they fuse in such a manner as to separate one part of the fourth ventricle from the main cavity (Fig. 15-27). The giant nerve cells making up these masses are visceral motor types. In the region of immense development of these "nuclei" for the electric organs, the viscerosensory area appears to have been displaced far laterally.

In many bony fishes the medulla shows a great development of the viscerosensory zone, particularly in relation to nerves VII (facial), IX (glossopharyngeal), and X (vagal). In *Cyprinus,* the carp, for example, these parts are very evident on the dorsal surface of the brain. This hypertrophy is apparently related to the wide extent of the organs of taste sensation. Taste buds in these fishes are not confined to the oral cavity but

occur scattered over the entire upper surface of the body; they are innervated by branches of the cranial nerves VII, IX, and X.

The somatosensory zone lies dorsal and lateral to the viscerosensory zone. In fishes it consists of three functional parts: (1) general sensory of the skin, of which the trigeminal nucleus is the most important part, its spinal nucleus continuing into the substantia gelatinosa of the spinal cord; (2) the centers for the lateral line system of the head and trunk; and (3) the statoacoustic system for equilibrium and hearing. In fishes there is no division of the acoustic system within the medulla into dorsal and ventral parts. This occurs only **after regression of the lateral line system** when the cochlear paths become well developed.

A trapezoid body, as a compact mass of horizontally running (crossing) fibers, begins to be defined in the medulla of reptiles but is not clearly seen until the mammals. The fibers of the great trigeminal nerve enter the brain between the border of the pons and the rostral border of this trapezoid body.

The pyramidal tracts represent chiefly the masses of motor fibers from the neocortex of the cerebrum of mammals. They lie along the ventral portion of the midbrain, become diffuse in the pons, and are again very definite columns on the ventral side of the medulla. The extent of their development is correlated with that of the motor cortex. Thus in the hedgehog, with a poorly developed motor cortex, the pyramids are inconspicuous. In man, with his highly developed motor cortex, they are large, conspicuous structures.

The pyramids, the trapezoid body, and the pons are structures of later evolutionary development, added on to the older parts of the brain, the **paleoencephalon** of Edinger (1908). Their development appears to be related to the development of the wide areas of neocortex of the cerebral hemispheres and to the newer parts of the cerebellum, in particular the lateral lobes. These changes, unlike some of the adaptations seen in lower forms (e.g., the medulla oblongata of *Torpedo* and *Cyprinus*, are **qualitative** rather than simply quantitative.

Cerebellum

The cerebellum in its adult form in lower vertebrates, and during some stage of development in all vertebrates, consists of a transverse plate located at the anterior end of the fourth ventricle of the brain. The rostral surface of the plate is continuous, in the rhombomesencephalic fissure, with the midbrain. The caudal surface forms a part of the wall of the fourth ventricle. In this simplest form the cerebellum appears as a continuation of the dorsolateral wall of the medulla, the area in which are found nerve centers related to equilibrium and balance.

The cerebellar plate shows an outer, molecular layer with scattered cells and an inner, granular layer with a dense population of small neurons that have been derived by proliferation of the ventricular epithelium. These layers constitute the gray matter of the cerebellar cortex. In reptiles, birds, and mammals a third layer, the fibrous layer, is added internal to the granular layer. This is "white matter."

In phylogenetic development the plate or body of the cerebellum grows larger, and on each side of it the **cerebellar auricles** appear. These increase in size and are destined to be the equilibratory

parts of the cerebellum. The other great function of the cerebellum seems to be a control, by modulating procedures, of almost all kinds of skeletal muscle activity. Thus the cerebellum shows a varied development in different kinds of animals in relation to their types of motion and way of life. The degree of variation seems to be greater than for any other part of the brain.

CYCLOSTOMATA AND CHONDRICHTHYES

The myxinoids have almost nothing that can be called a cerebellum. *Petromyzon* has a very inconspicuous cerebellum, a transverse plate related to the equilibratory part of the medulla.

In elasmobranchs the cerebellum shows more variety in size and development than any other part of the brain (Fig. 15-28). It does, however, have a consistent basic plan—the simple median portion (body) and two caudal, ear-shaped swellings (pars auricularis). In *Carcharias* the auricular parts of the cerebellum show a great development and a number of deep furrows, giving this part of the brain a complicated aspect. In *Acanthias*, although the auricular part of the cerebellum is large, its surface is smooth.

The rays have a well-developed auricle. The *Torpedo* shows a relatively small degree of development. In these fishes the layer of Purkinje cells, large motor neurons, has appeared at the junction of the molecular and granular layers. These neurons serve as a final common path for all impulses leaving the cerebellar cortex.

OSTEICHTHYES

The cerebellum of bony fishes shows an even greater range of variation in relative size than that of the sharks, depending apparently on way of life. In the Mormyridae it is often so large as to cover almost all the rest of the brain! The mormyrids are an interesting group of bony fishes from the Nile river, which show many primitive characteristics in their main organ systems but whose sense of **hearing** is highly developed. Active fishes, such as the trout, often have a cerebellum as well developed as the birds'. In some bony fishes, it is small and platelike.

In all teleosts, however, the cerebellum consists of two parts: the **body** and the **valvula.** The body

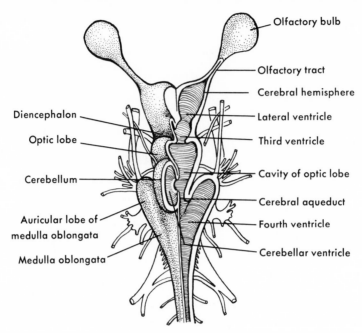

Fig. 15-28. Brain of dogfish shark, *Squalus acanthias.* Dorsal aspect is shown, with portion of wall removed to show ventricles of brain.

includes both of the parts that were mentioned above for the cartilaginous fishes (body and pars auricularis). The pars auricularis is only poorly defined in bony fishes. The valvula is special for the bony fishes. It seems to be due largely to development of a mesencephalocerebellar tract. Hypertrophy of this tract in turn appears dependent on a great development of the lateral-line organs. The body protrudes freely just posterior to the midbrain; the valvula pushes forward deeply into the ventricular cavity of the midbrain and hence is not visible in an undissected brain. The valvula is a more variable part of the teleost cerebellum than the body and is the portion that is so highly developed in mormyrids (Suzuki, 1932).

The efferent tracts of the cerebellum in the teleosts are generally similar to those of elasmobranchs. They arise chiefly from the cerebellar cortex, some going to subcerebellar nuclei and others traveling for longer distances. These tracts carry motor impulses that will move the animal or modify movement in accordance with the sensory impulses brought to the cerebellar cortex by vestibular lateral-line sensory stimuli. These

stimuli are "associated" in the cortex by means of synapses in the granular and molecular layers.

AMPHIBIA

The cerebellum of amphibians in general shows no advance over that of fishes. In fact, it is generally very small and simple, consisting of a transverse plate. The auricles are always clearly defined. The central portion lying between the two auricles may vanish almost completely, so that the auricles may comprise nearly the entire cerebellum. An interesting difference in this part of the brain exists between the tailed and tailless amphibians. The central portion, or **corpus cerebelli,** is much better developed in the tailless amphibians such as *Rana.* In the tailed forms the auricles reach a full development, and the central part is reduced to little more than a commissural role, that is, as a transverse connecting structure.

The functional explanation of this difference apparently lies in the diminution of the lateral line and vestibular systems in the tailless forms and in the greater size of spinocerebellar and tectocerebellar tracts. The latter tracts go primarily to the corpus cerebelli; the auricles are the centers for

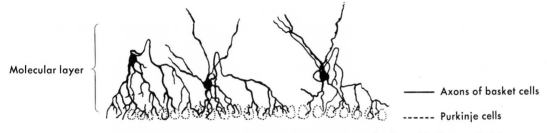

Fig. 15-29. Cerebellum of *Chameleo vulgaris.* Cells in black are basket cells that send their dendrites down to twine about the bodies of the Purkinje cells (outlined by dotted lines). (Redrawn from Ramón y Cajal, 1911.)

vestibular and lateral-line reception and integration.

The division of the frog cerebellum into the following layers is not as clear-cut as in the higher forms: (1) an outer molecular layer of stellate cells and unmyelinated fibers, (2) a middle layer of large Purkinje motor cells, and (3) an inner layer of granular cells, larger cells, and neuroglia.

REPTILIA

With the great differences in methods of locomotion among the reptiles, it is not surprising to find great variations in structure of the cerebellum. Such variation is seen not only among individual orders but also among suborders. The simpler form of cerebellum in reptiles is seen in *Hatteria* and *Varanus.* In the snakes the change from the simple form is seen only in a thickening of the cerebellar plate. In the Chelonia and Hydrosauria the dorsal and lateral parts of the organ grow considerably to form a rounded mass that contains its own ventricular cavity.

However, in all reptiles the corpus cerebelli (body) is much better developed than the auricles, a condition to be expected with the complete lack of lateral-line nerves and the well-developed spinocerebellar tracts. The terrestrial life with its new demands on the musculature of trunk and extremities requires fuller development of tracts between spinal cord and cerebellum.

In the Reptilia groups of nerve cells that have been subcerebellar in lower forms come to lie within the cerebellum. These **intracerebellar nuclei** consist of rather vaguely defined medial and lateral nuclei. Synapsing with these cells are the axons of many of the Purkinje cells, which are well developed in all reptiles (Fig. 15-29).

AVES

The cerebellum of birds shows a higher degree of surface differentiation than that of reptiles, with deep furrows dividing it into definite lobules. Its shape is generally similar to that found in the Crocodilia, and as in those reptiles, it contains its own ventricle. The basic plan of the bird cerebellum varies much less within the class than is true for either reptiles or mammals. Indeed, uniformity is more pronounced in the birds in the overall brain as well as in a number of other features of the body.

The cerebellum of birds corresponds chiefly to the **vermis** (corpus cerebelli, or middle portion) of the cerebellum of mammals. There are no lateral lobes. This central portion is relatively more massive and shows a more complex pattern of fissures than the cerebellum in reptiles. Its large size is due chiefly to further development of spinocerebellar fibers, which end only in this portion. There are nuclei in the white matter of the bird cerebellum, but strands of gray matter often still connect them to subcerebellar vestibular regions. Most of the cerebellifugal fibers (leading away from the cerebellum) arise from the intracerebellar nuclei, but some still come from the cerebellar cortex as in the lower forms. A pair of **flocculi** (auricles) are present, as in mammals. The flocculi vary considerably in size in different families of birds.

A sagittal section of the cerebellum of any bird shows the beautiful treelike appearance that in this class and in mammals was called the arbor vitae by earlier anatomists. Branches of white matter ramify out from the central mass and are covered by multitudes of small neurons, the "granules" of the cerebellum. The granular layer

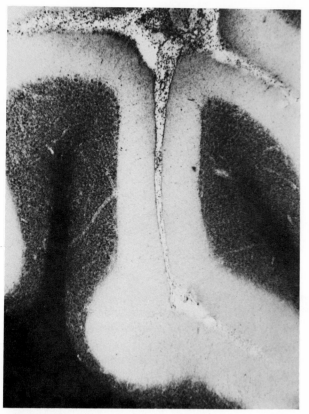

Fig. 15-30. Photomicrograph of cerebellar cortex. (×3.5.)

has a reddish cast in the fresh condition. With the more common stains, it appears as a great mass of deeply basophilic spheroidal nuclei. Outside the granular layer are the Purkinje cells with clear nuclei, conspicuous nucleoli, and large flakes of Nissl material. It is difficult to homologize the parts of the cerebellum among the various classes of vertebrates. However, it seems probable that the lingula and pars auricularis are ancient components. It seems likely also that the lingula corresponds with the valvula of teleost fishes, and the pars auricularis with the part by the same name in elasmobranchs. The large dome-shaped portion rising up between these parts, then, would be the more recently developed part.

MAMMALIA

The cerebellum of the mammal (Fig. 15-30) is much like the birds, although its ventricular cavity is greatly restricted. The surface has folds

arranged parallel to the main fissures, so that a sagittal section gives the characteristic arbor vitae appearance.

It is only in the mammals that the so-called **neocerebellum** of Edinger is developed. It consists of massive additions, the **lateral lobes,** which carry impulses from the neocortex of the **cerebrum,** relayed through the numerous nuclei of the pons. The lateral lobes are more highly developed as the cerebrum takes on more significance in the more advanced forms. The paleocerebellum now is restricted to a medial position and is known as the **vermis.** The pars auricularis (flocculus) is a well-defined, but relatively small, portion of the mammalian cerebellum and, with the vermis, is a part of the paleocerebellum.

The cerebellum in mammals shows many variations in general size, shape, and complexity of folding. A discussion of these would involve too much of comparative gross anatomy. Indeed, the microscopic structure of the cortex, even in the strictly mammalian neocerebellum, is remarkably similar to that of birds and varies little in the different orders of mammals. Of the three layers of the cerebellar cortex, the outer one has some small nerve cells, the middle has the large Purkinje cells with extensive branching processes of the dendrites (Fig. 15-31), and the innermost layer has numerous small nerve cell bodies. Axons from the Purkinje cells arise from the side opposite to that of the dendrites, acquire myelin sheaths, and give off collaterals that pass in a vertical direction to contribute to a basketlike network around the Purkinje cells. The cerebellum is concerned with coordination, posture, and equilibrium and with movements of striated muscles.

Nuclei, collections of gray matter in the central white core of the cerebellum, are well defined in all mammals, as in birds. They apparently occur in some of the lower forms as well. The ratio of volume of nuclei to that of the total medullary (myelinated) mass varies in different mammals: mouse, 1 to 2; hamster, 1 to 2.5; rabbit, 1 to 3; dog, 1 to 4; and man, 1 to 15. It appears that the greater growth of the lateral hemispheres is more related to development of the medullary mass as a whole than of the intramedullary nuclei.

In mammals, as in birds, the spinocerebellar

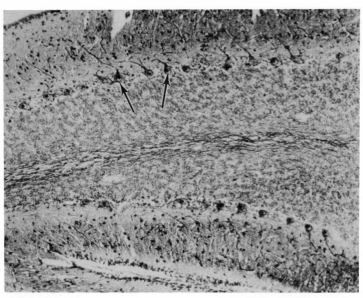

Fig. 15-31. Section through cerebellum of rabbit, showing Purkinje cells (arrows). Microscopic section here does not show their extensive branching. (×460.)

tracts end in the vermis. Other fibers from the medulla oblongata are distributed to both paleocerebellar and neocerebellar cortex. Most of the efferent tracts of the mammalian cerebellum arise from the nuclei in the white matter, but it is thought that the cortex of the flocculi (auricles) gives rise to some of them as well. The flocculus retains direct connections with the vestibular apparatus and appears to be related to the coordination of eye motions. An important pathway from the **dentate nucleus** of the cerebellum via the large **red nucleus** of the **midbrain** to the **thalamus** carries proprioceptive impulses that are relayed from the thalamus all the way to the cerebral cortex.

Midbrain

The midbrain, or mesencephalon, includes the **corpora quadrigemina** and the **cerebral peduncles.** The part of the ventricular system included in the midbrain is the **aqueduct of Sylvius,** which connects the third and fourth ventricles. The midbrain is the least modified part of the primitive neural tube remaining in the adult brain. The corpora quadrigemina (four-part bodies) consist of a pair of **superior colliculi** (Fig. 15-32), where reflex optic centers are located, and a pair of **inferior colliculi,** with reflex auditory centers. In many of

the lower vertebrates the superior colliculi, known as optic lobes, are very large and conspicuous structures (Fig. 15-28).

Fig. 15-33 illustrates a simple arrangement in which two different sensory paths discharge into a single center, which then serves as origin for a final common pathway in the *Necturus* midbrain. The upper portion of the roof of the midbrain receives fibers from the optic tracts; the lower part receives fibers from the tactile and primary acoustic centers. A single **intercalary neuron** is seen to send one dendrite up to receive visual stimuli and another dendrite down to receive tactile or acoustic stimuli. Thus visual and auditory stimuli can be received simultaneously by the one neuron. This neuron then discharges through its axon to the nucleus of nerve III or to some other motor center (Herrick, 1917).

The nuclei of the midbrain in mammals include the **red nucleus,** the **substantia nigra,** the nucleus of the dorsal longitudinal bundle, the interpeduncular nuclei, and the nuclei of nerves III (oculomotor) and IV (trochlear). In vertebrates with nonfunctional eyes, because of parasitic way of life *(Myxine glutinosa)* or cave habitat *(Proteus anguineus),* the motor nuclei of the eyes are lacking or very difficult to identify.

The red nucleus is found in fishes and all other

376 HISTOLOGY OF THE VERTEBRATES

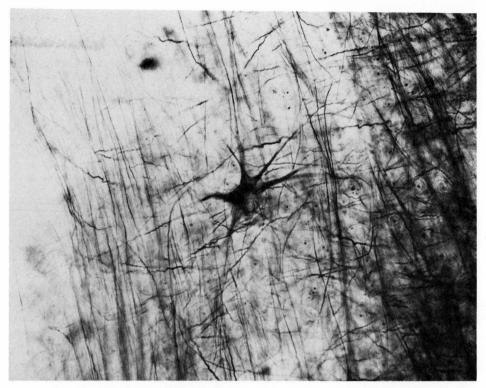

Fig. 15-32. Photomicrograph of neuron from superior colliculus of cat. (×250.)

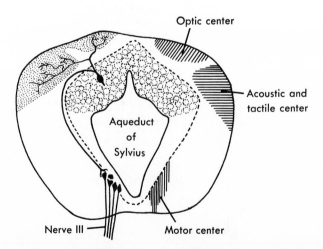

Fig. 15-33. Cross section of midbrain of *Necturus,* showing single correlation neuron of mid-brain roof. One dendrite of neuron ramifies in the optic center, where some fibers of optic tract are terminating; another dendrite ramifies in acoustic and tactile center. Axon, of which only a short portion is shown, descends to synapse with motor neurons of nucleus of oculo-motor nerve. (Redrawn from Herrick, 1917.)

classes of vertebrates. In mammals it consists of two portions, a phylogenetically older part ("large-celled" portion) and a newer ("small-celled" portion). The small-celled part is present **only in mammals.** The substantia nigra, a large nucleus consisting of neurons containing masses of **melanin pigment,** is seen only in mammals and is dependent, apparently, on the development of the cerebral peduncles. The nucleus of the longitudinal bundle is present in all classes of vertebrates. It is important for conjugate movement of the eyes. The interpeduncular nuclei appear to precede the development of peduncles, since they are found in some of the lowest vertebrates.

Thalamus

The thalamus **(diencephalon)** has been the subject of many careful anatomic and physiologic studies. It is present in all vertebrates (Fig. 15-34) but adds on new portions as the phylogenetic scale is ascended. In the human brain the thalamus consists of a dorsal part, having two main groups of sensory nuclei (particularly for touch and sight), and a ventral region **(subthalamus)** with motor nuclei. The lateral group of dorsal sensory nuclei is of most recent origin and may be called the new thalamus **(neothalamus);** the medial sensory nuclei and the motor nuclei constitute the ancient thalamus **(paleothalamus).** The basic thalamic associations important in all vertebrates occur in the paleothalamus. The neothalamus serves as a vestibule, or relay station, for messages on their way to the cerebral cortex where "higher" types of associations occur. It is important to note that much sensibility remains when the cerebral cortex of higher mammals, including man, is removed or damaged.

The nuclei of the neothalamus comprise by far the largest part of the thalamus in the human brain. They mediate general, cutaneous, and deep sensibility, receiving the fiber tracts known as the spinal, trigeminal, and medial lemnisci (Fig. 15-35). The nuclei themselves are named simply

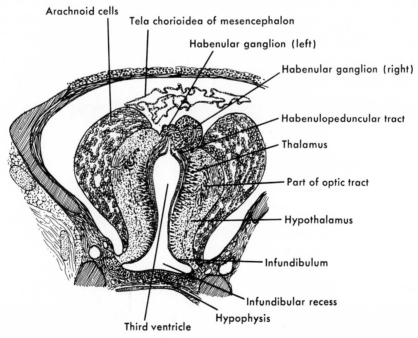

Arachnoid cells
Tela chorioidea of mesencephalon
Habenular ganglion (left)
Habenular ganglion (right)
Habenulopeduncular tract
Thalamus
Part of optic tract
Hypothalamus
Infundibulum
Infundibular recess
Hypophysis
Third ventricle

Fig. 15-34. Diencephalon of *Petromyzon*. Side walls are formed by the thalamus. Above it are the two habenular ganglia, of which the right is far larger than the left. (This is not a matter of obliquity of section but an interesting natural asymmetry of the nervous system.) Note great mass of arachnoid cells that surrounds much of the brain in *Petromyzon*. (Redrawn from Krause, 1923.)

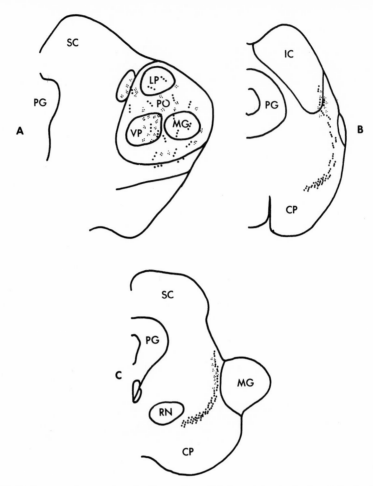

Fig. 15-35. Pathway and some terminations of medial lemniscus in brain of the hedgehog **(A)** and the slow loris **(B** and **C).** Degenerating fibers are shown as dots. *CP,* Cerebral peduncle; *MG,* medial geniculate body; *LP,* lateroposterior nucleus; *PG,* periaqueductal gray; *PO,* posterior nucleus; *RN,* red nucleus; *SC,* superior colliculus; *VP,* ventroposterior nucleus. (Redrawn from Schroeder and associates, 1968.)

according to position: lateral, ventral, and posterior.

Special parts of the thalamus are concerned with the optic pathways **(lateral geniculate bodies and the pulvinar)** and with the auditory pathways **(medial geniculate bodies).** The primary visual pathways, consisting of the axons of the large ganglion cells of the retina, have been studied recently by Laemle (1968) in two species of lemurs. Unilateral enucleation was performed, and the degenerative process of the fibers and chromatolytic changes in the nerve cells in the nuclei of the diencephalon and midbrain were

studied. The fibers could be traced through optic nerve, optic chiasma (Fig. 15-36), and optic tract to the nuclei of termination. By this method, the lateral geniculate body was shown to have a striking laminar structure (Fig. 15-37).

Belonging to the thalamus also are (1) the **hypothalamus,** which includes the **tuber cinereum** from which the **infundibulum** of the **hypophysis** (pituitary gland) arises and, posterior to it, the two **mammillary bodies;** (2) the **epithalamus,** which includes the **pineal body;** and (3) the roof of the third ventricle.

Just above the optic tracts on either side, in

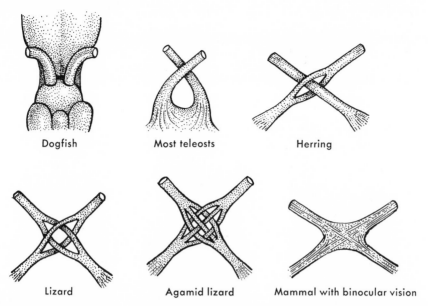

Dogfish   Most teleosts   Herring

Lizard   Agamid lizard   Mammal with binocular vision

Fig. 15-36. Some of the various types of optic chiasma in different vertebrates. (Redrawn with slight modification from Weichert, 1965; after Wiedersheim [herring and agamid lizard] and Kingsley.)

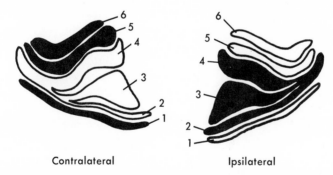

Contralateral   Ipsilateral

Fig. 15-37. Distribution of terminations of primary visual pathway in lateral geniculate body in lemurs. Successive laminae are indicated by numbers. Laminae shown in black have undergone transneuronal degeneration; that is, nerve cells in these laminae that synapse with axons from retina have degenerated. Note that specific laminae are allotted for the crossed (contralateral) and uncrossed (ipsilateral) fibers. (Courtesy Dr. L. K. Laemle, New York, N. Y.)

the substance of the hypothalamus are two significant nuclei, the **paraventricular** and the **supraoptic nuclei.** The cells of these nuclei, have some peculiar features, as seen in all classes of vertebrates: (1) a multinucleate condition, (2) presence of capillaries running through their cytoplasm, and (3) the presence of inclusions, some of which are secretion products. Scharrer and Scharrer (1940) showed striking evidence of

secretory activity in the hypothalamus of fishes and amphibians (Fig. 15-38). These nuclei are concerned with control of the secretion of the hypophysis and with the fundamental vegetative activity of the organism. It is interesting that they appear to be resistant, to some extent, to the degenerative changes that occur with advancing age in various other parts of the nervous system (Andrew, 1956). Buttlar-Brentano (1954) has

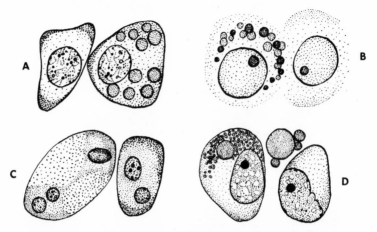

Fig. 15-38. Neurosecretory cells of the hypothalamus. **A,** Nucleus paraventricularis, banded watersnake *Lapemis hardwickii.* **B,** Nucleus preopticus of toad *Bufo arenarum.* **C,** Nucleus preopticus of marine fish *Tautoga onitis.* **D,** Nucleus preopticus of common killifish *Fundulus heteroclitus.* (Redrawn from Andrew, 1959; after Scharrer and Scharrer, 1940.)

shown that **reactive** changes occur in cells of the hypothalamus of man in old age. Such alterations are interpreted as possibly counteracting the tendency to degenerative change. It would be of interest to know whether similar reactive changes take place in other vertebrates.

Corpus striatum and cerebrum

The pathways that lie **beyond** the thalamus running to and from the **cerebral cortex** constitute the **corpus striatum.** It is a large mass of striated-appearing nervous tissue, consisting of alternating gray and white matter. The white matter includes the conspicuous **internal capsule.** The gray matter is composed of the **lenticular nuclei** (including the **putamen** and **globus pallidus)** and the **caudate nuclei.** In higher vertebrates these nuclei, known collectively as the **basal ganglia,** are centers of extrapyramidal motor control, the source of motor impulses not originating in the cerebral cortex. Disease affecting these ganglia results in symptoms such as rigidity, tremor, and bizarre involuntary movements.

A striking enlargement and differentiation of the corpus striatum is seen in phylogenetic development. In cyclostomes (Fig. 15-39) it is poorly developed. It seems to begin as a local thickening of the ventrolateral wall of the anterior telencephalon, perhaps as an olfactory re-

flex center. In elasmobranchs it appears as a markedly bulging part. In amphibians it is seen as part of the lateral wall of the cerebral hemisphere, connected with the thalamus by a mass of white matter. The great development of these fiber tracts going to and from the cerebral cortex accounts for the striated appearance seen in higher vertebrates.

The development of the corpus striatum and the cerebral cortex, or **pallium,** brings about a great modification of the form of the brain as a whole. This once inconspicuous area of the fish brain has so developed as to overshadow all other portions of the brain (Figs. 15-40 and 15-41).

Histologic examination of the frog brain shows that the walls of each hemisphere may be divided into a dorsal pallium, median-ventral septum, and lateral-ventral striata. Cell bodies are seen around the ventricles in several layers (Fig. 15-42); this is considered a primitive arrangement of nerve cell bodies. Most cells are pyramid-shaped. As always, the gray matter of the cerebral hemispheres consists of cell bodies, unmyelinated nerve fibers, and neuroglia cells; the white matter is made up of myelinated nerve fibers and neuroglia processes. Two enlargements at the anterior end of the hemispheres are the olfactory lobes from which olfactory nerves arise.

The cerebral hemispheres of reptiles are larger than those of the frog. A new associative area,

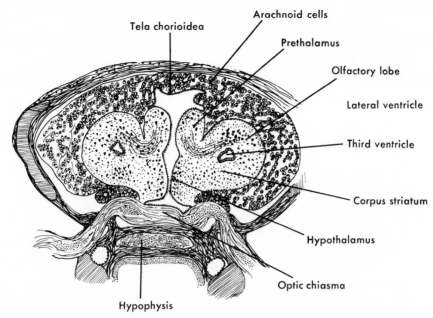

Fig. 15-39. Forebrain of *Petromyzon*. Section passes through olfactory lobes and optic chiasma. Note that ventricle of diencephalon (third ventricle) is shown along with the olfactory lobes, indicating how little developed is the forebrain. (Redrawn from Krause, 1923.)

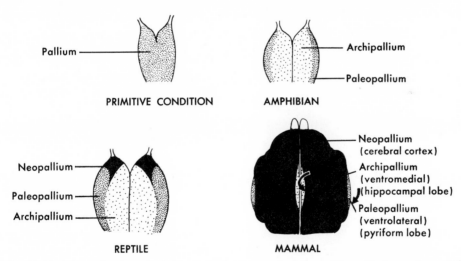

Fig. 15-40. Schematic view of phylogenetic development of the pallium. (Redrawn from Weichert, 1965.)

the neopallium, appears for the first time in the hemispheres. The cortex is enlarged and conceals the corpus striatum (basal ganglia) lying below it. A large corpus striatum is a characteristic feature of the brain of many reptiles. The olfactory lobes are distinct from the cerebral hemispheres.

Many birds show an enlargement of the forebrain that may indicate high intelligence. The cerebral hemispheres are much larger than any other part of the brain, but their surfaces lack the deep fissures found in the mammalian brain. Much of their bulk is made up of the corpus striatum, but dorsally there is a thick cortex de-

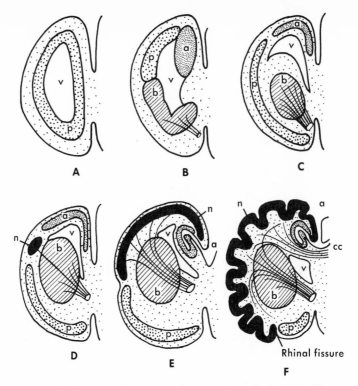

Fig. 15-41. Stages in development of corpus striatum and cerebral cortex. Transverse sections. **A,** Primitive stage. **B,** Amphibian. **C,** Primitive reptile. **D,** Advanced reptile. **E,** Primitive mammal. **F,** Advanced mammal. *a,* Archipallium; *b,* basal nuclei; *cc,* corpus callosum; *n,* neopallium; *p,* paleopallium; *v,* ventricle. (Redrawn from Carter, 1967.)

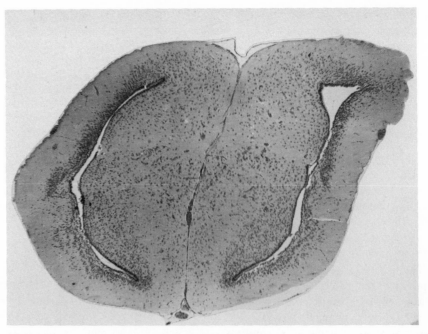

Fig. 15-42. Cross section through the cerebral hemispheres of the frog. Note the slitlike lateral ventricles. (×52.)

rived chiefly from the neopallium of which there is only a trace in reptiles. It is now thought that the corpus striatum is the actual location of complex integrating reactions in birds (Stettner and Matyniak, 1968).

The cerebrum reaches its greatest development in the mammals, its peak in man. It is the center of the highly integrative and associative neural interactions that typify the mammal, particularly the human being.

The cerebral cortex consists of numerous **gyri,** folds of gray matter with cores of white matter. Between these gyri are **sulci,** depressions into which the pia and arachnoid membranes dip. The white matter is made up of tremendous numbers of myelinated nerve fibers—some ascending to the cortex and some descending from it. There are also association fibers running from gyrus to gyrus—some linking together neighboring gyri and others travelling to distant ones. Some fibers travel from lobe to lobe. In man and many other mammals the cerebral hemispheres themselves are connected to one another by fibers making up the **corpus callosum.**

Six layers of cells are commonly recognized in the gray matter of the cerebral cortex in higher primates and man. Most of the cells are pyramidal, stellate (granular cells), and spindle-shaped. The molecular layer is the outermost and has few cells. It is made up mostly of fibers of underlying cells running in various directions but chiefly parallel to the surface. Apical dendrites of the deeper pyramidal cells ascend into this layer where they branch and form synapses with cells and fibers. The external granular layer lies next and has small triangular nerve cell bodies. The third layer is the pyramidal cell layer of large pyramid and granule cells. The fourth is the inner granule layer of small stellate cells. Axons of the larger stellate cells extend into the core of white matter, whereas axons of smaller cells remain in this layer. The fifth layer is called the internal pyramidal layer because of its large and medium-sized pyramid-shaped cells. Both small and large pyramidal cells have an axon arising at the base of the cell. The axon descends into the white matter where it becomes myelinated. The deepest layer is the polymorphic cell layer of cells of many shapes. These six layers are not sharply

demarcated and blend into each other. All the layers contain neuroglia cells, and the thickness of the layers varies according to the functions of different regions.

The white matter, which underlies the gray cortex, consists of bundles of myelinated fibers passing in all directions and supported by neuroglia cells. Such fibers belong to one of the following: association fibers, connecting different parts of the same hemisphere; commissural fibers, connecting areas of one hemisphere with areas of the other hemisphere; and projection fibers, connecting the cortex with lower centers.

References

Allen, W. F. 1917. Distribution of spinal nerves in *Polistotrema* and some special studies on the development of the spinal nerves. J. Comp. Neurol. **28:**137-214.

Andrew, W. 1956. Structural alterations with aging in the nervous system. Proc. Assoc. Res. Nerv. Ment. Dis. **35:**129-170.

Andrew, W. 1959. Textbook of comparative histology. Oxford University Press, New York.

Ariëns Kappers, C. U. 1929. The evolution of the nervous system in invertebrates, vertebrates and man. De Erven F. Bohn, N. V., Haarlem, Netherlands. 335 pp.

Ariëns Kappers, C. U., G. C. Huber, and E. Crosby. 1936. The comparative anatomy of the nervous system of vertebrates, including man. Vols. 1 and 2. The Macmillan Co., New York.

Bolk, L., E. Göppert, Z. Kollius, and W. Lubarsch. 1938. Handbuch der vergleichende Anatomie der Wirbeltiere. Urban & Schwarzenberg, Vienna.

Buttlar-Brentano, K. von. 1954. Zur Lebensgeschichte des Nucleus basalis, tuberomammillaris, supraopticus und paraventricularis unter normalen und pathogenen Bedingungen. J. Hirnforsch. **1:**337-419.

Carter, G. S. 1967. Structure and habit in vertebrate evolution. University of Washington Press, Seattle.

Dempsey, E. W., and S. Luse. 1958. Fine structure of the neuropil in relation to neuroglia cells. Pages 99-129 in W. F. Windle, ed. Symposia on the biology of neuroglia. Charles C Thomas, Publisher, Springfield, Ill.

Edinger, L. 1908. The relations of comparative anatomy to comparative psychology. J. Comp. Neurol. **18:**437-457.

Herrick, C. J. 1917. The internal structure of the midbrain and thalamus of *Necturus*. J. Comp. Neurol. **28:**215-348.

Herrick, C. J. 1930. An introduction to neurology. W. B. Saunders Co., Philadelphia.

Kappers, C. V. A., G. C. Huber, and E. C. Crosby. 1936. The comparative anatomy of the nervous system of vertebrates, including man, vol. 1. The Macmillan Co., New York.

Kingsley, J. S. 1912. Comparative anatomy of vertebrates. P. Blakiston's Son & Co., Philadelphia.

Krause, R. 1923. Mikroskopische Anatomie der Wirbeltiere in Einzeldarstellungen. IV. Walter de Gruyter & Co., New York.

Laemle, L. K. 1968. Visual pathways of the lemurs. Anat. Rec. **160**(2):380-381. (Abstr.)

Palay, S. L. 1944. The histology of the meninges of the toad *(Bufo)*. Anat. Rec. **88**:257-270.

Pesetsky, I. 1969. Carbonic anhydrase activity in ependymoglial cells of lower vertebrates. Histochemie **19**:281-288.

Ramón y Cajal, S. 1911. Histologie du système nerveux de l'homme et des vertébrés. Norbert Maloine, Paris.

Schroeder, D. M., D. Yashon, D. P. Becker, and J. A. Jane. 1968. The evolution of the primate medial lemniscus. Anat. Rec. **160**:424. (Abstr.)

Schultz, R. L., E. A. Maynard, and D. C. Pease. 1957. Electron microscopy of neurons and neuroglia of cerebral cortex and corpus callosum. Am. J. Anat. **100**:369-427.

Scharrer, E., and B. Scharrer. 1940. Secretory cells within the hypothalamus. Proc. Assoc. Res. Nerv. Ment. Dis. **20**:170-194.

Speidel, C. C. 1919. Gland cells of internal secretion in the spinal cord of the skates. Publ. 281. Carnegie Institute, Washington, D. C.

Stettner, L. J., and K. A. Matyniak. 1968. The brains of birds. Sci. Am. **218**:64-76 (June).

Suzuki, N. 1932. A contribution to the study of the Mormyrid cerebellum. Ann. Zool. Jap. **13**:503-524.

Tretjakoff, D. 1909. Das Nervensystem von Ammocoetes. I. Das Rückenmark. Arch. Mikrosk. Anat. **73**:607-680.

Van der Horst, C. J. 1918. Die motorischen Kerne und Bahnen in dem Gehirn der Fische. Tijdschr. Ned. Dierk., vol. 16.

Weiss, P., and H. B. Hiscoe. 1948. Experiments on the mechanism of nerve growth. J. Exp. Zool. **107**:315-395.

Weiss, P., and H. Wang. 1935. Neurofibrils in living ganglion cells of the chick, cultivated in vitro. Anat. Rec. **67**:105-117.

Wiedersheim, R. 1909. Vergleichende Anatomie der Wirbeltiere, 7th ed. Verlag Gustav Fischer, Jena.

16 SENSE ORGANS

THE NATURE OF SENSATION

One of the general properties of all living substance is **irritability,** the capacity to respond to a stimulus (Hoar, 1966). This property is really the starting point of the nervous system and also the beginning of the sense organs, which may be considered a vital part of the basic integration that is characteristic of nervous phenomena. By division of labor, this capacity to respond to environmental stimuli has led to individual cells' being specialized to pick up specific environmental changes and to initiate nervous impulses to be sent to some integration, or nervous, center (Cordier, 1964). In response, this center will accordingly give appropriate reactions. A sense organ therefore is a structure adapted to respond to a specific environmental variable to which it is selectively hypersensitive.

The stimuli themselves do not reach the brain; only the nervous impulses are carried there, and an interpretation is made about the nature of the environmental change (sensation). The sense organs sort out the information they receive and relay it to the brain centers. The source of all this information is both the external and internal environment of the organism. Every organism is constantly bombarded by environmental disturbances, and the receptors are selectively stimulated by different kinds of stimuli such as pressure, heat, light, sound, and touch and are connected by chains of neurons to different parts of the brain (Case, 1966). The basis for interpreting different kinds of sensations depends on the presence of several different kinds of receptors in the body. A receptor is highly sensitive to a particular kind of stimulus and its range of energy change (Prosser and Brown, 1961). Receptors concerned with heat, for example, must come in contact with heat to institute a nervous impulse. Those concerned with smell react only to odoriferous substances.

Our study of the senses is chiefly subjective, especially so with regard to what other animals experience (Gordon, 1968). Anatomic investigations indicate that some animals have senses or ranges of sense perception for which man has no counterpart. Bats and others are known to respond to supersonic waves far beyond man's capacity of sound perception. The pit viper has facial pits extremely sensitive to radiant heat, but these are absent in man. Nor can physiologists be sure that the nature or intensity of a sensation experienced by an animal are the same as those of his own. Animals often have great development of one type of receptor and a relative neglect of another. Dogs have keen sense of smell but relatively poor eyesight. Birds have the best organs of vision in the animal kingdom, but their sense of smell is extremely poor. Although other animals may appear to have about the same sensory equipment as man, it is often quite erroneous to make generalizations about their sensory capabilities (Grinnell, 1968).

CLASSIFICATION OF RECEPTORS

Sense organs can be classified in various ways because there are many morphologic types at different levels of organization. Not all are built on the same plan, and physiologists are not sure about the kind of sensations they arouse. This is

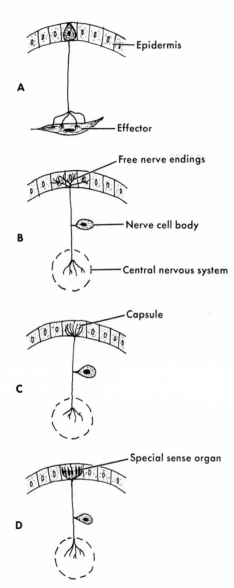

A

— Epidermis

— Effector

— Free nerve endings

B

— Nerve cell body

— Central nervous system

C

— Capsule

D

— Special sense organ

Fig. 16-1. Evolution of the receptors. **A,** Primitive neuro-sensory cell whose cell body is the receptor. **B,** Primary receptor with nerve cell body deeper and with naked nerve endings as primary sense organs. **C,** Secondary sense organ with special terminal capsule about nerve endings. **D,** Tertiary or special sense organ with specialized epithelial sensory cells in addition to the deeper conductive nerve cells.

especially the case with general receptors that have a wide distribution over the surface of the body or throughout the visceral organs. Special receptors are those that have a restricted distribution, and their functions appear to be more definite. Although sensitivity in all vertebrates in general is thought to vary mostly in range and not in kind, it is very difficult to determine minute variations in animals other than man and especially so with those sensory perceptions with which we are not familiar.

The evolution of receptors has occurred in various steps showing a gradation from simple to complex (Fig. 16-1). The first was a primitive **neurosensory** cell whose cell body was hypersensitive to stimuli of certain kinds. This type was followed by a **primary** receptor scheme in which the naked neuron terminals (or free nerve endings) serve as the receptor instead of the cell. As further modification occurred, the primary receptor became invested with specialized supporting cells that were adapted for increasing the sensitivity of the receptor (the **secondary receptor).** The climax was reached with the **tertiary** receptors in which accessory epithelial sensory cells become sensitive to stimuli and are the actual receptors, while sensory neuron terminals carry impulses to the central nervous system. This last type is represented by the great sense organs of eye, ear, and some others. A modification of this last type occurs in the primitive, but specialized and complex, olfactory organ in which the bipolar neurosensory cells with supporting cells are the actual receptors having nerve fibers carrying impulses to the central nervous system. In the olfactory organ, the receptors are not the terminal fibers nor separate sensory cells (as in the eye and ear) but are the actual nerve cells interspersed among non-sensory supportive cells.

On the basis of source of stimuli, receptors may be classified as follows:

1. **Exteroceptors.** Receptive to external environment stimuli; includes all special sense receptors (touch, pressure, pain, temperature).
2. **Interoceptors.** Receptive to internal environmental stimuli such as those from the visceral organs (excretion, digestion, and circulation).

3. **Proprioceptors.** Receptive to stimuli from within the body wall, giving an awareness of degree of tension of muscle and position of joint components; all are primary or secondary and include joint receptors and muscle spindle receptors.

The exteroceptors represent all morphologic types; the others, primary or secondary types.

On the basis of the kind of stimulus received or the stimulus to which they react, receptors may be classified as follows:

1. **Chemoreceptors.** Sensitive to chemical stimuli; include olfactory, Jacobson's organ, and taste buds.
2. **Mechanoreceptors.** Sensitive to mechanical stimuli; include touch, pressure, and low-frequency vibration (sound).
3. **Radioreceptors.** Sensitive to high-frequency vibrations (temperature and light); include organ of vision (eye) and thermoreceptor organs (pit organs, ampullar organs).

GENERAL SENSIBILITY RECEPTORS

The receptors of general sensibility are in the form of free nerve endings or encapsulated nerve endings.

The skin is the most extensive sense organ of the body (Sanders, 1947). In all vertebrates there are sensory endings (**free** nerve endings) that ramify among the epidermal cells and are stimulated whenever the skin comes in contact with other objects. These free nerve endings are among the oldest general exteroceptors among vertebrates, and they are the only general sensory endings in the skin of cyclostomes, on which they are scattered on the body surface and the gills. Here they are restricted to a primitive sensibility of awareness to environmental stimuli (protopathic sensibility) that does not discriminate differences in the kind of stimuli (epicritic sensibility).

The naked nerve endings are often found entwined about hair follicles or other objects, or they sometimes form a terminal aborization. The free endings may show small, knoblike enlargements at their distal ends. They receive the sensation of light touch or of pain. They are distributed in the deeper layers of the epidermis or in the connective tissue.

In addition to free nerve endings (primary sense receptors) many vertebrates have **encapsulated** nerve endings, often referred to as corpuscles or secondary sense receptors or organs. Examples are the corpuscles of Meissner, end bulbs of Krause, pacinian corpuscles, genital corpuscles, and Merkel's corpuscles. They are found mostly in amniotes where the distribution of the various kinds is also spotty. The different corpuscles vary in their makeup. They usually consist of a central elongated granular mass of epithelioid cells covered with many concentric, thin layers of connective tissue. Sometimes a little connective tissue and some capillaries lie between adjacent sheets of such cells. At one pole of the corpuscle the nerve fiber enters and terminates in a small swelling, usually near the region of the granular mass. The neurolemma and the endoneurium of the nerve fiber blend in with the connective tissue capsule. Such a corpuscle is the **pacinian corpuscle,** which is found in the mesenteries and connective tissue and beneath the skin or mucous membrane (Loewenstein, 1960). This corpuscle is sensitive to pressure, and its nerve ending is supposed to be depolarized by the stretching of the nerve fiber and its axolemma under squeezing pressure—the axolemma becomes more permeable to sodium ions, which causes depolarization.

In **Merkel's corpuscle,** each branch of a terminal arborization ends in a shallow, disklike plate that is closely applied to a modified epithelial cell. This type of receptor is really a modification of free nerve endings, for it lacks a capsule. Merkel's corpuscles are common in the submucosa of the tongue and in the stratified epithelium of the oral cavity. When stimulated, they arouse the sensation of touch as does the stimulation of **Meissner's corpuscle.** The latter corpuscle is ovoid and consists of a central mass of flattened epithelioid cells that are penetrated by curved nerve endings, the whole being enclosed in a many-layered connective tissue capsule.

Other interesting capsular corpuscles are found among vertebrates. One of these is **Krause's end bulb** of spherical shape found in the conjunctiva, skin of lips, and the glans penis. This corpuscle consists of a thin connective tissue capsule enclosing a mass of nerve fibers formed from the breaking up of the supplying nerve into interlacing fibrils. These corpuscles are sensitive to thermal stimuli. The **genital corpuscles** in the glans penis of man are of similar structure. **Grandry's corpuscle** is found in the skin on the bill of the duck and other waterfowl. In this corpuscle two or more epithelioid cells are enclosed in a connective tissue capsule and supplied by a nerve fiber as a tactile disk. **Herbst's corpuscle** is also found in waterfowl, where they are mostly restricted to the region around the month. It has a capsule almost like that of a pacinian corpuscle. The entering nerve lies in a single layer of heavy, cubical cells that enclose it like a cylinder. This corpuscle is also one of touch.

In the muscles and tendons of vertebrates are found neuromuscular and neurotendinous organs of touch. These sensory organs are often called spindles because they enclose spindle-shaped areas of the muscle or tendon. They record the pressures to which muscles and tendons are subjected (proprioception). Their structure varies with different groups, but in general the nerve fiber branches into many fibrils surrounded by connective tissue. These fibrils terminate by branching and spiraling around the muscle and tendon tissue. They are best demonstrated in mammals.

CHEMORECEPTORS

Chemoreception is often considered to be a universal sense. Vertebrate cells (the same would be true of invertebrates) are coordinated, or controlled, by some form of chemical substance. Hormones, carbon dioxide, oxygen, and chemical transmitters of nerve terminals afford a chemical communication that is involved in every aspect of body functions. There is reason to believe that such chemical communication evolved before the nervous system which relies on chemical substances in the performance of its functions.

Chemoreceptors in animals serve to provide information about the environment by detecting chemical substances coming from various sources (Benjamin and associates, 1965). Such stimuli may be potential food, mating reactions, potential dangers, and others. Reactions to these were early embodied into reflexes, so that stimulation of such receptors was quickly transformed into appropriate behavior patterns. Even the simplest of receptors, the free nerve endings of the skin and mucous membranes, are responsive to chemical irritants, but two sets of specialized receptors, the olfactory organ and the taste buds, are the chief ones emphasized by vertebrates. Smell and taste have been intimately associated throughout vertebrate evolution, and their close association has often resulted in confusion regarding the physiologic distinction between the two. For man, and probably for other animals as well, the flavor of foods may be a combination of taste and smell. However, they differ in many ways (Haagen-Smit, 1952). Smell is a function of the olfactory organ, and substances to be smelled must be volatile; taste is a function of taste buds, and a substance to be tasted must be dissolved in a fluid. The nerves involved in olfaction are those in the olfactory cranial nerve; those in taste are from the facial and glossopharyngeal cranial nerves. Olfactory organs have a wider range of sensitivity; taste buds are capable of giving four basic sensations: sweet, bitter, salt, and sour (Beidler, 1965).

Olfactory organs. These organs are the most ancient of the sense organs of vertebrates (Allison, 1953). Olfactory lobes make up a large part of the fish's brain. The olfactory organ in most fishes consists of the olfactory membrane in the blind nasal pits; in higher vertebrates it is the olfactory membrane of the nasal cavity. Olfactory organs were in existence before taste buds, but both function by sampling fluid-borne chemicals. Even all airborne substances must enter into solution on the nasal epithelium in order to be detected. In vertebrates the olfactory receptors are not in the chief respiratory stream as the animal breathes (Grinnell, 1968). Only a small fraction of inspired air normally passes over the olfactory receptors. This is why animals must sniff to increase the sampling of the air over the olfactory membrane. Cell bodies of most other afferent neurons have migrated for better protection to a more central position in the cerebrospinal ganglia, but the cell bodies of olfactory afferent neurons are placed at the surface in the olfactory areas of the epithelium of the mucous membrane (Fig. 16-2). Olfactory cells are both receptors and conductors.

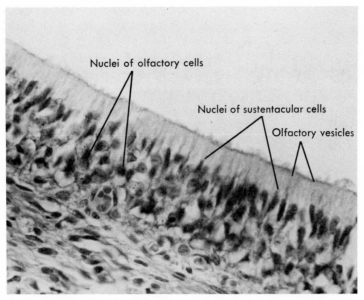

Fig. 16-2. Section of olfactory epithelium, which is found in the upper part of both lateral and medial walls of the nasal cavity. Note the olfactory vesicles extending along the free surface of the epithelial cells. Each vesicle contains a number of stereocilia (nonmotile cilia) and represents a single dendritic process from the bipolar neuron (sensory receptor). The sustentacular cells are most numerous and have oval nuclei and striated free borders. The olfactory cells are spindle-shaped, bipolar neurons with round nuclei toward the middle and deeper areas of the cells. (×840.)

The evolution of the olfactory organs has been a long one with many variations. On cyclostomes, and possibly in the extinct ostracoderms, the olfactory sacs have fused into a single olfactory sac with a single external nostril opening, but with paired nerves. The olfactory sac is blind, but there is a canal extending from the base of the sac in lampreys. In the hagfishes this canal extends into the pharynx as a nasopharyngeal canal and carries water to the gills. Other fishes have paired olfactory sacs. In elasmobranchs a superficial groove runs from each nasal orifice to the corresponding corner of the mouth. The olfactory membrane in the sac is much folded to increase the area of the epithelium. Each naris, or opening into the nasal cavity, is divided into an incurrent and an excurrent aperture. As the animal moves, water is forced in one aperture and out the other.

The lungfishes (order Dipnoi) were the first to introduce a connection between the nasal pit, or blind olfactory sac, and the pharynx. They were the first to have internal nares (choanae) and thus have the first internal oronasal canals. This arrangement preadapted these vertebrates for air breathing. In the amphibians the nasal passages open at the anterior end of the palate, posterior to the lateral extensions of the vomer bone. In the upper part of the nasal passages the olfactory epithelium or organs are located (Smith, 1951). Terrestrial forms have increased the epithelial surfaces by folds and have mucous glands to keep the epithelium moist. In reptiles, birds, and mammals the nasal passages are long, and the internal nares open far back in the pharynx because they have a false or secondary palate. In mammals the olfactory organs have reached their greatest efficiency. Extensive convolutions of the nasal epithelium overlaying the turbinated bones have greatly increased the epithelial surface. In man the olfactory area extends from the roof of the nasal cavity and down the nasal septum onto the superior nasal concha. Its total area is not extensive.

The microscopic structure of the olfactory organ is rather complicated (Frisch, 1967). It con-

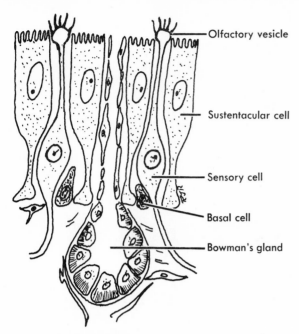

Fig. 16-3. Olfactory sustentacular cells.

sists of a mucous membrane with a thick pseudo-stratified epithelium and a lamina propria. There are no goblet cells in the epithelium. The epithelium is made up of two kinds of cells: the supporting (sustentacular) cells and the olfactory (sensory) cells (Fig. 16-2). A third type of cell (basal) may serve as reserve cells for forming new sustentacular cells (Fig. 16-3). The **sustentacular cells** are tall and bear microvilli (with a granular cover) that give the appearance of a striated border. They have an ovoid and lightly stained nucleus and give a brownish yellow color to the epithelium. Their nuclei are placed nearer the cell top than those of olfactory cells. The **olfactory cells** are bipolar with a distal fiber (dendron) and a proximal fiber (axon). Their cell bodies fit into the region where the sustentacular cells have become narrow. Their dendrons are found in the crevices between the supporting cells. Their nuclei are round and usually more darkly stained than those of the sustentacular cells. At the tip of the dendritic process is the **olfactory vesicle,** which is a solid bulblike mass of cytoplasm that bulges through the surface of the epithelium (Figs. 16-2 and 16-3). The vesicle bears on its surface a number of nonmotile cilia (stereo-

cilia) (Reese, 1965). The shafts of these cilia have the typical 2 + 9 arrangement of tubules found in the cilia, but in their more distal portions they have only two tubules. The lamina propria has secretory cells (the glands of Bowman) that have ducts to the surface of the epithelium and constantly bathe the olfactory cilia with a mucuslike substance. In this fluid the gases responsible for odors may dissolve (Moulton and Beidler, 1967).

The sense of smell may be extremely sensitive, and animals are able to distinguish many different odors. It is thought that olfactory cells are specialized for certain basic odors and that the receptors for these basic types are not distributed evenly through the olfactory epithelium (Amoore and associates, 1964). A variety of odors might be distinguished by using various combinations of the different receptors. There is at present no way to discriminate differences among olfactory cells with respect to their specialization for different kinds of odors (Lettvin and Gesteland, 1965).

The olfactory organ relays its nervous impulses directly to the brain by fibrous extensions of the cells, the fibers of which make up collectively the olfactory nerve. The brains of birds have a very small olfactory center, which accounts for their poor sense of smell.

Jacobson's organ. The olfactory epithelium is actually deployed in many vertebrates into two regions. One is the nasal olfactory organ just described and the other is a vomeronasal organ called Jacobson's organ (Smith, 1960). This organ is found exclusively in certain tetrapods, being best developed in snakes and lizards (Grinnell, 1968) although it is found in the embryos of all tetrapods (except perhaps the birds). Adult crocodiles and chelonians lack them, as do the higher mammals. They are rudimentary in adult anurans, certain urodeles, and many mammals. Monotremes, marsupials, and some primitive placentals among mammals have the best developed ones.

Jacobson's organs are paired blind cavities opening into the anterior tip of the roof of the mouth. The organ is best understood in its simplest form in urodeles (tailed amphibians) where it consists of a groove in the ventrolateral floor of the nasal sac opening into the roof of the mouth. In other animals where it is present, each organ is mostly, or entirely, separated from the nasal sac and appears as a blind sac connected with the mouth. It is in the shape of a curved blind tube that opens through the roof of the mouth cavity in front of the internal nares on each side. In lizards and snakes Jacobson's organ serves to detect the odor of substances taken into the mouth by picking up and sampling the odoriferous particles flicked into these pockets by the forked tip of the tongue.

The sensory epithelium lining the pits or pockets of Jacobson's organ are also made up of sustentacular and sensory cells. The sustentacular cells form the outer nuclear layer of the epithelium and have their apices joined together as terminal bars to form an outer nonciliated membrane. Unlike olfactory epithelium, the sensory cells are placed at the basal layer of the epithelium. Their oval nuclei are distributed at several levels. The sensory cells send processes to the outer membrane formed by the sustentacular cells and form terminal bulbs for detecting odors. These processes do not reach the lumen and do not have cilia.

Taste buds. Taste buds are found in all vertebrates. In fishes they have a wide distribution over the body, being found over the surface of the head, trunk, and mouth regions. The barbels of the catfish have many of them. They are widely distributed in the floor of the mouth, the side walls, and the pharynx of fishes. They are especially numerous in bottom-feeding fish, where taste buds may be found on the entire body surface even to the tips of the tail. In amphibians taste buds are distributed within the mouth and pharynx and often on the head regions. They occur on the tongue of most tetrapods except birds. Reptiles and birds tend to concentrate them in the pharynx. In mammals they are more common on the tongue where they are associated with elevations called papillae (Fig. 16-4). They are most abundant in mammals and least common in birds. As the individual ages, there are fewer taste buds. Man may have some taste buds on the hard palate, inside the cheek, and in the pharynx, but the child has far more of them in those places.

There are supposed to be four basic tastes in

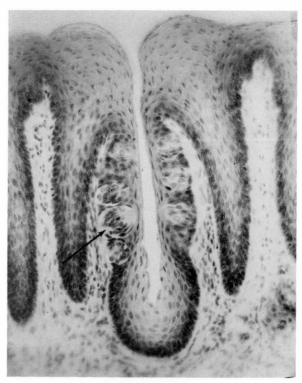

Fig. 16-4. Taste buds in the epithelium of the side of a foliate papilla of rabbit. (×760.)

man, acid (or sour), bitter, sweet, and salt. Buds are usually grouped into four functional categories with reference to these basic tastes. It is possible for drugs to inhibit one taste and not another. These taste buds are not distributed evenly, so that some tastes are more easily detected in some parts of the tongue than others. Some taste buds are nonfunctional. Many complex tastes may depend on smell, as vapors rise from the food and stimulate the olfactory nerve endings. Some authorities also add to the four basic tastes those of an alkaline and a metallic taste.

Histologically, a taste bud is an oval or barrel-shaped cluster of elongated cells within an epithelial covering (Fig. 16-4). The vallate papillae of the tongue have more cells than other parts of the tongue. Fewer are found in the moat or trench around the papilla. Some are also found in the fungiform and foliate papillae. There are two types of cells in the taste buds: **taste-receptor** cells and **supporting** cells. The supporting cells are narrower at each of their ends than in the middle and are shaped like barrel staves. At the end of the barrel-shaped structure near the surface, they are so arranged that they surround a little depression, the **pit,** which communicates with the surface by means of a small canal called the **inner taste pore** that is open to the outside of the covering epithelium. The taste (neuroepithelial) cells are long, narrow cells that are intermingled with the supporting cells in the central part of the bud. They average about a dozen to each bud. Each cell gives rise to a short apical process that extends into the pit. Taste cells have elongated nuclei and microvilli on the apical process, which project into the inner taste pore.

For a substance to be tasted, it must become dissolved in saliva and pass by the taste pore into the pit where the apical processes are affected; then nervous impulses convey the sensation over branches of cranial nerves VII, IX, and or X, depending on the position of the taste bud. Some fibers from these nerves branch, form a plexus on the taste bud, and with their knobbed endings (boutons) make contact with the surface of the bud.

MECHANORECEPTORS

Mechanoreceptors deal with mechanical irritability, which includes many forms of stimuli. The varieties of receptors that are so stimulated include nearly all types from free nerve endings to specialized tertiary organs. The stimulus usually acts on the sensory nerve terminal itself, although such terminals are associated with other tissues (hair, cilia, bulbs, etc) that may support or augment the stimuli involved. A receptor's specificity to a particular stimulus is found chiefly in its location and its mechanical reaction to environmental forces. A specific receptor may be responsive in a certain way to a brief and slight stimulus and in a different way when subjected to an intense irritation. When a mere **touch** stimulus is prolonged, it may be interpreted as **pressure** or even **pain** when the stimulation is intense. Mechanoreceptors provide the central nervous system with information about certain changes in the physical nature of the environment, such as vibration or sound, pressure, gravity, stretch, flow of a fluid, orientation or equilibrium, and acceleration. Many of their stimuli and involved

receptors overlap those described under general sensibility in a previous section, as for example, certain encapsulated corpuscles. Many of the mechanoreceptors are located in the dermis of the skin, tendons, periosteum, and mesenteries, but they also include the specialized tertiary organs of the **lateral line system** and the **ear.**

Lateral line system. This is represented by a specialized series of organs located down each side of the body and usually making a pattern over the head, especially around the eye and on the lower jaw (Wright, 1951) (Fig. 16-5). They are cutaneous receptor organs and are rows of pits, depressions, or grooves in which are clustered sensory cells, each provided with a sensory hair from its free surface and innervated by a nerve fiber. These clusters of sensory or hair cells are commonly called **neuromasts** (Fig. 16-6) and arise as epidermal placodes (thickenings).

Lateral line organs are present in cyclostomes, both bony and cartilaginous fishes, aquatic larval stages of amphibians, and adult aquatic amphibians. They are absent in reptiles, birds, and mammals, although a few vestiges may be found in some reptiles. Although neuromasts vary in their arrangement in different groups, they are all fluid-filled pits, cups, or ampullae. They may open by pores onto the surface of the animal, or they may be found under the skin in a closed canal that may have pores along its length. The primitive arrangement was to have the neuromasts in discontinuous pits that made up a system consisting of linear rows of the sense organs lying in individual shallow pits. Later, the organs were protected by the formation of a continuous shallow groove, a condition found today in certain primitive fish (order Holocephali). A further advancement in the system was the closing of the grooves into canals or tunnels. Cyclostomes and amphibians tend to have the primitive arrangement, or a slight modification of it, for they have only free neuromasts. In elasmobranchs and some teleosts the neuromasts are placed in tubes near the surface of the body, with occasional pores to the outside. The common arrangement among fishes is that of a main canal running along each side of the trunk and tail, with a complicated pattern of branching canals, on the sides and tops of the head. This cranial series of canals includes

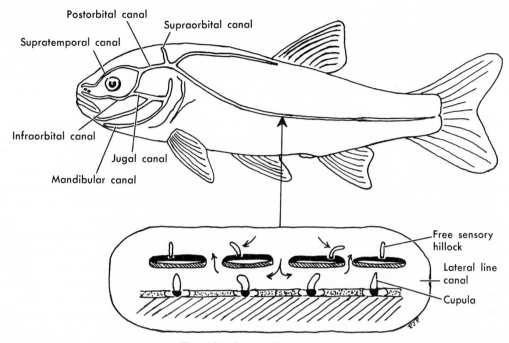

Fig. 16-5. Lateral line system.

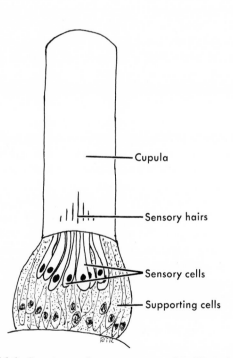

Fig. 16-6. Structure of a neuromast. (Modified from Ballard, 1964.)

supratemporal, supraorbital, infraorbital, mandibular, and hyomandibular canals.

In the head region of elasmobranch fishes the lateral line system has evolved a special type of structure in which modified sensory cells are sunk deeply into flask-shaped pits and embedded in a gelatinous substance. These are the ampullae of Lorenzini.

Microscopically, a neuromast (Fig. 16-6) is arranged between a basement membrane and a semigelatinous mucous substance through which stimuli are transmitted. This fluid may also be water (in the open groove type) on the outside of the animal. The neuromast is a cluster of receptor, or sensory, cells supported by sustentacular cells, and possibly by special basal cells attached to the basement membrane, as well as by mantle cells that separate the neuromast from the surrounding epithelium. Each sensory cell bears a single hair that is made up of many short stereocilia and one long kinocilium. The column-shaped receptors send delicate protoplasmic filaments into the fluid, protected by a jellylike coating called the **cupula,** which often takes the form of an elongated dome. Each hair may be enclosed in a tube within the cupula. Three types of these apical projections from the receptor are recognized: the motile kinocilium, the immotile stereocilia, and fine microvilli.

The results of extensive investigations on the functions of the lateral line organs have produced many theories of their functions (Dijkgraaf, 1963). A common theory proposes that they are sensitive to the vibrations of the surrounding water movements over their body. The stimulating effects may be related to the asymmetry of the fine structure of the sensory hairs (as revealed by the electron microscope), which have a cluster of stereocilia, together with a kinocilium at one side, or edge, on the apical surface of the sensory cell. There are two kinds of sensory hairs, depending on which side of the cluster of stereocilia the single kinocilium is placed. Movement of a fluid (water) over a neuromast will cause a displacement of the cupula. Reception or excitation will occur only when the sense hair is bent in a plane that goes through the kinocilium located on the far side of the patch of stereocilia; bending to the opposite side will cause inhibition. The stimulating effect from one part of the sensory cells may dominate the inhibiting stimulation of the remaining sensory cells. Together, the two kinds of sensory hairs makes possible a detection of movement in either direction.

The ampullae of Lorenzini are known to be sensitive to changes in hydrostatic pressure (formerly thought to be sensitive to temperature) and to weak electric fields (Murray, 1965). Recent experiments indicate that they may be involved in salinity measurements, since they respond to small changes in the percentage of salt in sea water.

The lateral line organs are connected with branches of cranial nerves VII (facial), IX (glossopharyngeal), and X (vagus).

The ear. The inner ear, or membranous labyrinth, of vertebrates is the organ of equilibrium and hearing. Wherever found, the outer and middle ears are merely accessory organs for receiving, amplifying, and transmitting the waves of sound. Only the inner ear is present in all vertebrates; the middle ear appears first in amphibians, and the outer ear is found only in amniotes (Katsuki, 1963). The equilibrium function

of the inner ear appeared in evolution before the sense of hearing, and its structure is about the same in all vertebrates. On the other hand, the structures pertaining to hearing, or audition, have undergone striking evolutionary changes from the lowest to the highest of the vertebrate groups.

The origin of the paired inner ear (or membranous labyrinth, as it is commonly called) arose first in fishes as an ectodermal placode directly in line with the placodes of the lateral line or neuromast system just described. This placode becomes the **otic** vesicle of the embryonic condition. The labyrinth may be considered as an enlarged anterior region of the lateral line system. It consists of a series of semicircular canals (usually three in number) and two chambers or sacs, an upper **utriculus** and a lower **sacculus.** In its development the otic vesicle sinks into the mesenchyme to form the slender **endolymphatic duct,** which remains in communication with the surface in some, but is closed off in others. As the vesicle enlarges, it becomes divided into the utriculus and the sacculus. From these two sacs all the sensory elements of the membranous labyrinth arise. The semicircular canals are connected at both ends with the utriculus. These semicircular canals are arranged perpendicular to each other, two being vertical and one horizontal, thus conforming to the three planes of space. (Figs. 16-7 and 16-8). With the exception of the lampreys (which have two) and the hagfishes (which have one), all living vertebrates have three semicircular canals.

At the end of each canal there is an enlargement, the **ampulla,** in which there is a **crista,** or crest, made up of rows of hair cells with cilia embedded in a gelatinous cupula (Fig. 16-9). From the floor of the sacculus there is an evagination, the **lagena,** which becomes longer and actually coiled (in mammals) to form the **cochlea.** There are also three patches of sensory epithelium in the utriculus and sacculus, called **maculae.** One of these is in the wall of the utriculus (**utricular macula**); another is in the floor of the sacculus (**saccular macula**); and a third is in the lagena (**lagenar macula**). They bear sensory hair cells that project into a gelatinous mass containing mineralized concretions, or **otoliths.** These otoliths are absent in birds, mammals and most

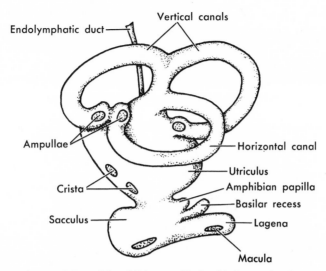

Fig. 16-7. Dissected ear of frog. "Amphibian papilla," with its patch of sense cells, is found only in this class of vertebrates and may be one of the chief hearing organs.

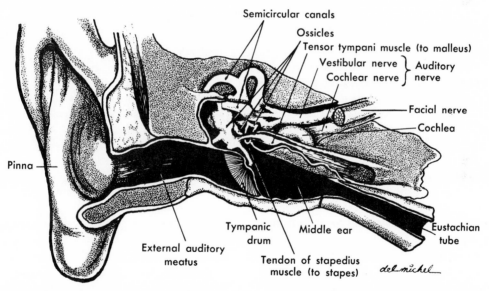

Fig. 16-8. Human ear, showing relations of the external, middle, and inner ear. Note the tensor tympani muscle.

reptiles. Rotational changes in the head or changes in linear velocity cause the otoliths to move, stimulating the sensory hairs and giving information about the posture of the body. This information is carried by afferent nerve fibers (of the VIII cranial, or auditory, nerve) to the brainstem and cerebellum where appropriate motor reflex actions are initiated. The fluid in the semicircular canals stimulates the sensory cells of the crista when displaced by turning movements in the three planes of space. All these functions of the membranous labyrinth are concerned with the sense of equilibrium.

The entire membranous labyrinth of semicircular canals—utriculus, sacculus, and lagena—is enclosed in cartilage or bone that conforms to

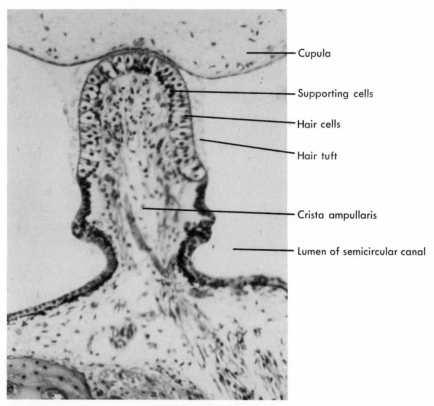

        — Cupula

        — Supporting cells

        — Hair cells

        — Hair tuft

        — Crista ampullaris

        — Lumen of semicircular canal

Fig. 16-9. Crista ampullaris in the ampulla of a semicircular canal. Epithelium over crista is modified into sustentacular and hair cells. Resting on the crista can be seen a small portion of the cupula, a gelatinous mass, into which hair cells project. It is thought that the cupula is formed by coagulation of the gelatinous material about the hairs by the fixative reagents. Hair arrangement can be revealed by the electron microscope. (×350.)

the sacs and canals of the labyrinth, the so-called **skeletal labyrinth.** This skeletal labyrinth is lined with membranous periosteum or perichondrium, and the space between the membranous labyrinth and the skeletal labyrinth is filled with **perilymph** from the cerebrospinal fluid. This fluid acts as a cushion and protection for the delicate membranous labyrinth. Within the labyrinth the viscous **endolymph** occurs and is vitally concerned with the functions of equilibrium as we have seen.

The second great function of the ear is that of **hearing,** or **audition.** This part of the ear has undergone many evolutionary changes in its development from fishes to mammals. Although the sacculus plays an obscure role in equilibration functions, the depression in its floor has evolved into an elongated duct known as the **cochlear duct,** which is filled with endolymph. As was

pointed out earlier, the lagena is provided with a **lagenar macula,** which persists in all vertebrates except viviparous mammals. The lagenar macula, along with the utricular macula and the saccular macula, is concerned with equilibrium or static position. A second sensory area, the **basilar macula,** or **papilla,** also appears in the lagena. This macula is sensitive to sound waves. Hearing is very restricted in fishes. It is thought that the saccular macula may respond to water vibrations of low frequency. Several fishes are known to make noises, and some make use of a chain of weberian ossicles to carry vibrations to their ears from their air bladder (Lanyon and Tavolga, 1960). As the cochlear duct evolves, so does the basilar macula, which differentiates into the **organ of Corti** running the entire length of the cochlear duct (Figs. 16-10 to 16-12). This organ of Corti

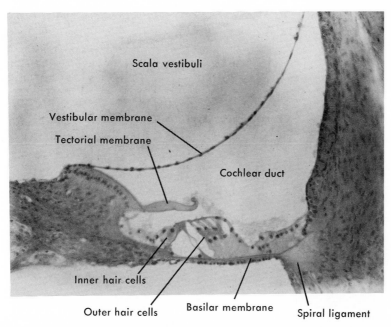

Scala vestibuli

Vestibular membrane

Tectorial membrane

Cochlear duct

Inner hair cells

Outer hair cells

Basilar membrane

Spiral ligament

Fig. 16-10. Organ of Corti (guinea pig). Note that basilar membrane supports neuroepithelial structure of the cochlear duct (scala media, or the organ of Corti), the structure concerned with hearing. Preparation is a vertical section of cochlear duct and related structures. (×260.)

contains receptor hair cells that are specialized to pick up sound waves of certain frequencies or tones along its course. The perilymphatic space that surrounds the membranous labyrinth is also found around the lagena, and as the lagena develops, it forms a tube (the **perilymphatic duct**), in which lies the cochlear duct. During development, the perilymphatic duct is actually divided into two separate ducts, a **scala vestibuli** lying above the cochlear duct and a **scala tympani** below it (Fig. 16-11). The total cochlea is thus made up of these three parallel tubes, only one of which, the median cochlear duct, is a part of the membranous labyrinth. The scala vestibuli and the scala tympani come together (in the **helicotrema,** Fig. 16-12) at the apex of the cochlea, but the cochlear duct, or scala media, ends blindly there. As already mentioned, the acoustic sensory cells making up the **organ of Corti** are located on the floor (basilar membrane) of the cochlear duct. The organ of Corti consists of a band or row of receptive cells with their free ends terminating in a tuft of hairs (Fig. 16-10). A **tectorial membrane** of gelatinous and noncellular substance rests over the sensory cells.

The receptor cells are connected at their bases with fibers from the larger division of the auditory, or acoustic, nerve (cranial nerve VII). The supporting cells of the organ of Corti are tall, columnar cells that are found in various groups and given special names. Within the organ of Corti a tunnel extends through its entire length. It is triangular in cross section and bounded at its base by the basilar membrane and at its sides by the inner and outer pillar cells. The hair cells, or receptors, are cylindrical in shape with basal nuclei and rounded bases. Their bases do not rest on the basilar membrane, but their rounded ends project into the cytoplasm of special supporting cells (phalangeal cells).

The cochlea of birds is not coiled but is elongated and curved. In most mammals it is coiled into a snail-shape with the coils varying from one (sloth) to five (guinea pig). The lagena is very short in amphibians and longer in reptiles. Amphibians also have, in addition to the lagena, an **amphibian papilla,** which is a small dorsal pocket from the sacculus (Fig. 16-7). It possesses sensory hairs that may have sensory responses.

The inner ear just described represents the

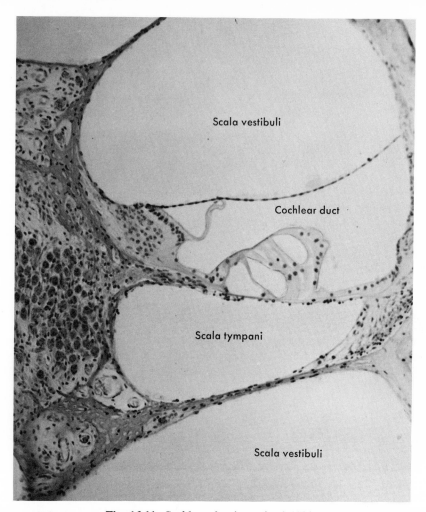

Fig. 16-11. Cochlea of guinea pig. (×100.)

receptor centers for responding to sound waves. But accessory organs have evolved to aid in hearing by localizing and picking up sound waves and conducting them to the sensory receptors in the inner ear. These accessory devices, **outer** and **middle** ears, are therefore specialized to bring sound waves from the environment to the receptor center (organ of Corti) in the inner ear (Fig. 16-8). Sound waves in the aqueous environment of fishes that can hear are transmitted to the air bladder (a resonator) from which the waves are sent by way of the weberian chain of bones to the perilymph and endolymph of the membranous labyrinth where neuroepithelial cells of the sacculus are stimulated. Tetrapods in an air environment have elaborated on this bone-conducting method with the outer and inner ears. This development has been a gradual one in the evolution of tetrapods. The middle ear came first in the salientians (anurans) of the Amphibia. It is a chamber derived from an extension of the first pharyngeal pouch and communicates with the pharynx by the eustachean tube. From the outside, the middle ear chamber is separated by a **tympanic membrane,** or eardrum, formed by the contact of endoderm with ectoderm. The tympanum is the place where the gill slit first opened to the exterior. It is flush with the surface of the head in frogs and toads, for no outer ear exists in amphibians. In all salamanders and caecilians (Apoda), a few salientians and lizards, and all snakes the middle ear has been lost. In the

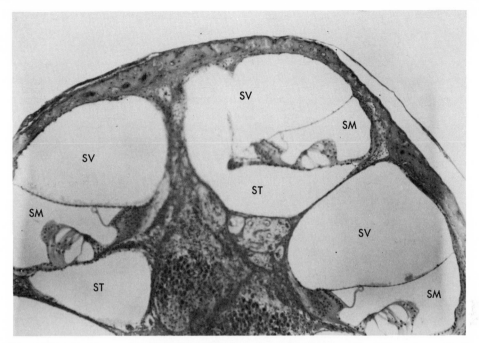

Fig. 16-12. Cochlea and helicotrema of guinea pig. *SM,* Scala media; *SV,* scala vestibuli; *ST,* scala tympani. The cochlea has been cut in such a way that the different spaces (scala) may be represented by more than one. (×160.)

middle ear of salientians, a slender bone (the **columella**) extends from the tympanum, or eardrum, across the middle ear to the fenestra ovalis, or oval window, a membrane covering the **scala vestibuli** of the organ of Corti. This bone (or cartilage) is derived from the hyomandibular bone of fishes where it serves to anchor the upper jaw to the skull and is no longer necessary when the jaw and cranium are firmly fused together (tetrapods). Any vibrations produced by sound waves on the eardrum are transmitted to the oval window by the columella. Reptiles and birds have about the same arrangement (a columella plus a small hinge bone). Mammals have three middle ear bones (ossicles): **malleus, incus,** and **stapes.** The malleus is derived from the articular bone of the lower jaw of vertebrates below mammals, and the incus from the quadrate bone of the upper jaw. The stapes is homologous to the columella. All these bones function in transmitting vibrations of the tympanic membrane to the oval window of the scala vestibuli where lymph carries the vibrations to the hair cells of the organ of Corti. The round window of the scala tympani bulges out to adjust to the pressure of the lymph produced by the incoming vibrations to the oval window.

Snakes and urodeles, which have no middle ear cavity and no eardrum, have a columella that stretches between the oval window and squamosal bone (urodeles) or quadrate bone (snakes). These animals are deaf to airborne sounds but can pick up sounds carried through the ground.

There is no outer ear in amphibians, but in some reptiles the tympanic membrane is sunk below the skin surface at the bottom of a shallow canal, the **external auditory meatus.** This canal is deeper in birds and mammals, and in the latter group, it is bounded externally by the cartilaginous **pinna** for collecting sound waves.

Katsuki (1963) has given a very extensive review of the comparative physiology of hearing.

The range of hearing varies among animals. Some have ears far more sensitive than man's, such as certain bats that make use of ultrasonic vibrations in their method of **echolocation** (Griffin, 1958).

RADIORECEPTORS

Radioreceptive organs are responsive to high frequency vibrations of temperature and light. Although there are many types of energy to which living organisms are exposed (Steven, 1963), there appear to be only two special kinds of receptors in the animal body that are sensitive to high frequency vibration, **thermoreceptors** and **photoreceptors.** Little is known about thermoreceptors, in spite of the concern animals have about temperature-sensing. It must be remembered that the surface of the earth is constantly bombarded with electromagnetic energy in the form of ultraviolet rays, visible light, infrared rays or heat, x rays, and radio waves. This energy in the form of little packets is what keeps life going. The wavelengths are inversely proportional in length to the energy in the packets. However, there is a restriction to the usefulness of the energy in the energy spectrum (McElroy and Glass, 1961). If the wavelengths are very short, they are destructive to living matter that absorbs them by breaking up molecules into ions; if too long, they are too weak to affect molecular energy. Within the range from 300 or 400 to 800 nanometers (nm) animals have carotenoid pigments that absorb these wavelengths (light) and excite photoreceptors of vision. As a result there has evolved in higher forms one of the most highly specialized of sense organs—the eye.

Thermoreceptors. Vertebrates are highly sensitive to temperature changes and undoubtedly have many temperature receptors scattered over their bodies (Murray, 1962). But specificity is very difficult to establish because many other receptors not directly involved in temperature sensing are also affected in some ways by temperature changes. Such temperature receptors are confined mainly to the skin, and many thermoreceptors are thought to be unspecialized free nerve endings. There is some evidence that encapsulated or secondary receptors are involved in temperature-sensing. The lateral line system of fishes may have temperature-receptor functions. The cranial ampullae of Lorenzini of elasmobranchs, already described with the lateral line section, have been given a temperature function by some authorities.

Clearcut examples of thermoreceptors are the **labial pits** of pythons and the **pit organs** of pit vipers such as rattlesnakes (Fig. 16-13). These snakes are mostly active at night when seeking food, which is mammalian prey. Such snakes have compensated for the poor vision of night and a naturally restricted hearing (except through ground vibration) by developing a specialized temperature receptor in a small pit on each side of the head between the ear and the external naris (Bullock and Fox, 1957). This pit, which varies slightly with the size of the animals is about 5 mm deep and 5 mm wide at the base. A thin membrane extends across the pit near the bottom and is densely covered with many bulbous trigeminal nerve terminals (Fig. 16-13). Experimentally, it has been shown that slight raises in temperature (0.002° C) will cause sensory detection of a warm-blooded animal several feet away. By the use of the two pits, precise location of the prey is made possible. This pit sense organ, which is the most sensitive temperature receptor known, well deserves the rank of a specialized organ. The labial pits of the pythons are arranged in a row along the upper jaw but are far less sensitive than

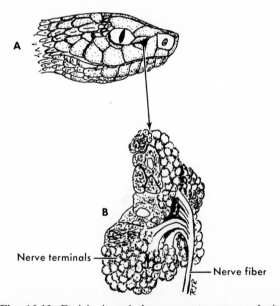

Fig. 16-13. Facial pit and thermoreceptor organ of pit viper. **A,** Location of pit between the eye and external nostril. **B,** Structure of a single nerve terminal in pit membrane, which is made up of many such terminals. The whole membrane may be called the thermoreceptor organ. (Modified from several sources.)

those found in pit vipers. Our knowledge about thermoreceptors has been reviewed recently by Murray (1962), who reveals how scanty is our present understanding of this field of sensory mechanisms.

The photoreceptor—the eye. Visual receptors of some kind are widespread throughout all vertebrates (Gregory, 1967). Fishes may have light sensitivity in their lateral line system as well as in scattered cells or clusters of cells over the body (Detwiler, 1943). The chief visual organs, however, are the highly specialized organs, the eyes. The basic plan of the eye shows remarkable uniformity throughout the history of vertebrates, including early fossil records of the phylum, insofar as it is possible to determine (Walls, 1942). Although all present vertebrates have only the paired eyes located between olfactory and auditory organs, the early vertebrate phylogeny indicates that there were at least two other visual organs, the **parietal** and the **pineal.** These were dorsal, median eyes that originated from median evaginations from the roof of the diencephalon of the brain. They were aranged in tandem, with the parietal being anterior and the pineal posterior. They may have developed in vertebrates about the same time as the bilateral eyes. These median eyes have had a variable history in the long phylogeny of vertebrates (Torrey, 1962). Some primitive vertebrates still living show median eyes with rudimentary rods, cones, and lens as well as other parts of a conventional eye (Eakin and Westfall, 1959). Among modern forms only the lamprey has median eyelike structures representative of both kinds of eyes, but the best developed median eye is the single parietal eye of the "living fossil" reptile, *Sphenodon,* although it is too rudimentary to be functional. The fossil record indicates that only one median eye was present, but some vertebrates seemed to have emphasized one kind of median eye and some the other. Other vertebrates have dorsal evaginations, usually the pineal, which is glandular in nature but forms no eye. Investigations seem to show that this gland may act as a sort of biological clock in controlling the reproductive cycles in mammals, but it may be responsible for pigment distribution in some animals (Breder and Rasquin, 1950).

Although the vertebrate eye has many variations in minor details, it has a uniformity in its structural plan (Smelser, 1961). However, its position, the orbit, shows many basic differences in the various groups of vertebrates. In most animals the eyes have become adapted to different ways of life, not by changed structural details but by changed location. In primates, including man, the eyes are placed forward so that the two visual fields overlap, which gives a sense of depth and distance. This type gives binocular vision. Predaceous animals in most major groups (e.g., cats and birds of prey) tend to have the same arrangement so that they have the overlap in the hunting direction. In bitterns (whose freezing stance with the bill canted up gives them binocular vision) and herons the orbits are found somewhat ventrally. Those that are preyed on, such as rabbits, have the eyes placed laterally (monocular vision) to get a wide angle of vision at the sacrifice of overlap. In the flatfishes, skates, rays, and some teleosts the orbits are dorsal and close together. In other flatfishes such as the sole, which lies on its side and has assumed an asymmetrical shape, the two orbits have come to lie on the same side. Other peculiar relations of the orbit are those in the hammerhead shark where they are placed far out on lateral processes or in those teleosts that have the eyes out at the ends of tentacles.

The eye is spherical with three concentric layers, the outer one being complete. The outer coat is the tough, fibrous **tunica fibrosa.** This coat is divided into two regions, the transparent **cornea** (Fig. 16-14) in front of the eye and the white **sclera** to which the extrinsic muscles are attached at the sides and back of the eyeball (Fig. 16-15). The second or middle coat (**tunica vasculosa,** or uvea) has three regions: a vascular, pigmented **choroid layer** closely applied to the sclera; a thickened, muscular **ciliary body** at the anterior end; and a colored disk-shaped continuation of the ciliary body (iris) (Figs. 16-16 and 16-17) with a central variable aperture, the **pupil.** The third and innermost coat is the **retina** (Fig. 16-20), which lines the tunica vasculosa throughout its extent. It is divided into an outer **pigmented layer** lying against the choroid coat and an inner **photosensory layer** containing the light receptors, the rods and cones. Light receptors are found only

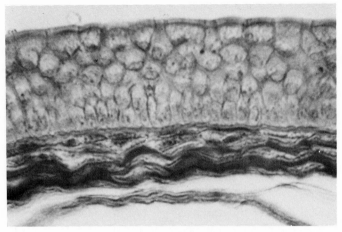

Fig. 16-14. Cornea of lamprey. (×75.)

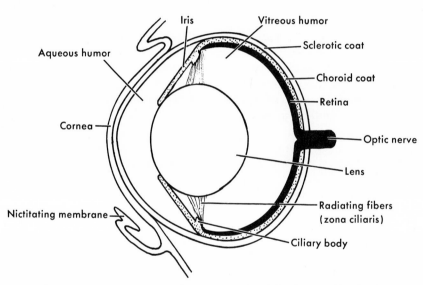

Fig. 16-15. Diagrammatic section through optical axis of frog's eye. The frog's eye at rest is farsighted, but accommodation for near vision is produced by moving the lens forward by means of the protractor muscles (not shown).

in the retina at the back and sides of the eye. The anterior part of the retina is nonsensory.

The eyeball is protected in front by the eyelids, movable folds of skin that are lined on the inside by a mucous membrane, the **conjunctiva.** A lacrimal apparatus of lacrimal gland (in upper lateral corner of the orbit) and ducts that drain excess tears into the nasal cavity keeps the cornea moist. Inserted on the outer coat, or tunica fibrosa, are the extrinsic muscles of the eye, the rectus and

oblique muscles, which are usually six in number among vertebrates.

In the interior of the eye just behind the iris is the biconvex lens (Fig. 16-18), held in place by a fibrous **suspensory ligament** that is attached to the ciliary body at the base of the iris. The lens has an outer elastic capsule around it, and from this capsule radial fibers (ciliary zonule) run out to make the suspensory ligament. The anterior portion of the lens may be covered with cuboidal epithelium (Freeman, 1964) (Fig. 16-19). Behind the lens is the large **vitreous body** of semisolid or gelatinous jelly. It helps transmit light rays, and its bulk tends to keep the lightly attached retina in place. The space between the cornea and the iris is the **anterior chamber;** that between the lens (and suspensory ligament) and the iris is the **posterior chamber.** These two chambers communicate through the pupil and are filled by the **aqueous humor,** a clear, watery fluid that is secreted by the ciliary body.

The retina (Figs. 16-20 and 16-21) consists of two strata, the outer **pigmented layer** against the choroid coat and an inner **sensory layer** containing two types of radially oriented sensory cells: (1) **rods,** which are long and narrow, and (2) **cones,** which are shorter and stout and shaped like long-necked flasks (Polyak, 1941). These rods and cones represent the actual receptors of the eye and are sensitive to light stimuli. The rods are sensitive to faint light, and the cones are sensitive to bright light and can make color discrimination (Hecht, 1937). Most strictly diurnal animals have only cones in the retina, or else the cones may greatly outnumber the rods (Wolken, 1961). Nocturnal animals tend to stress the number of rods. Most vertebrates have both rods and cones, but the relative differences in number of the two components may vary greatly. Man has a cone-rod ratio of about 1 to 20.

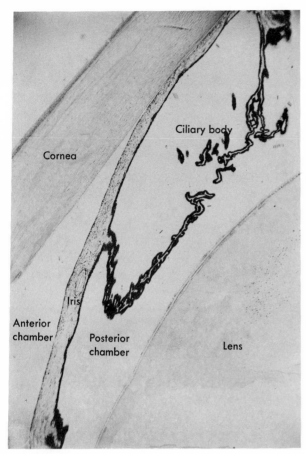

Fig. 16-16. Median section through rabbit's eye, showing iris and related parts. (×56.)

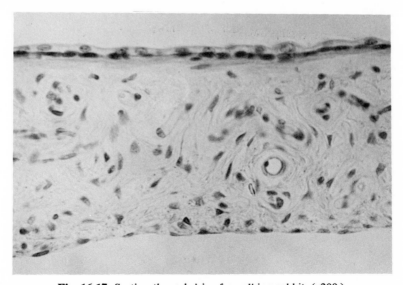

Fig. 16-17. Section through iris of an albino rabbit. (×200.)

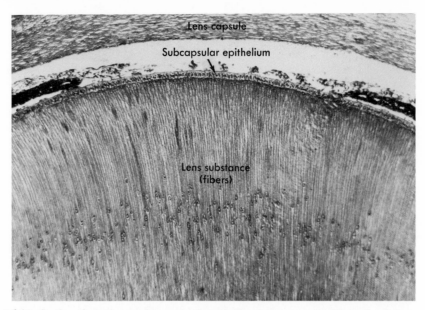

Fig. 16-18. Section through anterior part of crystalline lens of a pup. Note the lens capsule (partially torn), the single layer of cuboid cells on the anterior surface of the lens, and the lens substance of fibers that are modifications of epithelial cells. There is a middle zone of nuclei in a young animal, but these disappear later. (×290.)

Fig. 16-19. Section through anterior surface of the crystalline lens of *Anolis* eye, showing dorsal view of epithelial cells. In the vertebrate lens a single-layered epithelium (low columnar or squamous) is restricted to the anterior surface; the posterior epithelial cells are modified into concentrically arranged lens fibers making up the bulk of the lens. (×860.)

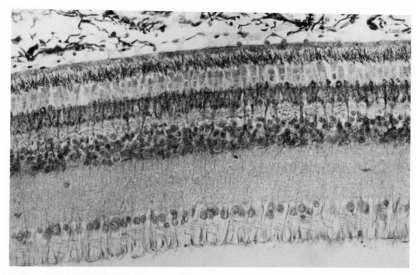

Fig. 16-20. Retina of turtle as seen with light microscope. This is the "classic view," ordinarily seen with routine staining. Staining by Golgi's method shows, in addition, an outer limiting membrane and an inner limiting membrane. Compare with Fig. 16-22. (×290.)

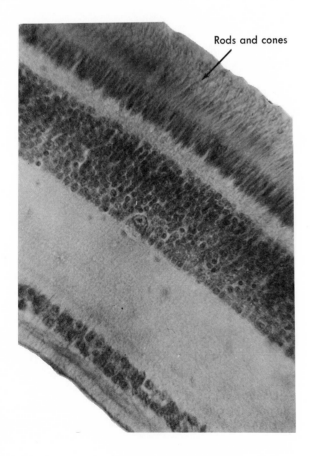

Rods and cones

Fig. 16-21. Section through retina of bird. Retina of birds is much thicker than that of mammals because of dense concentration of cones (especially in diurnal forms) and the high ratio of optic nerve fibers to visual cells. In some birds each cone may have its own bipolar and ganglion cells; man may have 125 visual cells to each bipolar and ganglion cell. Compare with Fig. 16-22. (×720.)

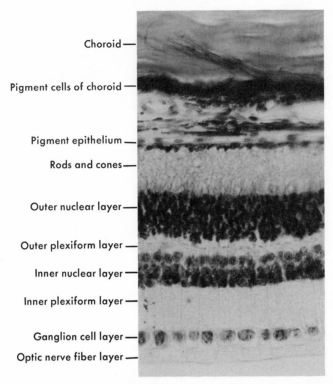

Choroid —

Pigment cells of choroid —

Pigment epithelium —
Rods and cones —

Outer nuclear layer —

Outer plexiform layer —

Inner nuclear layer —

Inner plexiform layer —

Ganglion cell layer —
Optic nerve fiber layer —

Fig. 16-22. Photomicrograph of rabbit retina, showing its chief layers. Outer nuclear, inner nuclear, and cell ganglion layers are the respective cell bodies of three sets of neurons that make up retina. The two sets of synapses form plexiform layers. (×550.)

The free ends of the rods and cones are oriented toward the pigmented, non-sensory outer layer. The rods and cones are actually specialized ends of neurons. The cell bodies of these neurons form a distinct layer internal to the rods and cones. Processes of these neurons then synapse with bipolar neurons, which in turn make synapses with the dendrites of other nerve ganglion cells (the **ganglionic layer**). Thus the retina is three neurons deep, which results altogether in ten distinguishable layers (Fig. 16-22). The axons of the nerve ganglion cells sweep over the inner surface of the retina and converge back of the eye to form the **optic nerve.** Where this optic nerve leaves the eye, there are no rods and cones—the **blind spot** (Fig. 16-23). The optic nerve passes to the brain through the three chief layers, retina, choroid, and sclera. The retina, at its front portion, loses its rods and cones and passes forward to cover the posterior surface of the ciliary body and iris as a pigmented layer. The retina receives blood from an artery which comes in at the optic nerve and sends a network of branches in the outer layer. Other blood vessels also enter the eye at various points.

Where the direct optic axis (a straight line through the center of the cornea, the pupil, the lens, and vitreous body) strikes the retina, there is a small yellow spot, the **macula lutea,** in the center of which is a tiny, shallow pit, called the **fovea centralis** (Mountcastle, 1974) (Fig. 16-24). This fovea contains only slender cones, closely packed together, and is the region of greatest visual acuity. Only the images formed here are interpreted sharply by the brain.

The fine structure of rods and cones (as revealed by the electron microscope) has been one of the most important developments in the study of the retina (Fig. 16-25). It is this fine structure that gives rise to the ten distinguishable layers previously noted. Cones and rods consist essentially of the same parts, but they differ in detail

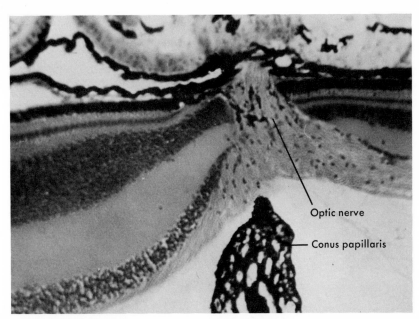

Optic nerve

Conus papillaris

Fig. 16-23. Section through retina of *Anolis,* showing point of entrance of optic nerve through retina (blind spot) and part of conus papillaris. The highly vascular and pigmented conus papillaris is supplied by an artery and vein from the optic nerve region and is purely a nutritive organ for inner layers of the retina. Some authorities consider it homologous to pecten of the bird's eye. (×100.)

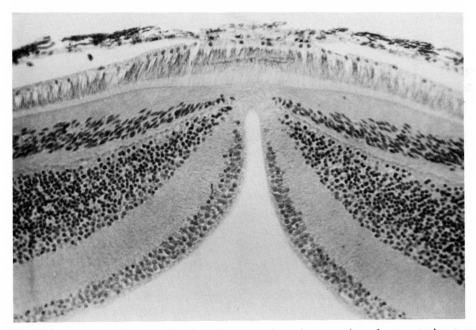

Fig. 16-24. Fovea centralis of lizard. This is a region of aggregation of cones to increase threshold of stimulation. Steep sides of fovea magnify slight movements and necessitate precision of focusing. Depth of the fovea in lizards equals or exceeds that of birds, reaching its greatest development in strictly diurnal forms. (×100.)

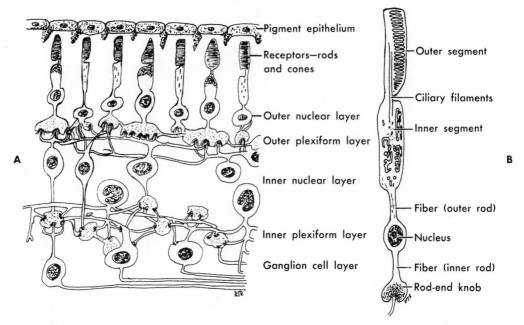

Fig. 16-25. **A,** Diagram of vertebrate retina, based upon a reconstruction from electron microscopy. The most revealing features are found in fine structure of rods and cones. **B,** Single rod (greatly enlarged). (Modified from many sources.)

(Villegas, 1964). Both are modified neurons, both have outer and inner segments, and both vary to some extent in different parts of the retina. Only that part of a photoreceptor (rod or cone) farthest from the center of the eye is involved in the reception of light. This part of the photoreceptor is called the **outer segment,** and its tip is surrounded by pigment of the pigment epithelium (Wolken, 1961). The electron microscope reveals that this outer membrane consists of double flat membrane disks stacked on each other at right angles to the long axis of the receptor. Located on or between these membrane disks are the photosensitive pigments. These disks are separated from each other by spaces, and they are solid in cones, hollow in rods. In rods this segment bears the photopigment rhodopsin (visual purple), whose biochemical change initiates a stimulation of the rod. Next to the outer segment is the inner segment, which is somewhat thicker and is connected to the outer segment by a narrow neck region containing a cilium-like structure attached to a centriole in the inner segment. This cilium, or fiber apparatus, may provide the outer segment with rigidity. The inner segment has an

abundance of longitudinally oriented mitochondria and rough-surfaced endoplasmic reticulum (Hollenberg and Bernstein, 1966).

The rest of the photoreceptor (rod or cone) is made up of an outer fiber between the inner segment and the cell body (with its nucleus) and an inner fiber between the cell body and the termination of the neuron in the plexiform layer, one of the ten layers of the typical retina. In the rod, the outer and inner fibers are narrow; in the cone they are somewhat short and thick.

Variations in both structure and function of rods and cones occur among vertebrates. In many submammalian vertebrates the rods and cones are contractile in the inner rod segment. When illuminated, this segment relaxes and pushes the photosensitive segment into the pigment epithelium. In dim light the inner segment contracts and pulls the outer segment out of the pigment epithelium and exposes it to more light.

There are striking similarities between the working of a camera and the mechanism of light reception in the eye. Each has a lens for focusing an inverted image of its visual field on a photosensitive surface. In the mammalian eye the bending and focusing of the rays are brought about by the cornea, aqueous humor, lens, and vitreous humor—all of which are refracting surfaces. In spherical aberration the periphery of the lens tends to focus more sharply than the center, and the eye corrects this by reducing the curvature of the cornea at its margins and increasing the density of the lens at its center. In chromatic aberration the shorter wavelengths are refracted more sharply than long ones and focus nearer the lens; the eye tends to filter out the shorter wavelengths by the yellow lens and yellow macula lutea. Accommodation to nearby objects is brought about in the mammalian eye by altering the shape of the lens. Ciliary muscles, which have their origin on the inside of the sclerotic coat near the cornea and are inserted on the ciliary body, contract to pull the ciliary body forward and inward. This decreases the tension on the suspensory ligament holding the lens, and the elasticity of the lens causes it to become nearly spherical. The thicker lens bends the rays of light and brings them to a proper focus for nearby objects. This method is used by the amniotes (except snakes). Accommodation may be accomplished in animals below amniotes by changing the position of the lens forward or backward. Lampreys and teleosts usually keep the lens in a forward position for near vision and move it back by special muscles for distant vision. Elasmobranchs and amphibians keep the lens in a rear position for distant vision and move it forward for near vision. Snakes have about the same method but use vitreous pressure to push the lens forward.

The pupil of the eye shows some variations in vertebrates. In some fishes the iris remains in a fixed position with the pupil unchanged, but most irises have dilator and sphincter muscles (under autonomic nervous control) for enlarging or contracting the size of the pupil in dim or bright light (Richardson, 1965). Some nocturnal animals (e.g., the cat) protect their eyes against bright light by having the pupils contracted into a slit. In many mammals there is a highly reflective layer, the **tapetum lucidum,** in the choroid layer back of the rods and cones; this causes light to be reflected back through the retina instead of being absorbed in the typical pigmented choroid (Bernstein and Pease, 1959). The tapetum lucidum in-

creases the effectiveness of light so necessary for nocturnal animals. Nocturnal animals often have large protruding corneas to collect as much light as possible.

The third eyelid, or **nictitating membrane,** arises as a vertical fold of the conjunctiva on the nasal side of the eye, covered by the upper and lower eyelids. It is transparent and can sweep the cornea to clean and moisten it without eliminating the light. The nictitating membrane is well seen in the frog and bird, although many other animals have it. Snakes and lizards have an unmoving transparent window formed by the fusion of eyelids. When the animal molts, this window is shed with its skin. Eyelids are characteristic of amniotes, but urodeles have no eyelids. Except for snakes, reptiles have movable eyelids, and usually the lower is more movable and larger.

Birds' eyes are interesting in many ways. The eyes are usually disproportionately large. Their foveal depression in the retina (some have two) is usually very deep for precise focusing and higher visual acuity, especially in diurnal birds. The peculiar cone-shaped pecten is restricted to birds and some reptiles, and no theory at present explains its function. The eyes of birds of prey have binocular vision.

Dermal light reactions, or responses to illumination, are due to photoreceptors widely scattered over the body. These responses may involve chromatophores, orientation movements, withdrawal reflexes, and other reactions. Such photoreceptors are most common in aquatic forms that have nonwaterproof skin but are rare in amniotes and certain terrestrial forms. They are mostly absent on animals that have well-developed eyes, but some have both eyes and dermal photoreceptors. Although photoreceptors for dermal light reactions have photochemical responses similar to those with optic responses, the sensitivity of eyes is far greater. Much is still to be learned about the dermal light sense. Steven (1963) has reviewed the subject recently.

References

Allison, A. C. 1953. The morphology of the olfactory system in the vertebrates. Biol. Rev. **28**:195-244.

Amoore, J. E., J. W. Johnson, Jr., and M. Rubin. 1964. The stereochemical theory of odor. Sci. Am. **210**: 42-49 (Feb.).

Ballard, W. W. 1964. Comparative anatomy and embryology. The Ronald Press Co., New York.

Beidler, L. M. 1965. Comparison of gustatory receptors, olfactory receptors, and free nerve endings. Symp. Quant. Biol. **30**:191-200.

Benjamin, R. M., B. P. Halpern, D. G. Moulton, et al. 1965. The chemical senses. Annu. Rev. Physiol. **16**:381-416.

Bernstein, U. H., and D. C. Pease. 1959. Electron microscopy of the tapetum lucidum of the cat. J. Biophys. Biochem. Cytol. **5**:35-39.

Breder, C. M., and P. Rasquin. 1950. A preliminary report on the role of the pineal organ in the control of pigment cells and light reactions in recent teleost fishes. Science **111**:10-11.

Bullock, T. H., and W. Fox. 1957. The anatomy of the infrared sense organ in the facial pit of the pit vipers. Q. J. Microsc. Sci. **98**:219-234.

Case, J. 1966. Sensory mechanisms. The Macmillan Co., New York.

Cordier, R. 1964. Sensory cells. *In* J. Brachet and A. E. Mirsky, eds. The cell, vol. 6. Academic Press, Inc., New York.

Detwiler, S. R. 1943. Vertebrate photoreceptors. The Macmillan Co., New York.

Dijkgraaf, S. 1963. The functioning and significance of the lateral line organs. Biol. Rev. **38**:51-106.

Eakin, R. M., and J. A. Westfall. 1959. Fine structure of the retina of the reptilian third eye. J. Biophys. Biochem. Cytol. **6**:133-134.

Freeman, J. A. 1964. Cellular fine structure. McGraw-Hill Book Co., New York.

Frisch, D. 1967. Ultrastructure of mouse olfactory mucosa. Am. J. Anat. **121**:87-120.

Gordon, M. S., ed. 1968. Animal function: principles and adaptations. The Macmillan Co., New York.

Gregory, R. I. 1967. Origin of eyes and brains. Nature **213**:369-372.

Griffin, D. R. 1958. Listening in the dark. Yale University Press, New Haven, Conn.

Grinnell, A. D. 1968. Sensory physiology. *In* M. S. Gordon, ed. Animal function: principles and adaptations. The Macmillan Co., New York.

Haagen-Smit, A. J. 1952. Smell and taste. Sci. Am. **186**:28-32 (March).

Hecht, S. 1937. Rods, cones, and the chemical basis of vision. Physiol. Rev. **17**:239-290.

Hoar, W. S. 1966. General and comparative physiology. Prentice-Hall, Inc., Englewood Cliffs, N. J.

Hollenberg, J. J., and M. H. Bernstein. 1966. Fine structure of the photoreceptor cells of the ground squirrel. Am. J. Anat. **118**:359-372.

Katsuki, Y. 1963. Comparative physiology of hearing. Physiol. Rev. **45**:380-423.

Lanyon, W. E., and W. N. Tavolga. 1957. Animal sounds and communication. Am. Inst. Biol. Sci. **7**:21-24.

Lettvin, J. Y., and R. C. Gesteland. 1965. Speculations on smell. Symp. Quant. Biol. **30**:217-225.

Loewenstein, W. R. 1960. Biological transducers. Sci. Am. **203**:98-108 (Aug.).

McElroy, W. D., and B. Glass, eds. 1961. Light and life. The Johns Hopkins University Press, Baltimore.

Moulton, D. G., and L. M. Beidler. 1967. Structure and function in the peripheral olfactory system. Physiol. Rev. **47**:1-52.

Mountcastle, V., ed. 1974. Medical physiology, 13th ed. The C. V. Mosby Co., St. Louis.

Murray, R. W. 1962. Temperature receptors in animals. Symp. Soc. Exp. Biol. **16**:245-266.

Murray, R. W. 1965. Receptor mechanisms in the ampullae of Lorenzini of elasmobranch fishes. Symp. Quant. Biol. **30**:233-243.

Polyak, S. L. 1941. The retina. University of Chicago Press, Chicago.

Prosser, C. L., and F. A. Brown, Jr. 1961. Comparative animal physiology, 2nd ed. W. B. Saunders Co., Philadelphia.

Reese, T. S. 1965. Olfactory cilia in the frog. J. Cell Biol. **25**:209-230.

Richardson, K. C. 1965. The fine structure of the albino rabbit iris with special reference to the identification of adrenergic and cholinergic nerves and nerve endings in the intrinsic muscles. Am. J. Anat. **114**:173-181.

Sanders, F. K. 1947. Special senses, cutaneous sensation. Annu. Rev. Physiol. **9**:553-568.

Smelser, G. K., ed. 1961. The structure of the eye. Academic Press, Inc., New York.

Smith, C. G. 1951. Regeneration of olfactory epithelium and nerves in adult frogs. Anat. Rec. **109**:661-668.

Smith, H. M. 1960. Evolution of chordate structure. Holt, Rinehart & Winston, Inc., New York.

Steven, D. M. 1963. The dermal light sense. Biol. Rev. **38**:204-240.

Torrey, T. W. 1962. Morphogenesis of the vertebrates. John Wiley & Sons, Inc., New York.

Villegas, G. M. 1964. Ultrastructure of the human retina. J. Anat. **98**:501-513.

Walls, G. L. 1942. The vertebrate eye and its adaptive radiation. Bloomfield Hills. Cranbrook Institute of Science Bulletin 19.

Wolken, J. J. 1961. The photoreceptor structures. Int. Rev. Cytol. **11**:

Wright, M. R. 1951. The lateral line system of sense organs. Q. Rev. Biol. **26**:264-288.

GLOSSARY

afferent Conveying, conducting, or leading inward or toward; as a blood vessel, duct, or nerve.

aglomerular Without glomeruli; characteristic of certain teleost fish.

analogous Similar in function and superficial structure but not necessarily in fundamental origin; e.g., the wing of the bird and the wing of the insect are analogous.

anastomosis The union of two or more arteries or veins or other vessels.

anterior Pertaining to the front or head end; cephalic, cranial. In erect animals such as man superior might be used.

aponeurosis Broad, fibrous sheet of tissue investing muscle.

archenteron Primitive digestive cavity of a metazoan embryo; it is formed by gastrulation.

argentaffin cell A type of cell, filled with argyrophilic granules, between epithelial cells in the upper part of the fundic gland of the stomach; may secrete serotonin, a vasoconstrictor substance.

arrectores pilorum Certain bands of smooth muscle cells that bring about erection of individual hairs.

atresia Degeneration of primary follicles in the mammalian ovary.

azurophilic Having an affinity for the methylene azure of blood stains.

blastopore The opening into a gastrula.

bouton The end bulbs of axons at a synapse; they are closely applied to the cell body or dendrites of the receiving cell.

brachial Pertaining to an arm or armlike process.

branchial Pertaining to gills.

brown fat Fat in which droplets remain more or less separate from each other; often called hibernating glands. Found in various mammals and in the human fetus but not in the adult.

brush border Brushlike appearance of free edge of certain epithelial cells, especially absorptive and gland cells; actually due to presence of microvilli.

bulbus arteriosus The bulb at the proximal end of the aorta; contains no cardiac muscle and is a part of the aorta (the conus is part of the heart).

bursa of Fabricius In birds, a lymphoglandular sac near the cloaca; may be involved in antibody formation.

caecum A saclike extension on the digestive tract, blind at the outer end.

caudad Toward the tail or posterior end.

cephalic Pertaining to the anterior end or head end.

chorion The outer double membrane around the embryo of certain vertebrates.

collagen Fibrous protein material in connective tissue; the most common protein in the body.

countercurrent osmotic multiplier system Found only in animals (mammals and birds) that have the loop of Henle which can excrete urine hypertonic to their blood.

cranial Pertaining to the head end; cephalic; the opposite of caudal.

cyanosis Blueness of skin and mucous membrane due to lack of hemoglobin in capillaries.

Dahlgren cells The neurosecretory cells of the urohypophysis; located along the posterior region of the cord in sharks.

distal Pertaining to free end of limb; located relatively farther away from the base or point of attachment, as opposed to proximal.

efferent Conveying, conducting, or leading outward or away from; as a blood vessel, duct, or nerve.

enteron A name for the digestive cavity, especially the part lined with endoderm.

epididymis A coiled duct that receives sperm from the seminiferous tubules and transmits them to the vas deferens.

epiphysis (pineal body) An endocrine gland on the dorsal part of the brain; may be a structure of photoreception. In the frog it forms the brown spot on the dorsal side of the head in front of the eyes.

extravascular Outside the blood system; in contrast to intravascular (inside the blood system).

fibrin Needlelike crystals that form the main support of the blood clot.

ganglion A group of nerve cell bodies acting as a center of nervous influence, outside central nervous system.

glomus A structure formed by the fusion of glomeruli into one mass; usually provided with a single afferent

411

arteriole and several efferent arterioles, in some species of lampreys.

gubernaculum A cord extending between epididymis and scrotum wall.

Hassall corpuscle Found in the medulla of the thymus and made up of concentrically arranged remnants of epithelial cells of the primordium of the thymus.

haversian system The arrangement of a haversian canal and its concentric lamellae to form bony structure.

hemopoiesis Tissue that forms blood corpuscles, white or red, or both.

homologous Basically similar as a result of similarity in embryonic origin and development; e.g., the forelimbs of the bird, bat, turtle, and horse are all homologous.

intercalated disks A cell junction at the end of each fiber in cardiac muscle and often arranged in a zigzag manner.

internal elastic lamina A layer of elastic connective tissue found in the arteries setting off the intima from the media.

isotonic Pertaining to static relationship between two solutions with osmotically equal concentrations of solutes when they are separated by a selectively permeable membrane.

juxtaglomerular cells Cells that are in close contact and communication with the macula densa cells, may secrete a hypertensive factor, renin, when blood pressure falls. They are found only in the afferent arterioles and occur in all mesonephric and metanephric kidneys.

loop of Henle A U-shaped loop found in the nephron and necessary for the production of hypertonic urine.

maturation Final stages in preparation of sex cells for fertilization.

medial Internal, as opposed to lateral (external); toward the midline of the body.

meninges The special connective tissue that protects the delicate nervous tissue of the brain and spinal cord.

mesenchyme Embryonic or unspecialized connective tissue arising from mesoderm and giving rise to many different type of cells; the nonepithelial portions of the mesoderm.

microvilli Numerous minute, fingerlike, irregular protrusions from cells of some tissues; visible with the electron microscope, and in large cells, like those of amphibians, with oil-immersion objective of light microscope.

motor plate The complex found at the junction of muscle and nerve fiber where the naked axon breaks up into a number of terminal ramifications.

myofilament One of the fibrils parallel to others, representing the contractile element of the muscle cell; in striated muscle tissue the fibrils have cross striations.

neuroglia Supporting and nutritive cells of the nervous system and consisting of three types: astrocytes, microglia, and oligodendroglia.

neurotubules Hollow tubules in the fine structure of the nerve cell; making up a large part of the neurofibrils.

Nissl bodies Regions of rough endoplasmic reticulum; related to protein synthesis in the nerve cell. They are present in cell body and dendrites, but not in axons.

papillary muscle Extensions or irregular projections of the myocardium of the heart; prevent the eversion of the valves by their chordae tendineae.

pelagic Pertaining to the open sea, or away from the shore.

perikaryon The cell body of the neuron.

periosteum The fibrous membrane surrounding bone and serving for the attachment of muscles and tendons.

peripheral Pertaining to the external surface.

pheromone A chemical signal (transmitted by a hormone secreted externally) between members of the same species.

plexus A network of anastomosing nerves or blood vessels.

polymorphonuclear Referring to a kind of leukocyte with a nucleus of two or more lobes.

posterior Pertaining to the tail or hind end, as opposed to anterior. In erect animals such as man inferior might be used.

primordial germ cells The first or beginning germ cells; they usually arise in extraembryonic regions (endodermal) and then migrate to the genital regions by transportation in the blood or by amoeboid movement.

proximal Located relatively nearer to the point of attachment or center of body, as opposed to distal.

renal Pertaining to the kidney.

rete mirabile A small network of blood vessels formed by the breaking up of a blood vessel into smaller branches; words mean "wonderful net."

sarcoplasmic reticulum A continuous system of membrane-limited tubules that form a meshwork around each myofibril; the counterpart of the smooth-surfaced endoplasmic reticulum of other kinds of cells.

sagittal Pertaining to the median vertical longitudinal plane dividing the animal into right and left halves.

Sertoli cells Cells interspersed among the spermatogonia in vertebrate testis to which spermatids may be attached for nutritive and endocrine functions.

stereocilia Nonmotile cilia composed of long, slender microvilli-like structures.

syrinx The sound organ of a bird; composed of modified tracheal and bronchial rings.

urohypophysis An enlargement at the posterior end of the spinal cord of elasmobranchs and teleosts containing secretions by neurosecretory cells of the cord.

villus A projection of the mucous membrane with cores of lamina propria over and between the folds of the intestine.

INDEX

Boldface numbers refer to illustration pages.

Male ducts, 229-230
Male reproductive system, 211-215
  of Amphibia, 222-223
  of Aves, 225-226
  of Chondrichthyes, 218-219
  of Mammalia, 228-232
  of Reptilia, 224-225
Malleus, 399
Mallory's connective tissue stain, 66, 68
Malpighian corpuscle, 150, 183; *see also* Renal corpuscle
Mammalia, 8-9
  adaptive nature of tissue of, 9
  adrenal gland of, 341-343
  blood cells in, 176-179
  blood vessels of, 144-148
  brain of, 8
  cartilage in fetuses of, 70
  cerebellum of, 374-375
  cerebrum of, 383
  ear of, 397, 399
  epidermal derivatives in, 56-57
  esophagus of, 269-271
  eye of, 409
  heart of, 141-144
  immune mechanisms of, 162
  integuments of, 54-61
  intestines of, 291-295
  liver of, 304-310
  lymphatic circulation and lymphoid tissue of, 154-158
  mesonephros in embryonic forms of, 184
  metanephros in, 184
  olfactory organs in, 389
  oral cavity and pharynx of, 249-254
  ossification in, 84
  pancreas of, 313-315
  parathyroid gland in, 338
  pineal body of, **344**
  pituitary gland of, 323-325
  reproductive systems of, 228-241
   ovaries in, 216
   sperm in, **215**
  respiratory systems of, 127-132
  spinal cord in, 367-368
  stomach of, 271-275
  striated muscle fibers of, **95**
  taste buds in, 391
  thymus bodies of, 162-163
  thyroid gland of, 333-337
  urinary systems of, 199-207
Mammary glands, 57, **59,** 61
Mammillary bodies, 378
Man; *see also* Mammalia
  cerebellum of, 374
  ear of, 395

Man—cont'd
  esophagus of, 270
  eyes of, 401
  lymph nodes in, 155, **156**
  medulla oblongata in, 370
  pharynx of, 254
  salivary glands of, 253-254
  sperm of, **215**
  spinal cord in, 368
  stomach of, 271-272
  taste buds in, 391
  teeth in, 253
  tonsils in, 253
Mancini, R. E., 62
Mark, J. S. T., 93
Marmot, 336
Marrow
  of guinea pig, **180**
  hemopoiesis and, 157-158
  red, 77, **87**
  yellow, 77
Marrow cavity, 83, 86, 87
Marshall, A. J., 125
Marshall, E. K., Jr., 199
Marsupials
  blood cells in, 177
  immunity and, 162
  Jacobson's organ in, 390
  pouched, 8
  spinal cord in, 368
  stomach of, 275
Masson's trichrome stain, 68
Mast cells, 65
Mathews, J. L., 79
Matrix, 18
  bone, 70, 79-80
  cartilage, 70, 71
  of chromosome, 21
  of connective tissue, 62, 67
  cytoplasmic, 19
Matyniak, K. A., 383
Mauthner's fibers, 363
Maximow, A., 64
McElroy, W. D., 400
Mechanoreceptors, 387, 392-399
  ear in, 394-399
  lateral line system in, 392-394
Media, 133
  of lymphatic vessels, 154
  tunica, of arteries, 140
Mediastinal lymph node, **156**
Mediastinum, 142
Mediastinum testis, 228
Medulla
  of adrenal gland, 339, 342, **343**
  of hair, 56
  of kidney
   of Aves, 197

Medulla—cont'd
  of Mammalia, 199
  of lymph node tissue, 154-155
  of ovary, 210, 215
   in Mammalia, 232, 235
Medulla oblongata, 369-371
Megakaryocytes
  in myeloid tissue, 77
  red marrow and, 157
Megaloblasts, 171
Meissner corpuscles, 356, **357,** 387
Melanins, 51
  in Mammalia, 56
  in midbrain, 377
Melanocytes
  in Aves, 51
  in connective tissue, 65
  in Mammalia, 56
Melanocyte-stimulating hormone, 41, 319
Melanophores
  in Amphibia, 46, 49
  in Aves, 51
  in dogfish shark, 41
  in Reptilia, 50
  in veins, 141
Melanophoric pericytes, 141
Melatonin, 344
Membrane granulosa, 228
Membranes
  basement, 28
  cell, **12**
   of cardiac fibers, 101
   of skeletal muscle, 93
  nictitating, 409
  nuclear, 12, 17, **20**
  tectorial, 397
  tympanic, 398
  tympaniform, 125
  Z, 93
Membranous labyrinth, 394
Membranous urethra, 206, 229
Meninges, 359-360
Menschik, Z., 69
Merkel's corpuscles, 387
Mesencephalon, 375
Mesenchymal cells
  connective tissue and, 62, 68
  undifferentiated, 63
Mesentery, ovarian, 235
Mesoderm, 26
  connective tissue and, 62
Mesomere, 182
Mesonephric ducts, 182, 183; *see also* Ureters
  in reptiles, 224
Mesonephros, 183, **184**